Millimeter-Wave and Infrared Multisensor Design and Signal Processing

For a complete listing of the *Artech House Radar Library,*
turn to the back of this book.

Millimeter-Wave and Infrared Multisensor Design and Signal Processing

Lawrence A. Klein

Artech House, Inc.
Boston • London

Library of Congress Cataloging-in-Publication Data
Klein, Lawrence A.
Millimeter-wave and infrared multisensor design and signal processing / Lawrence A. Klein.
p. cm.
Includes bibliographical references and index.
ISBN 0-89006-764-3 (alk. paper)
1.Multisensor data fusion. 2. Infrared detectors. 3. Millimeter wave devices. 4. Precision guided munitions. 5. Military surveillance.
I. Title.
TA1632.K555 1997
623'.042—dc21 97-12574
CIP

British Library Cataloguing in Publication Data
Klein, Lawrence
Millimeter-wave and infrared multisensor design and signal processing
1. Multisensor data fusion 2. Millimeter wave devices 3. Infrared detectors
I. Title
621.3'822
ISBN 0-89006-764-3

Cover design by Jennifer Makower

685 Canton Street
Norwood, MA 02062

International Standard Book Number: 0-89006-764-3
Library of Congress Catalog Card Number: 97-12574
10 9 8 7 6 5 4 3 2 1

To Jonathan, Amy, and Gregory

Contents

Preface

The topics presented in this book were developed as part of a course describing advanced sensors for modern defense applications. Although the subject matter originally addressed military smart weapons requirements, most of the subject matter (such as electromagnetic sensor operation in noise and ground clutter, atmospheric and weather limitations on sensor performance, individual sensor signal processing options, and data fusion) was eventually applied to other applications as well. Modifications to the original syllabus easily incorporated additional sensor applications, including space-borne sensors for Earth resource monitoring and weather forecasting and sensors for management of vehicular traffic. Still later incarnations of the course addressed fusion of multispectral and hyperspectral data.

The major premise used in synthesizing this material was that multisensor architectures could be designed and produced cost effectively to enhance the detection of military targets and other observables under a diverse set of operating conditions such as high clutter, inclement weather, and countermeasures. The selected topics were drawn from more than twenty years of experience in designing space-based sensors and sensors for military weapon systems. Experience had shown that the design of these sensor systems could not proceed without knowledge of passive and active sensor operation, atmospheric and weather effects, detection theory, clutter decorrelation, measures of optical image quality, signal processing algorithms, and data fusion. Therefore, many aspects of these subjects were compiled for inclusion in this volume. There were other subjects to be mastered as well, such as signal processor throughput requirements, warhead capabilities, missile guidance, and benefit-to-cost ratio analysis. However, a discussion of these latter topics is beyond the scope of this book.

In reviewing the history of the publication cycle with Artech House, I found that many members of its editorial staff participated in bringing *Millimeter-Wave and Infrared Multisensor Design and Signal Processing* to fruition. They include acquisitions editor Mark Walsh and current and past members of his staff, Alexia Rosoff, Kimberly Collignon, and Rebecca Warren. The efforts of executive editor Judi Stone and senior production editor Kate Feininger and their staff in having galleys prepared in a timely manner and in providing editorial assistance were invaluable in launching the book. I thank these and many other friends, students, and colleagues for their conversation and comments over the years. The book is stronger for their insights. Its faults are of course my own.

Lawrence A. Klein
Placentia, California
June 1997

CHAPTER 1

Introduction

1.1 THE NEED FOR SENSORS IN MODERN WEAPON SYSTEMS

Although microwave and infrared sensors have historically been used for military applications, weather forecasting, and earth resource monitoring, their development accelerated as they were incorporated into high volume military projects. Eventually, the microwave sensor was augmented by sensors operating in the millimeter-wave region so that size and weight could be reduced. Alone or in combination, millimeter-wave and infrared sensors were integrated into weapon systems to increase accuracy and performance in inclement weather and in countermeasure environments. The development of these smart munitions began in the 1970s as a response to the proliferation of tank and artillery systems by the Warsaw Pact nations. The United States and NATO believed that the requisite volume of fire from their conventional weapons was not adequate to defeat a massive assault on Western Europe by the Warsaw Pact. It was also argued that the development of smart weapons that operated in a "fire-and-forget" mode would raise the nuclear threshold by offering alternative weapon systems to defeat an assault by Warsaw Pact forces. The change in weapon system development and acquisition philosophy to incorporate sensors for target detection, tracking, and weapon guidance marked a transition from a "think many" to a "think smart" mentality.

In addition to the development of munitions that incorporated sensors, the western allies pursued negotiations to reduce the number of conventional weapons that were available to both sides. These efforts, which began in the 1980s, culminated in the Conventional Forces Europe (CFE) Treaty signed in November 1990. The CFE Treaty established equal amounts of conventional armament in Europe for both NATO and the Warsaw Pact countries as shown in Table 1.1 [1].

Based on force levels in 1989, the Soviets would make a 60% cut, while NATO would reduce its armament by less than 5%. But during two years of talks, the Soviets moved large amounts of weaponry beyond the Urals to Siberia, and thus beyond the geographic boundaries affected by the treaty. As a result of these

Table 1.1
Comparison of NATO and Warsaw Pact Conventional Forces at the Time of the CFE Treaty

	Before Treaty		*After Treaty*	
Weapon	*NATO*	*Warsaw Pact*	*NATO*	*Warsaw Pact*
Main battle tanks	22,230	36,798	20,000	20,000
Artillery	18,500	32,000	20,000	20,000
Armored combat vehicles	30,000	47,220	30,000	30,000
Fixed wing aircraft	5,700	9,790	6,800	6,800
Helicopters	2,235	3,550	2,000	2,000

Source: Toth, R. C., "22 Nations Sign Pact to Cut Arsenals," *Los Angeles Times,* Nov. 20, 1990, p. 1.

transfers, only about half the armament once projected for destruction was actually destroyed or converted to peaceful uses. Even though conventional forces were reduced as a result of the treaty, work continued on the development of smart weapons for two primary reasons: (1) fewer numbers of smart munitions were required to neutralize a target, and thus offered a cost-effective alternative to conventional munitions, and (2) the survivability of military personnel increased since fire-and-forget artillery and missiles were launched from more protected locations or allowed the soldier to withdraw immediately after firing his weapon.

1.2 INCREASING WEAPON SYSTEM EFFECTIVENESS WITH SENSORS

Increased weapon system effectiveness may be achieved in several complementary ways, including the use of integrated sensors and munitions. Several of these concepts were addressed by Norman Augustine [2]. His ideas are summarized below:

- Optimize the weapon delivery system and the ordnance it delivers as an entire weapon system. One example of this approach is to equip modern, high-performance aircraft with laser guided and other smart munitions, rather than conventional unguided ordnance, to reduce the number of sorties required to neutralize a target. The strike precision afforded by the smart munitions not only reduces the amount of ordnance that must be delivered, but also increases the survival rate of the pilot and aircraft since fewer missions need be flown over hostile territory.
- Match the weapon systems to the scenario. Data from a British exercise show that helicopters are effective attacking tanks on the move, attaining a 7 to 1 kill ratio in favor of the helicopters. However, they are not effective when used against tanks in static phases of combat where the tanks are

protected by air defenses. In the latter case, the kill ratio changed by an order of magnitude to favor the tanks.

- Develop all-weather systems. The value of this approach is to extend the window of opportunity for launching a counter-attack. For instance, in a European battle scenario, a visual attack is only possible 12 to 40% of the time depending on the specific battle location.
- Design relatively uncomplicated systems that can be used in several applications, for example, a laser-guided missile that can be launched from aircraft as well as by ground forces.
- Use the ground soldier as the eyes and ears of the combat forces, rather than as the principal assault and defense element. Implementation of this idea leads to strategies such as equipping the soldier with a laser target designator and miniature radio, and relying on modern fire support systems for the actual fighting.

These concepts guided the implementation of modern weapon systems that incorporate sensors as an integral part of the system. The sensors increase the probabilities of target detection, target classification, and target identification. As a result, fewer munitions are required to destroy a target and, simultaneously, the safety of combat personnel is increased by removing them from active participation in the final stages of the target engagement.

Figure 1.1 illustrates the target detection, classification, and identification concepts in a "fire-and-forget" weapon system. The first row of Figure 1.1 describes target detection by showing a long-range view of military vehicles on a road as might be observed by an approaching missile. The weapon system performs automatic target recognition by translating object-like shapes into target detections by analyzing the objects for target-related signal strength and features. As the missile draws closer to the intended targets, as in the second row, additional target features are observable. Proper signal processing algorithm design allows the added information to be used to classify the vehicles as a tank, armored personnel carrier (APC), and truck. At even closer range, depicted in the third row, it may be possible to identify the specific models that correspond to each class of vehicle, for example, an M-60 tank. Improving the detection, classification, and identification probabilities of a weapon system addresses one of the top concerns of the military, that of finding bona fide targets.

The challenge for a combined sensor/munition system is to locate and track a target, and then either fire a warhead projectile at the target from a standoff distance, or guide a missile directly into the target itself. Effective sensors must function in potentially inclement weather, high clutter, and countermeasure backgrounds illustrated in Figure 1.2.

Inclement weather elements include clouds, rain, and fog that can limit the visibility of sensors that are part of an integrated munition system. These weather conditions can restrict the effective range of sensors in the visible and infrared (IR) spectral bands and also those that operate at millimeter-wave (MMW) frequencies

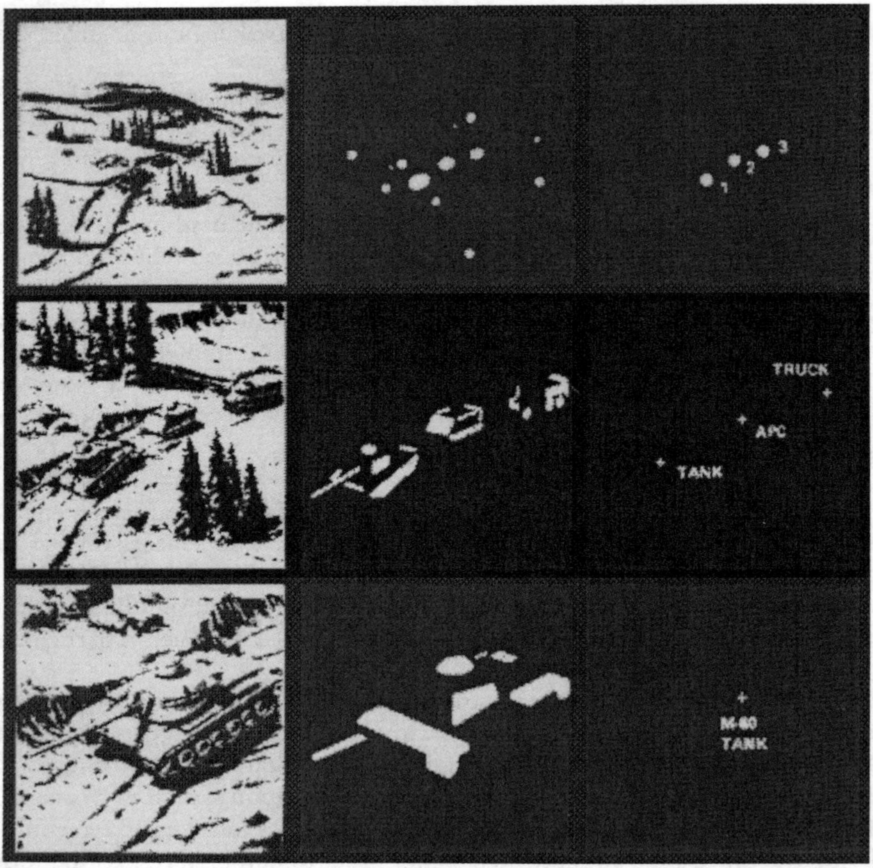

Figure 1.1 Target detection (first row), classification (second row), and identification (third row) as might be performed by a sensor in a "fire-and-forget" weapon system.

at extended ranges of tens of kilometers. The high clutter backgrounds shown in Figure 1.2 are represented by snow, trees, and plowed fields. These backgrounds can produce high clutter at millimeter wavelengths, making target detection difficult for sensors that do not transmit complex waveforms or use sophisticated signal processing. Standing water also provides a challenge for passive MMW sensors because it sometimes mimics the signature of armored vehicles. Countermeasures are represented in Figure 1.2 by smoke, dust, corner reflectors, and active jamming. Combinations of these may be deployed at all frequencies in the electromagnetic spectrum.

Even if a sensor is successful in detecting objects in this challenging environment, it must accurately differentiate military vehicles from commercial and civilian buildings and vehicles. This involves the target classification process. Still further refinement in target selection may be desirable if multiple types of military vehicles are present. Under these conditions, it may be necessary for the sensor to identify each specific target type and then seek out the target with the highest priority.

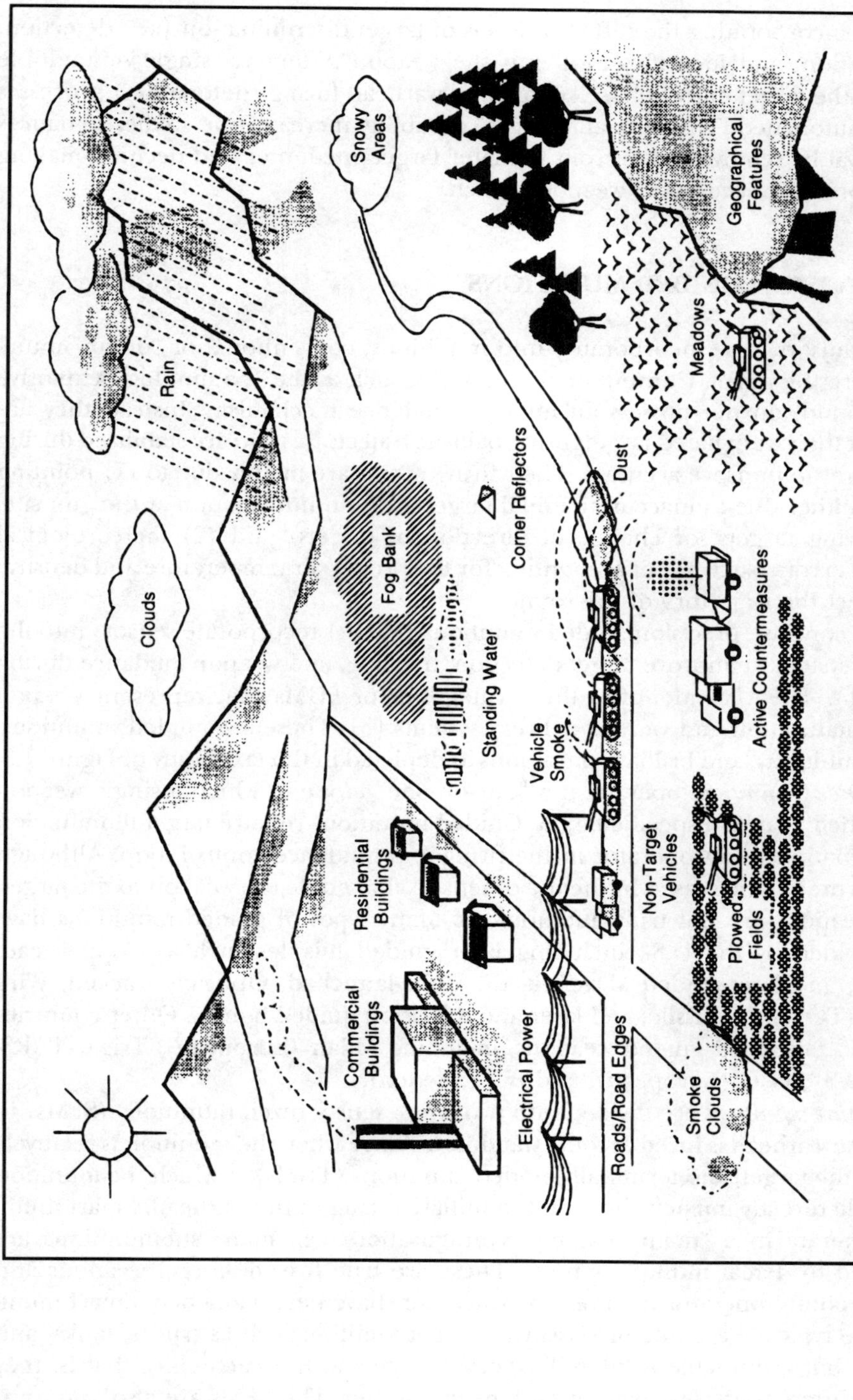

Figure 1.2 Inclement weather, background clutter, and countermeasure elements that hinder sensor performance.

By incorporating the different levels of target discrimination (i.e., detection, classification, and identification) into the weapon system via single or multiple sensors, the target detection, selection, and warhead fuzing (detonation) processes can be automated. This "fire-and-forget" capability increases the soldier's chances of survival by removing him from the later target encounter and decision-making operations that occur after weapon launch.

1.3 PRECISION GUIDED MUNITIONS

Before sensors were incorporated into munitions, conventional or "dumb" munitions were the norm. Conventional munitions, such as the 155 mm high explosive (HE) round, do not possess an internal guidance mechanism. Instead they are aimed at their designated targets using ballistic trajectories and are stabilized during their flight to improve accuracy. Their firing errors are mainly due to (1) pointing errors (either due to inaccurate initial target location information at the gun site, fast moving targets, or change in direction of targets) and (2) meteorological induced errors (such as not accounting for the proper air temperature and density) that affect the trajectory of the round.

By contrast, precision guided munitions (PGMs) incorporate sensors into the weapon system to improve target detection, tracking, and weapon guidance during the flight. The U.S. identifies three categories of PGMs that represent weapon system maturation and various delivery systems [3]. These are guided munitions, smart munitions, and brilliant munitions as depicted in the taxonomy of Figure 1.3.

Guided munitions operate in a "one-on-one" mode in which a single weapon is launched against a specific target. Guided munitions require target illumination or designation by an operator in the firing and guidance control loop. Although their accuracy is increased by the use of sensors that guide the weapon to the target, these weapons are not truly autonomous. Many types of guided munitions have been fielded by the U.S., including laser guided missiles such as Copperhead, Hellfire, and laser-guided Maverick; the Tube-launched, Optically tracked, Wire guided (TOW) 2A missile; and laser and television guided bombs. Other countries have also produced guided weapons, including Milan (European), Trigat/PARS-3/ATGW3/AC3G (European), and Bill (Sweden).

Smart munitions are divided into two types: sensor-fuzed munitions (SFMs) in which the warhead is fuzed at some standoff distance when the munition is relatively close to the target, and terminally-guided munitions (TGMs) in which the munition or missile directly impacts the target to initiate fuzing of the warhead. Smart munitions operate in a "many-on-many" configuration, i.e., many submunitions are launched to defeat multiple targets. These are true fire-and-forget weapons and do not require operator intervention once they have been launched. Smart munitions are typically capable of classifying target families, such as trucks, tanks, and bridges, but are unable to identify specific targets within each class, that is, they do not differentiate one type of tank from another. The SFMs are also known as

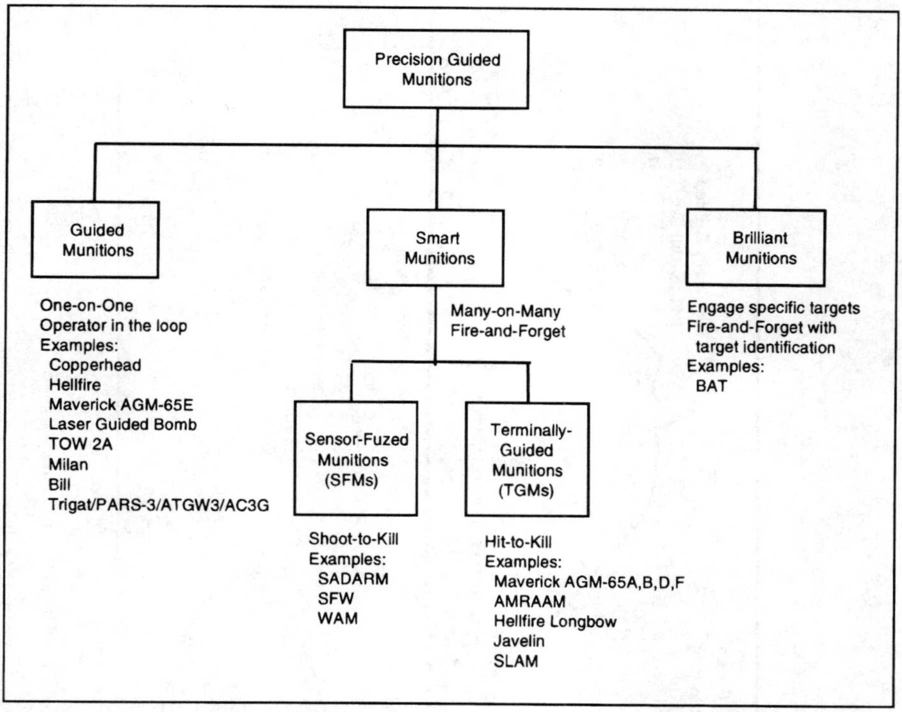

Figure 1.3 Precision guided munition classification.

shoot-to-kill weapons since they fire their warhead at the target from a standoff distance. Similarly, TGMs are called hit-to-kill weapons since they fuze their warhead by direct target impact, as shown in Figure 1.4.

The SFM is delivered to the target area in a carrier such as a 155 mm round, rocket, or large missile. Once over the targets, the SFMs are ejected and slowed in their descent using components such as a ram air-inflatable device (RAID). At a preset altitude, a parachute is deployed to stabilize and provide a scan mechanism for the sensor that is built into the munition. Once a target is detected and is within firing range, the warhead (or lethal mechanism) is fuzed and fired at the target as shown on the left of Figure 1.4. Examples of SFMs are Sense and Destroy Armor (SADARM), Sensor Fuzed Weapon (SFW), and the Wide Area Mine (WAM). TGMs may be delivered to the target area in larger rockets or missiles or may be launched one at a time. In the final stages of engagement shown on the right of Figure 1.4, the TGM is guided into the target where warhead fuzing is initiated. Examples of these systems are television-guided and imaging infrared (IIR)-guided versions of Maverick, Advanced Medium Range Air-to-Air Missile (AMRAAM), Hellfire Longbow with an all-weather MMW sensor, Javelin, and the Standoff Land Attack Missile (SLAM). A comparison of SFM and TGM attributes is presented in Table 1.2.

Brilliant munitions operate in an autonomous fire-and-forget mode with the capability to identify as well as classify targets. This feature allows them to differentiate

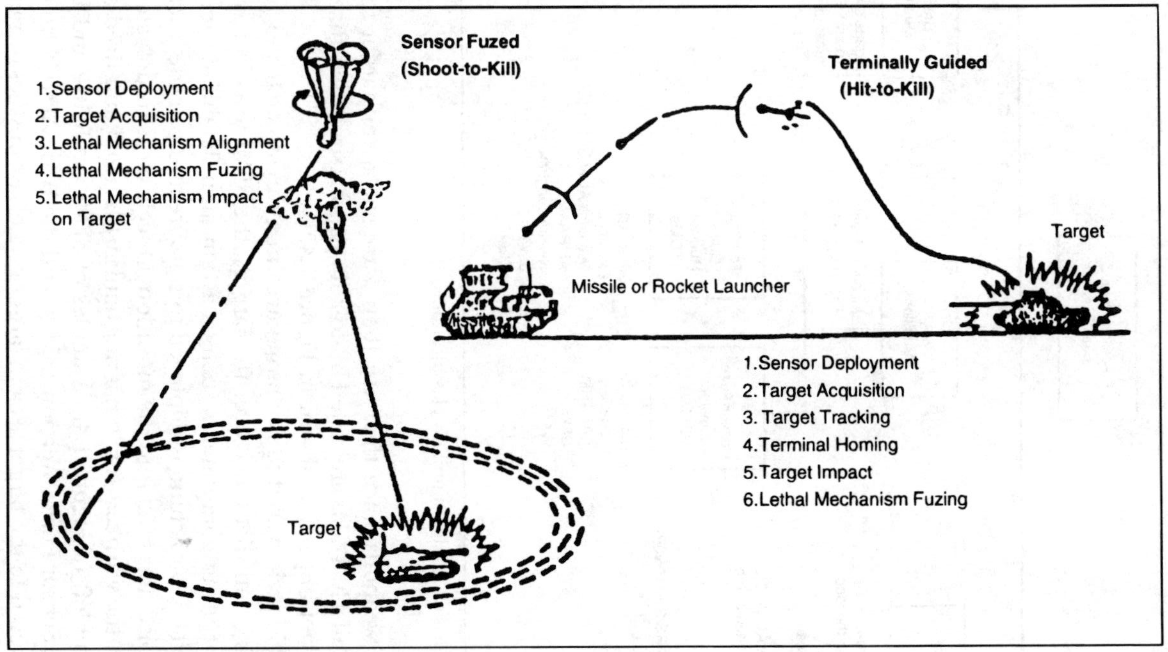

Figure 1.4 SFM (shoot-to-kill) and TGM (hit-to-kill) warhead fuzing concepts.

Table 1.2
Attributes of Sensor-Fuzed Munitions and Terminally Guided Munitions

Munition Type	*Best Suited With*	*Other Attributes*
Sensor-fuzed	• Small target location errors • Small delivery errors • Stationary and moving targets • Counterfire role (defeat of enemy artillery) • Softer targets and top attack against harder targets	• Top attack where armor is thinner • Does not maneuver to target • Smaller target search area • Lower complexity • Explosively formed penetrator (EFP) warhead • Lower cost
Terminally guided	• Larger target location errors • Larger delivery errors • Moving and stationary targets • Antiarmor and deep battle roles • Harder targets • Low and high value targets	• More powerful warhead to penetrate frontal armor • Maneuvers and tracks targets with gimbals, scanning mirrors, or electronically scanning antennas • Larger target search area • Higher complexity • Shaped charge warhead • Higher cost

and select specific target models within a family. Thus, they can differentiate between an M-1 tank and a T-62 or T-72 tank. Brilliant munitions currently being developed include the Brilliant Antiarmor Submunition (BAT), which employs acoustic and infrared sensors and, in an advanced version, adds a MMW sensor as well.

The types of sensors used in a variety of precision guided munitions are listed in Table 1.3. Their production costs vary from ten thousand dollars or less for the sensor-fuzed munitions, tens of thousands of dollars for the air-to-ground terminally guided munitions, and hundreds of thousands of dollars for the air-to-air terminally guided munitions.

1.4 MISSION-BASED WEAPON SELECTION

As the signal processing capability or intelligence built into the munition increases, so does its ability to correctly recognize valid targets. This allows fewer precision guided munitions to be used to defeat a target. While each precision guided munitions may cost more than a conventional munition, the total cost of the munitions and risk to the delivery system (e.g., aircraft) decrease since fewer munitions are required to neutralize a target. In addition, the risk to combat personnel is reduced. These ideas are illustrated conceptually in the cost per kill versus intelligence curve of Figure 1.5. The types of precision guided munitions are shown along the curve,

Table 1.3
Sensors Used in Precision Guided Munitions

Munition Category	*Example*	*Sensor System*
Guided	Copperhead	Laser guided
	Hellfire	Laser guided
	Maverick AGM-65E	Laser guided
	Guided bomb	Laser & television guided
	TOW 2A	Wire guided
	Milan	Wire guided
	Trigat/PARS-3/ATGW3/ AC3G medium range (2000 m)	CO_2 laser guided
	Bill	Wire guided
Sensor Fuzed	SADARM	IR, active & passive MMW
	SFW	Two-color IR*
	WAM	Seismic, acoustic, two-color IR
Terminally Guided	Maverick AGM-65A,B,D,F	A & B: TV guided; D & F: IIR guided
	AMRAAM	Microwave monopulse radar
	Hellfire Longbow	MMW
	Javelin	Long wavelength band IIR
	SLAM	Long wavelength band IIR
	Trigat/PARS-3/ATGW3/ AC3G long range (4500 m)	Long wavelength band IIR
Brilliant	BAT	Acoustic, IR, MMW

*Detects energy emitted in two distinct wavelength bands.

although their exact placement is mission dependent. On the left hand side of Figure 1.5 are the conventional munitions, many of which are required to defeat a target. On the right are the precision guided munitions. The TGMs are used in a one-on-one scenario, while the SFMs are used in many-on-many missions.

One factor in selecting a weapon for a mission is the available alternatives. For example, in a one-on-one scenario, guided missiles may be the weapon system of choice. Older models of guided munitions are locked onto the target before launch, while newer models have the ability to lock onto a target after launch, further increasing their effectiveness and operator survivability. In a many-on-many scenario that involves an array of tanks or artillery targets, it may be more expedient and less costly to deploy sensor-fuzed munitions from large caliber guns or from rockets. Two other issues that affect munition selection are target location errors and delivery system accuracy. These help define the search area size and the false

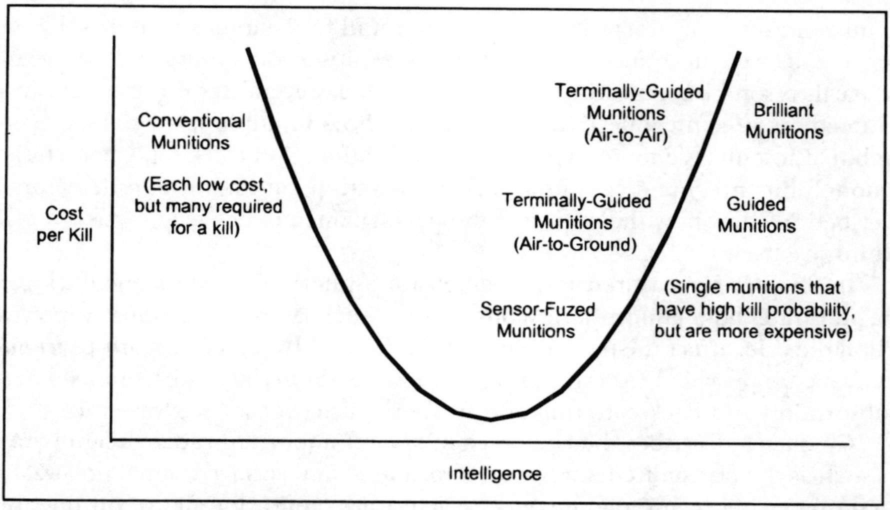

Figure 1.5 National cost per kill versus weapon intelligence for conventional, smart (sensor-fuzed and terminally guided), and brilliant munitions.

target and clutter rejection capability that are needed by the weapon's integrated sensor system. Target hardness contributes to the specification of warhead requirements. Harder targets may require a hit-to-kill weapon, while targets which deploy countermeasures, such as reactive armor that impedes the formation of the shaped charge warhead, may require a missile with a dual warhead. The first warhead detonates the reactive armor, allowing the second warhead to enter the target in the area cleared by the first. Finally, the cost of the weapon versus the cost of the target must be considered. Weapons employed against higher value targets are usually allocated a higher cost goal than those used against counterfire batteries.

1.5 BOOK CONTENTS

This book describes the operation, tradeoffs, clutter scenarios, performance measures, and signal processing techniques applicable to passive and active millimeter-wave and infrared sensors that find use in precision guided weapon systems and surveillance systems. Some weapon requirements are achieved using a single, high performance sensor, while others require multiple, lower performance sensors that are matched to the specific requirements of the mission. By properly selecting the sensors and combining their data, a wide range of signature generation phenomena can be exploited to improve target detection probability and identify weather, countermeasure, or land-based features of interest.

Chapter 2 discusses the potential benefits of a multisensor suite to improve weapon system performance in inclement weather, clutter, and countermeasures. It describes how a sensor's operating frequency or wavelength may be adapted to the mission, target characteristics, and clutter and false target signatures that arise from the background and the atmosphere. Multisensor applications to weather and earth resource monitoring are also presented. Several sensor and data fusion architectures are introduced to demonstrate how multiple sensor data may be combined to enhance target detection, classification, and tracking. Characteristics of monolithic integrated circuits and thermal and quantum infrared detectors are described to show how these govern the performance of active and passive MMW and infrared sensors.

In Chapter 3, measured and modeled atmospheric absorption and backscatter data illustrate how atmospheric constituents, such as oxygen, water vapor, and carbon dioxide, affect the performance of MMW and IR sensors at various frequencies. Data are provided for rain, snow, and fog weather-related obscurants that add to absorption and backscattering and further restrict sensor performance.

Chapter 4 describes the relevance of several important radar design parameters such as the transmitted waveform, transmitted and received signal polarization, signal processing technique, mixer characteristics, and noise figure on the ability of the radar to detect and classify objects. This is followed by a discussion of the operation of pulse, frequency-modulated continuous wave (FMCW), CW Doppler, pulse Doppler, and synthetic aperture radars. A central theme of the chapter is detection theory and how it is applied to optimize target detection probability in regions of high ground clutter. The influence of antenna beamwidth and target radar cross section fluctuation on the angle tracking accuracy of a radar is discussed. Measured backscatter data for forest and snow high clutter backgrounds are presented to show their potential for masking desired target signatures. Several parametric and nonparametric feature-based algorithms are described that can be applied to time domain and frequency domain data to detect and classify targets.

Chapter 5 discusses how to process information contained in the intermediate frequency (IF) signal at the output of a mixer in an FMCW radar. The IF amplitude and phase data allow improved target characterization as compared with the detection of only video information. The ability of an FMCW radar to calculate range using ground clutter (e.g., in an altimeter application) or to detect targets in clutter motivates a discussion of how to increase the number of independent samples of data processed by the radar. The probability of target detection can be increased by increasing the number of samples and by reducing the effects of noise and clutter. The independent samples are increased through diversity gain via temporal and frequency decorrelation, while the noise and clutter effects are limited through integration gain. The radar range equation allows the received power from either the ground clutter or target to be calculated, and the detection probability to be computed. Fractal modeling of high resolution IF range data provides simulated imagery to develop prototype algorithms that can later be used to analyze real range data acquired from radar testing against specific targets and backgrounds.

Chapter 6 describes the operation and applications of passive MMW radiometer sensors. Radiative transfer theory is introduced to show how atmospheric constituents or objects located in the atmosphere and on the ground give rise to the signatures that are detected by radiometers. The effects of finite gain antenna patterns, lossy receiver components, and impedance mismatches between components are analyzed to show their impact on the detection of emitted and reflected energy from objects in the sensor's field of view. Properties of total power, Dicke, and noise injection radiometers are presented, along with their applications to precision guided munitions, weather forecasting, and measurement of background and man-made material properties. Passive signature data for various types of ground terrain, water, and high clutter snow backgrounds are given to show their effects on target detection and material characterization. The relation between dielectric constant and emissivity is derived to illustrate their linkage and the effect these important material properties have on the passive signature.

The theory of operation, design characteristics, and performance measures for passive infrared sensors are discussed in Chapter 7. Forward looking infrared (FLIR), infrared search and track (IRST), and IR radiometer sensors are addressed. The influence of the sensor performance goals and operating environment on establishing the sensor design are described. The number of pixels (picture elements) that must be resolved on a target to meet target detection, orientation, classification, or identification requirements is given for first and second generation FLIRs. Performance measures (such as signal-to-noise ratio, noise-equivalent temperature difference, minimum resolvable temperature difference, etc.) are defined and applied to design examples for imaging and nonimaging sensors. Passive IR sensors do not possess a mixer as do MMW radars and laser radars. Since the mixer is the dominant noise source in active sensors, the noise statistics that affect the passive sensor are different. Therefore, a discussion of the probability of detection that applies to direct detection of photon energy for signal photon-limited and background-limited cases is provided. The small signal and small noise case (i.e., detection of photon energy in quantum noise) is also discussed. The chapter concludes with a discussion of signal processing techniques and algorithms that can be applied to multi-pixel imagery to achieve automatic target recognition.

Chapter 8 illustrates military applications of laser radars to target designation, ranging, and target classification, identification, and tracking. Solid state lasers are well-suited for these applications because of their small size, rugged design, and adequate output power. By extending the transmitted wavelength to greater than 1.4 μm, eye safe operation can be achieved. This has motivated the development of mid-infrared band solid state lasers. The radar range equation for lasers is derived for resolved targets and unresolved targets. Noise sources are discussed to show their effect on the signal-to-noise ratio for direct detection and heterodyne detection laser radars. The detection probability calculations that characterize laser radar performance in target detection are shown to be a function of the target type (resolved or unresolved), laser detection method (direct or heterodyne), noise- or clutter-limited operation, and target fluctuation characteristics. Finally, a comparison is presented of signal detection theory as applied to passive and active infrared and MMW sensors.

Properties of windows and domes used in dual-mode MMW and IR sensors are presented in Chapter 9. The need for such materials is demonstrated by outlining the general window and dome requirements of common aperture MMW and IR seekers that transmit or receive dual-wavelength radiation. Properties of monolithic and layered window and radome walls are included. Their physical properties are discussed in terms of the complex dielectric constant of the materials. The requirements to withstand heat, thermal shock, moisture, and rain erosion are illustrated through application-specific design parameters. Data that show the effects of temperature on the dielectric constant and loss tangent of dual-mode window materials are presented.

Chapter 10 incorporates many of the subjects discussed in earlier chapters by applying them to the design of a dual-mode MMW and IR seeker for a precision guided munition application. A dual sensor configuration is developed from the mission requirements for scan area; viewing geometry; missile diameter, velocity and altitude; and detection performance. MMW sensor design issues include radar and waveform type, operating frequency, gimbaled or electronically-scanned antenna, number of antenna beams, and application of coherent or noncoherent processing. The design of the IR sensor addresses imaging or nonimaging operation, wavelength band, detector material, optical aperture size, number of rows and columns of detectors, detector instantaneous field of view, spatial resolution, and signal processing algorithm selection. Seeker kinematic design parameters are also explored.

References

[1] Toth, R. C.,"22 Nations Sign Pact to Cut Arsenals," *Los Angeles Times*, Nov. 20, 1990, p. 1.

[2] Augustine, N. R., "Land Warfare," *IEEE Trans. on Aerospace and Electronic Systems*, AES-22, No. 5, Sept. 1986, pp. 486–512.

[3] *Smart Munitions: An Introduction to the Concepts and the Technologies—Primer*, U.S. Army Smart Munition Center, Redstone Arsenal, AL, Sept. 1987.

CHAPTER 2

Multiple Sensor System Benefits, Architectures, and Component Properties

Sensor systems composed of two or more sensors can be designed to operate in inclement weather, to support robust performance in high clutter, and to provide countermeasure resistance for a variety of military applications. Although millimeter-wave (MMW) and infrared (IR) sensors are the primary focus here, other sensors (such as magnetic, seismic, and acoustic) are often deployed for specific military applications. Multiple sensor systems are also used for civilian purposes such as weather forecasting, earth resource and crop management, and pollution monitoring. By properly selecting the sensor types, a wide range of signature generation phenomena, such as those listed in Table 2.1 [1], can be exploited to improve the detection and classification of observed objects.

The requirements of a particular application constrain sensor specifications such as weight, volume, power, cost, and detection and false alarm probabilities. Other mission-specific sensor specifications include the need for real-time signal processing; classification, identification, or tracking of objects; operating range; and sensor kinematics. This chapter discusses the benefits of a multisensor suite, the attributes and selection of active or passive modes of sensor operation, selection of the receiver frequency spectrum, techniques to combine or fuse sensor data, and the properties of critical MMW and IR sensor hardware components.

2.1 SENSOR CHARACTERISTICS AND APPLICATIONS ACROSS THE ELECTROMAGNETIC SPECTRUM

Multiple sensors that operate in the electromagnetic spectrum can maximize the independence of their acquired data in two ways. The first requires the operating

Table 2.1
Signature Generation Phenomena

Detectable Signature	*Signature Dependencies*
MMW radar cross section	Shape, material composition, surface smoothness and regularity, gaps, cavities, receiver polarization.
MMW apparent temperature	Emissivity and temperature of object, receiver polarization and incidence angle, surface roughness, weather, atmospheric conditions.
Infrared emission and reflection	Radiance produced from within the object, such as by engines, and radiance produced by natural sources, such as direct heating by the sun or by reflected radiation.
Visible reflection	Weather, atmospheric conditions, contrast with the background, visible emissions from exhausts.
Electronic emissions	Active sensor and transmitter sources such as communications equipment, navigation and guidance systems, fire control systems, electronic countermeasures, and, in general, any other source of electromagnetic radiation.
Acoustic	Engine noise, noise of object as it moves through air or moves on the ground surface such as produced by an airframe or ground vehicle.

frequencies to be selected from as wide an expanse across the electromagnetic spectrum as design constraints and scenario operation permit. The second requires the sensors to operate in both active (transmit and receive) and passive (receive only) modes as indicated in Figure 2.1. The emissive and atmospheric reflectance signatures that contribute to target detection by passive sensors are represented in the upper part of the figure. The portion of the transmitted energy that is scattered toward the receiver and leads to detection of targets by active sensors is depicted in the lower part of the figure. The received signal is a function of the sensor design and the target and background characteristics as summarized in Table 2.2. Active sensors are subject to design parameters that control the transmitted frequency and polarization, waveform shape, and power while passive sensors are not. The target shape, size, material, small-scale structure, orientation, and relative motion are some of the factors that influence the signature detected by active sensors. The emissivity, roughness, incidence angle, and temperature of the surface, as well as the receiver polarization, affect the passive signature. Background properties such as clutter type, weather and other atmospheric effects, and the presence of countermeasures affect the signatures presented to active and passive sensors by absorbing and scattering real target signature energy and by creating false target signatures.

Examples of active sensors are MMW and laser radars. Examples of passive sensors include MMW radiometers and IR radiometers, FLIR (forward looking infrared) sensors, and IRST (infrared search and track) sensors. In order to select the sensor operating frequencies (or equivalently wavelengths), tradeoffs are made based upon factors such as component size, resolution, weather, atmospheric gaseous constituents, clutter, countermeasures, and cost. For example, a

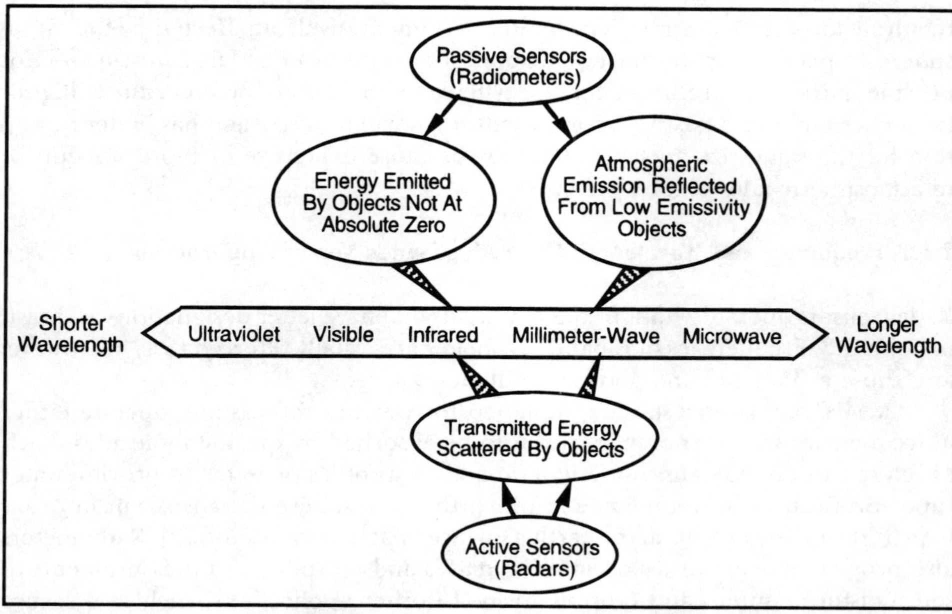

Figure 2.1 Signature-generation phenomena across the electromagnetic spectrum.

Table 2.2
Sensor, Target, and Background Parameters That Contribute to Signature Characterization

Sensor	*Target*	*Background*
• Spatial resolution	• Shape	• Clutter type
• Transmit and receive frequencies	• Overall physical size	• Clutter distribution
• Frequency stability	• Small-scale structure	• Clutter magnitude
• Polarization	• Gross and small-scale signature parameters	• Clutter decorrelation time
• Transmit waveform	• Orientation	• False targets and sun glint
• Transmit power	• Number and relative positions	• Jammers
• Scanning mechanism	• Material	• Rain
• Noise	• Temperature	• Smoke
• Sensitivity	• Relative motion	• Dust
• Bandwidth		• Haze
• Operating range		• Fog
• Data registration		• Clouds

relatively low frequency microwave radar is comparatively unaffected by the atmosphere (especially for shorter range applications), but can be large in size and not provide sufficient resolution for a particular application. By contrast, a higher frequency radar or a passive infrared sensor is smaller in size and has better resolution for the same size aperture, but may be more expensive or more susceptible to atmospheric and weather effects.

2.1.1 Frequency and Wavelength Operating Bands Versus Applications

Radar sensors operate within frequency bands that have letter designations as shown in Table 2.3. Frequencies in K band and below are usually referred to as microwave, and those at Ka band and above as millimeter-wave.

MMW radiometer sensors, designed for weather forecasting, operate either at frequencies where energy is known to be absorbed by specific molecules (such as oxygen to provide atmospheric temperature profiles or water to provide water vapor profiles) or at frequencies at which the atmosphere is transparent in order to provide measurements at the earth's surface or at lower altitudes [2]. Radiometers also provide imagery of ice-ocean boundaries and oil spills, and measurements of soil moisture content and crop maturity. In other applications, such as a secure communications system, frequency selection may again be at a strong atmospheric absorption band, such as the 60 GHz oxygen complex, to prevent transmission over long distances and to make interception of the information difficult.

IR sensors operate over spectral regions in the near-, mid-, and far-infrared wavelength spectral bands that correspond roughly to 0.77 to 1.5 micrometers

Table 2.3
Radar Spectrum Letter Designations

Designation	*Frequency (GHz)*	*Free Space Wavelength (mm)*
L	1.0 to 2.0	300 to 150
S	2.0 to 4.0	150 to 75.0
C	4.0 to 8.0	75.0 to 37.5
X	8.0 to 12.0	37.5 to 25.0
Ku	12.0 to 18.0	25.0 to 16.6
K	18.0 to 26.5	16.6 to 11.3
Ka	26.5 to 40.0	11.3 to 7.5
Q	33.0 to 50.0	9.1 to 6.0
U	40.0 to 60.0	7.5 to 5.0
V	50.0 to 75.0	6.0 to 4.0
E	60.0 to 90.0	5.0 to 3.3
W	75.0 to 110.0	4.0 to 2.7
F	90.0 to 140.0	3.3 to 2.1
D	110.0 to 170.0	2.7 to 1.8
G	140.0 to 220.0	2.1 to 1.4

(μm), 3 to 5 μm, and 8 to 40 μm, respectively. The band limits are frequently adjusted for specific applications. They may be further restricted with spectral filters to maximize the response to particular object signatures and to eliminate undesired returns from the atmosphere and surrounding background.

2.1.2 Monostatic and Bistatic Radar Operation

Active sensors such as MMW radars operate in monostatic and bistatic configurations. In the monostatic mode, the transmitter and receiver are collocated and the receiver processes energy that is backscattered from objects in the field of view. In the bistatic mode, shown in Figure 2.2, the transmitter and receiver are spatially separated. The receiver processes the transmitted energy that is scattered along the target echo path towards its antenna by objects in the field of view. When the bistatic angle β is zero, the configuration collapses to the monostatic case. Bistatic radars do not enjoy as many applications as monostatic radars. They do find use, however, in applications requiring detection and tracking of stealth targets, air-to-ground attack scenarios, satellite tracking, semiactive tracking of missiles, and passive situation assessment where they function as receivers [3].

2.1.3 Signature Generation Phenomena

In both the monostatic and bistatic MMW radar configurations, the received signal contains information about scatterer size and location as illustrated in the left panel

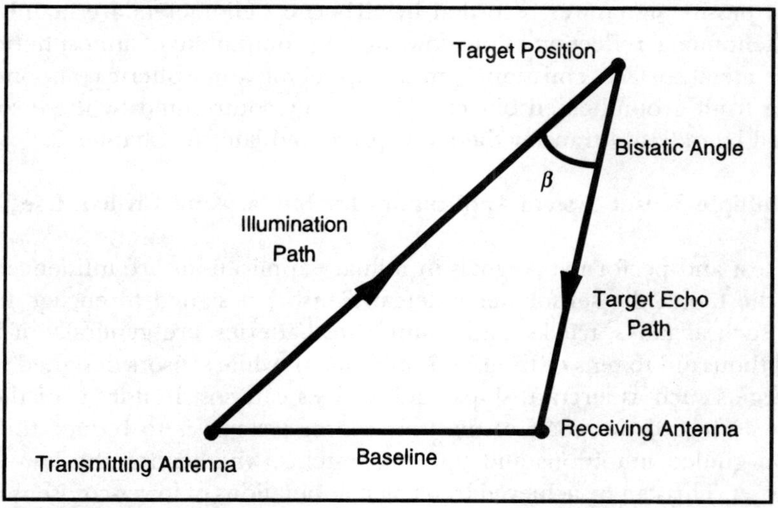

Figure 2.2 Bistatic radar geometry.

of Figure 2.3. IR laser radars, shown in the right panel of the figure, provide information similar to that from MMW radars, but at a higher resolution because of their shorter wavelength. However, they are subject to greater atmospheric degradation and an inability to search large areas in as short a time as the larger beamwidth MMW radars. In addition to scatterer size and location, the energy received by a laser radar is also responsive to the differences in reflectance between the objects and their backgrounds. This added discriminant has been shown to be useful in differentiating targets from backgrounds and other objects [4].

IR passive sensors respond to the apparent temperature difference between target and background. The apparent temperature is a function of the target emissivity in the IR spectral band of the sensor and the absolute temperature (in degrees Kelvin) of the observed targets and the atmospheric emission. Temperature sources within the sensor itself emit energy that also contributes to the apparent temperature. FLIR, IRST, and radiometer sensors are examples of passive IR devices. FLIRs are primarily used to provide high-resolution imagery of a scene, while IRSTs are primarily used to locate a "hot" area on an object and then track it. Radiometers are precisely calibrated instruments used to measure the emissivities of materials as a function of wavelength and incidence angle.

One method of obtaining two-dimensional imagery from an IR sensor is by scanning a one-dimensional linear detector array through the scene of interest as illustrated in the central panel of Figure 2.3. Other IR imaging methods include using a single detector element in a raster scan, or employing a two-dimensional staring array as will be discussed in Chapter 7.

Millimeter-wave radiometers, although not shown in Figure 2.3, respond to apparent temperature similarly to the IR passive sensors. However, their output is a function of the emissivity of the objects in the MMW spectrum of the receiver. Since metal objects have low emissivity, and hence high reflectivity at MMW frequencies, the passive signatures recorded by airborne radiometers are mainly due to three phenomena: reflection of the downwelling component of atmospheric energy from the metal surface, emission from the upwelling atmospheric component, and emission from ground-based objects. The energy contributed by these sources is described by radiative transfer theory as presented later in Chapter 6.

2.1.4 Multiple Sensor System Applications for Military and Civilian Use

Sensor cost and performance goals in military applications are influenced by the cost of the target the sensor helps defeat. Sensors designed to engage low-value targets, such as tanks, trucks, and counterfire batteries, are generally of low cost (several thousand to tens of thousands of dollars), while sensors designed for high-value targets such as aircraft, ships, and bridges can cost hundreds of thousands of dollars. One of the goals of multiple sensor systems is to reduce the cost of precision guided munitions and tracking systems, whether for the low- or high-value target. This can be achieved by using combinations of lower-cost sensors, each of which responds to different signature generation phenomena, to obtain target

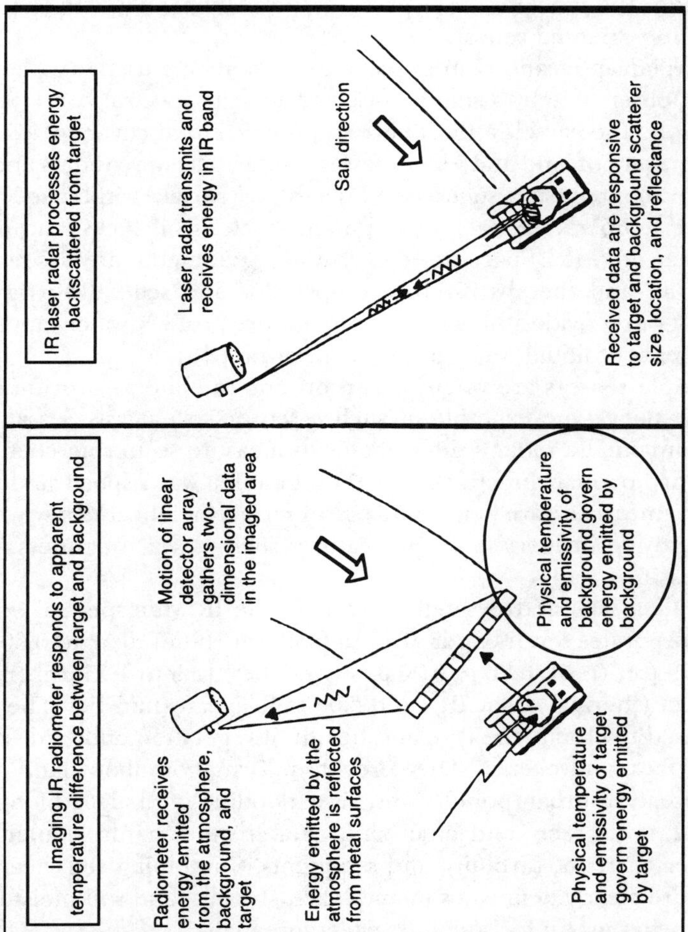

Figure 2.3 Target signatures generated by independent phenomena from active and passive sensors operating in different spectral regions.

classification and tracking data previously available only with expensive, single phenomenon-oriented sensors.

In civilian applications, multiple sensor systems are used to gather information that is not observable by a single sensor or a single spectral band. Space-based IR sensors probe the earth's atmosphere to provide cloud coverage information and thermal imagery of land and water masses, while microwave radars and radiometers supply data about sea state such as wave height, sea surface wind speed and direction, sea surface temperature, salinity, pollutants such as oil slicks, and delineation of the ice-ocean edge [2]. Space-based microwave radiometers are also used to measure soil moisture and thereby estimate crop yields and schedule irrigation, and in meteorology to provide atmospheric temperature profiles, water vapor profiles, and measurements of liquid water in the form of rain [5].

Multiple sensors are often placed on one satellite to provide multispectral, multimode data. One example of such a sensor system was Seasat, launched in 1978. It contained a radar scatterometer to measure scattering coefficients of the ocean in orthogonal directions and thereby infer wind speed and direction, an altimeter to measure wave height and ocean surface height, and a synthetic aperture radar to provide imagery of ocean features such as sea roughness and ocean-ice boundaries [6].

The Landsat-4 and -5 satellites contain the IR Multispectral Scanner (MSS) which incorporates seven bands: 0.45 to 0.52 μm (blue), 0.52 to 0.60 μm (green), 0.63 to 0.69 μm (red), 0.76 to 0.90 μm (near IR), 1.55 to 1.75 μm (mid IR), 10.40 to 12.50 μm (thermal or far IR), and 2.08 to 2.35 μm (mid IR). The multispectral data are used to classify the type and health of vegetation and to describe features of rivers, lakes, and oceans. They provide information about land and sea areas such as spread of urban populations, river flooding, and algae blooms [7,8,9,10]. The visible, red, green, and blue band data create a natural-looking image that shows water currents, turbidity, and sediments. Combining red, near IR, and mid IR data more clearly delineates inland lakes, streams, and soil moisture variations, the latter being useful for soil and vegetation analysis.

The Advanced Very High Resolution Radiometer (AVHRR) is currently carried by the National Oceanic and Atmospheric Administration (NOAA) NOAA-6 and seven satellites. It has five spectral bands, 0.580 to 0.680 μm, 0.725 to 0.1.100 μm, 3.550 to 3.930 μm, 10.50 to 11.30 μm, and 11.50 to 12.50 μm, that are used in weather forecasting, mapping ocean currents, and detecting cloud patterns. Edge detection algorithms can be applied to the mesoscale (50 to 300 km) AVHRR data to study ocean features, such as the Gulf stream and cold and warm eddies, and plot their boundaries [11]. Thermal or far-IR AVHRR data are used to measure sea-surface temperature. Accurate cloud detection by the AVHRR is necessary in this application since undetected clouds introduce errors into the surface temperature calculation, as their effect on the brightness temperature measurement is not accounted for if their presence is unknown. Accurate cloud detection is also important for global climate studies and earth radiation budget measurements since the presence or absence of clouds determines the amount of incident and reflected radiation from the Earth's surface. To this end, a number of cloud detection and classification techniques have been developed, among them being the application of the multilayer feed-forward neural network [12].

2.2 SENSOR AND DATA FUSION CONFIGURATIONS

Once the sensors and the required uses for the data (e.g., target detection, classification, identification, tracking, and aimpoint location for the military application) have been selected, methods for combining the sensor data in some optimal fashion must be chosen. Data fusion is the term usually applied to this process [13]. Data fusion configurations or architectures may be defined by the amount of signal processing that occurs in each individual sensor before the data are combined in a fusion processor. Architectures using this taxonomy include central-level fusion, also referred to as pre-individual sensor processing fusion or centralized fusion; sensor-level fusion, also referred to as post-individual sensor processing fusion or autonomous fusion; and combinations of the above, referred to as hybrid fusion. The resolution space and processing level of the data may also be employed to define another fusion lexicon. The nomenclature used here is pixel-level, feature-level, and decision-level fusion [14,15,16,17].

Before examining data fusion architectures in detail, it is worthwhile to introduce several terms that are used to describe portions of the data fusion process. These are alignment, association, correlation, estimation, classification, and cueing, as defined below [18]:

- *Alignment:* Referencing of sensor data to a common time and spatial origin;
- *Association:* Using a metric to compare tracks and data reports from different sensors to determine candidates for the fusion process;
- *Correlation:* Processing of the tracks and reports resulting from association to determine if they belong to a common object and thus aid in detecting, classifying, and tracking the objects of interest;
- *Estimation:* Predicting an object's future position by updating the state vector and error covariance matrix using the results of the correlation process;
- *Classification:* Assessing the tracks and object discrimination data to determine target type, lethality, and threat priority;
- *Cueing:* Feedback of threshold, integration time, and other data processing parameters or information about areas over which to conduct a more detailed search, based on the results of the fusion process.

2.2.1 Central-Level Fusion

Central-level fusion is applicable when the sensor outputs measure similar physical phenomena. Otherwise, sensor-level or decision-level fusion must be used. Each sensor in a central-level fusion architecture feeds minimally processed data into the fusion processor as shown in Figure 2.4. Minimal processing includes operations such as filtering and baseline estimation.

The major advantage of centralized fusion is that all the information in the raw data is available to the fusion algorithm. Hence, a fusion algorithm can potentially be designed to incorporate the sensor information before it is modified or corrupted by the processing that would otherwise occur in the individual sensors. However, the fusion algorithm is more complex than the type used with sensor-level fusion as it must operate on minimally analyzed data, and must also process data at higher rates. In the fusion processor, the input signals are examined for target features or attributes that aid in tracking and discriminating among objects. Central-level fusion is optimal for tracking objects, performing better than sensor-level fusion in estimating or predicting the future position of the object. Blackman [19] points out that the more accurate tracking is due to a combination of effects: (1) processing all the data in one place; (2) forming the initial tracks based on observations from more than one sensor, thus eliminating tracks established from partial data received by the individual sensors; (3) processing sensor data reports directly, eliminating difficulties associated with combining the sensor-level tracks produced by the individual sensors; and (4) facilitating multiple hypothesis tracking by having all data available in a central processor. Deficiencies of the method are reflected in the large amount of data that must be transferred in a timely manner to the central processor(s) for processing. Centralized fusion target tracking and discrimination

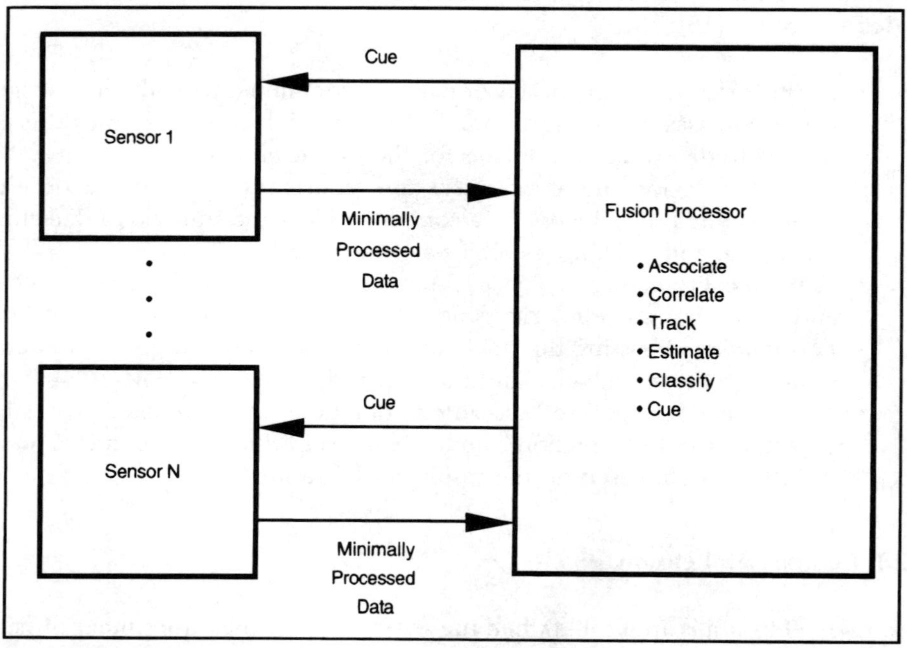

Figure 2.4 Central-level fusion.

algorithms can be written to tolerate lack of particular sensor inputs. A hybrid fusion algorithm may also be designed to combine both target tracks and unassociated data from multiple sensor inputs.

2.2.2 Sensor-Level Fusion

The architecture for sensor-level fusion is illustrated in Figure 2.5. It is optimal for classifying objects if the sensors use independent signature generation phenomena to develop information about the identity of observed objects, i.e., they derive object signatures from different physical processes and generally do not false alarm on the same artifacts [20].

Fully processed outputs from the sensors are input to a fusion processor when sensor-level fusion is utilized. For each sensor, the signal processing can be optimized to account for the sensor's resolution, frequency, field of regard, scan rate, and other attributes [21]. Time-domain processing can be used for one sensor, frequency-domain techniques with another, and multi-pixel image processing algorithms with a third. Two types of information must be input by each sensor to the fusion processor. These are: (1) how well, or with what confidence, the sensor has been able to detect, classify, or identify the objects in the field of regard, and (2) the position and accuracy to which the object has been located. With this information, a fusion algorithm can be designed to combine the sensor data and improve upon the classification and tracking decision made by any one sensor acting alone. In fact, the efficiency of sensor-level fusion for detecting, classifying, and identifying targets is equal to the efficiency of central-level fusion when the sensors derive their information from independent signature-generation processes [20].

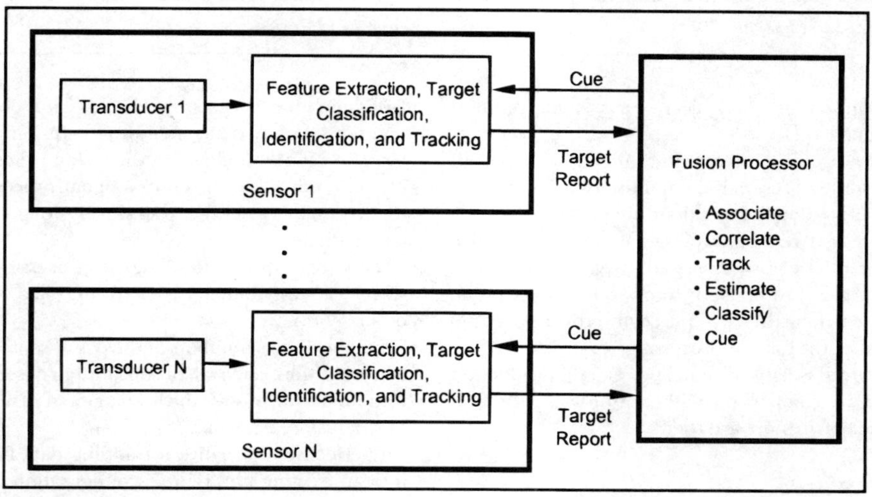

Figure 2.5 Sensor-level fusion.

Sensor-level fusion can provide cueing information to further optimize individual sensor performance. For example, if a region of high clutter is found, a signal may be sent to the appropriate sensor to increase the threshold setting. Alternatively, when the fusion processing identifies a decoy, a message describing the decoy's location may be sent to minimize target search-related signal processing in this region. Still another application of cueing is to initiate a search of a small, but high interest region using a sensor of limited field of regard having high resolution, such as a laser radar [22]. Central-level and sensor-level fusion are compared in Table 2.4.

2.2.3 Hybrid Fusion

In hybrid fusion (see Figure 2.6), the central-level fusion process is supplemented by individual sensor signal processing algorithms that may, in turn, provide inputs to a backup sensor-level fusion algorithm as well. Hybrid fusion allows the tracking benefits of central-level fusion to be realized using the minimally processed sensor data inputs and, in addition, allows fusing of target tracks as computed by the individual sensors in a sensor-level fusion process. Hybrid fusion can also be used to support target attribute classification when the sensor data are not truly generated by independent phenomena. The main disadvantage of hybrid fusion is the increased processing complexity and potentially increased data transmission requirements.

Table 2.4
Advantages of Central-Level and Sensor-Level Fusion

Central-Level Fusion	*Sensor-Level Fusion*
• Optimization of object track and position estimates • More accurate object discrimination than with sensor-level fusion if the multisensor data are not generated by independent phenomena • Increased reliability of signal processing hardware if fewer processors are used overall to support the fusion algorithms; reliability can be increased further, if required, by providing redundant paths for the processing • Reduced weight, volume, power, and production cost in comparison with sensor-level fusion if fewer processors are used	• Discrimination among potential targets or objects of interest before data entry into the fusion processor, to reduce the load on the fusion processor • Optimization of each sensor's signal processing to the nuances of the transducer design and kinematics • Cueing to adjust sensor signal processing or search area parameters based on data from other sensors • Flexibility in the numbers and types of sensors to allow addition, removal, or substitution of sensors without having to alter the fundamental structure of the fusion algorithm • Cost-effective alternative for adding data fusion into an existing multisensor configuration

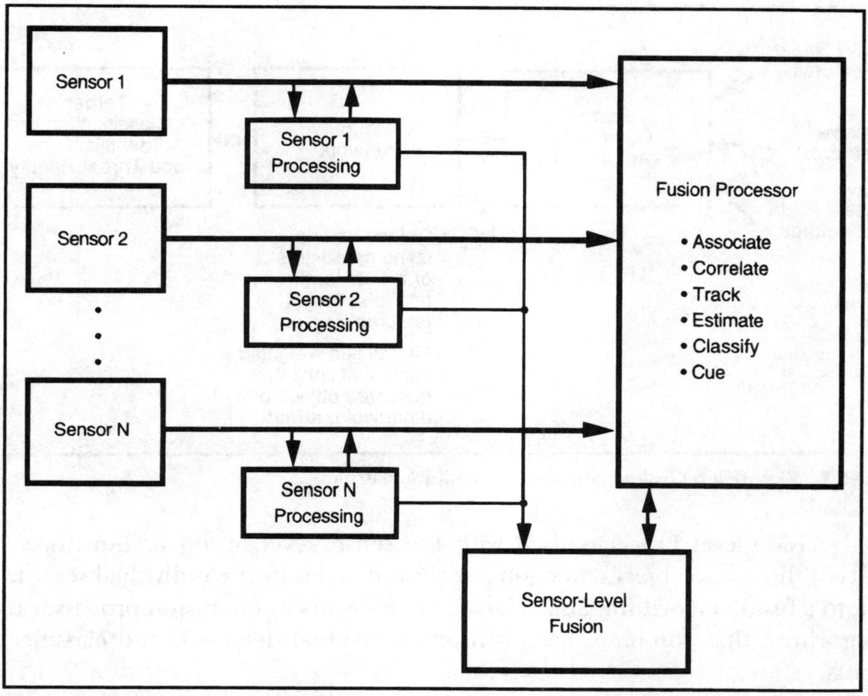

Figure 2.6 Hybrid fusion.

2.2.4 Pixel-, Feature-, and Decision-Level Fusion

In pixel-level fusion, minimally processed data are combined at the pixel or resolution cell level of the sensors and are used in a central-level fusion architecture. For example, pixel-level fusion is applied to Landsat imagery to detect diseased crops or identify a particular crop. Identification is not made using the individual spectral bands of data, but rather the information from all bands is combined in a pixel-level fusion process before the scene is identified.

In feature-level fusion, features that help discriminate among objects are extracted from each sensor's data and then are combined to form a composite feature vector representative of the object in each sensor's field of view. Feature-level fusion is found either in central-level or sensor-level fusion architectures. As an example of a composite feature, consider one constructed by stringing individual sensor feature vectors end-to-end (concatenation) to form a longer vector that serves as the input to a classifier. Classification may occur using microprocessor-based algorithms or in a multilayer neural network as shown in Figure 2.7 [23]. Here features from a millimeter-wave radar, passive infrared sensor, and laser radar are combined to form a composite vector that is input to the neural network. The network, trained off-line to recognize the objects of interest and differentiate them from false targets, produces a weighted value that links the observed object to a particular target with some probability, confidence, or threat priority.

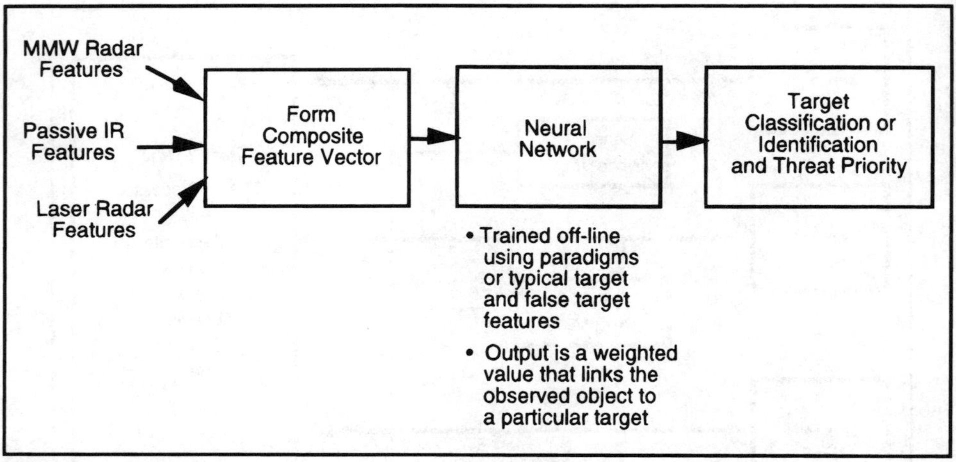

Figure 2.7 Feature-level fusion with a neural network classifier.

Decision-level fusion is used with the sensor-level fusion architecture. The results of the initial object detection and classification by the individual sensors are input to a fusion algorithm. Final classification occurs in the fusion processor using an algorithm that combines each sensor's individual detection and classification outputs.

2.2.5 Sensor-Level Fusion for Automatic Target Recognition

An example of a multiple sensor system configuration suitable for automatic target recognition in military applications is given in Figure 2.8. For illustration, MMW radar, MMW radiometer, and passive and active IR sensors are shown. In this sensor-level fusion architecture, each sensor processes its data with algorithms that are

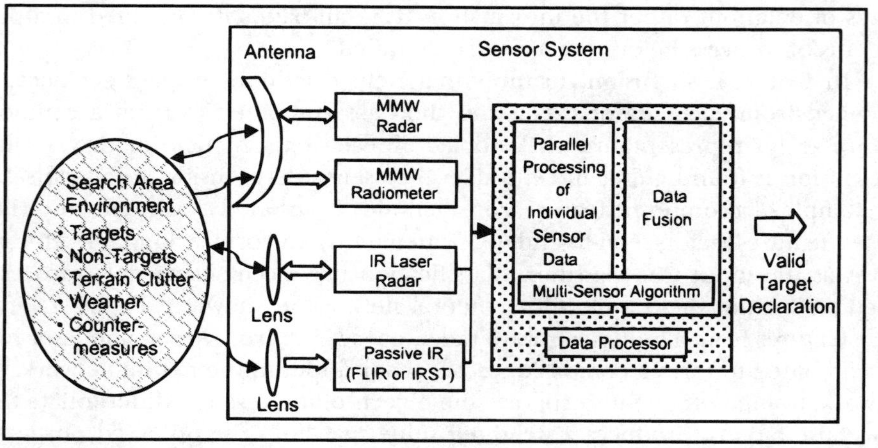

Figure 2.8 Sensor-level fusion concept for automatic recognition of objects using multiple-sensor data.

tailored and optimized to the received frequency band, active or passive operating mode of the sensor, sensor resolution and scanning characteristics, target and background signatures, etc. Results of the processing by the individual sensors are forwarded to a fusion processor where they are combined to produce a "validated target" or "no desired target encountered" decision. If track estimation is the desired output of the multiple sensor system, then centralized fusion may prove to be more optimal in producing accurate tracks in many applications. Those data that are correlated are used to form new tracks, update existing tracks, and estimate future positions of the objects [17].

2.3 BENEFITS OF MULTIMODE, MULTISPECTRAL SENSOR SYSTEMS

A quantitative argument can be made for the use of multiple sensor systems to improve the detection probability of suppressed signature targets as shown in Figure 2.9. The lower curve gives the detection probability for a single radar sensor, operating with a false alarm probability (P_{fa}) of 10^{-6}, as a function of signal-to-noise ratio. Detection probability, false alarm probability, and signal-to-noise ratio are interrelated as discussed further in Chapter 4.

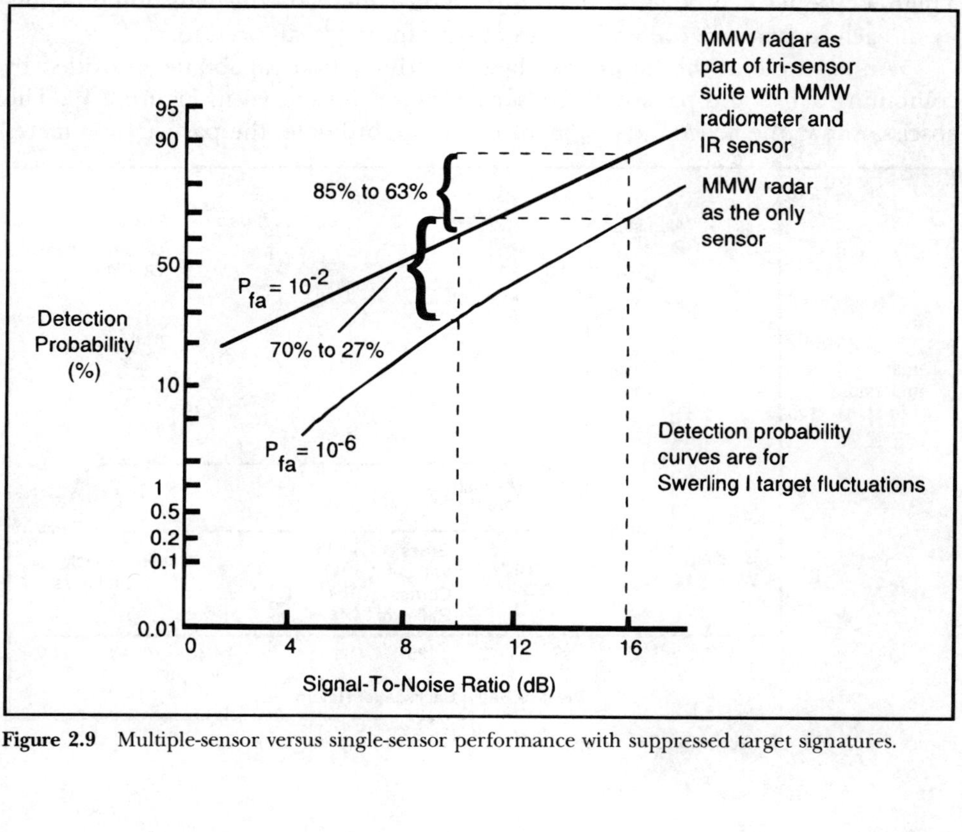

Figure 2.9 Multiple-sensor versus single-sensor performance with suppressed target signatures.

Assume that the detection probability of 0.7, obtained when the false alarm probability is 10^{-6} and the signal-to-noise is nominal at 16 dB, is adequate for the radar operation. When the target signature is reduced, however, and the signal-to-noise ratio decreases to 10 dB, the detection probability falls to 0.27, generally not acceptable for radar system performance. If, however, the radar is one of three sensors that detect the target, where each sensor responds to unique signature generation phenomena and does not generally false alarm on the same events as the others, then the false alarm rejection can be distributed among the three sensors. The system false alarm probability of 10^{-6} is recovered later in the fusion process when the data are combined, for example, through a voting fusion algorithm that combines outputs from the series and parallel combinations of the sensors [24]. When the false alarm rejection is able to be divided equally among the sensors, the radar performance shifts to the upper curve marked with the 10^{-2} false alarm probability. Now the nominal target signature (with a signal-to-noise ratio of 16 dB) yields a detection probability of 0.85, but even more importantly, the reduced-signature target (with a signal-to-noise ratio of 10 dB) yields a detection probability of 0.63, which is 2-1/3 times greater than before. Thus, the use of multiple sensors allows the false alarm rejection to be spread over the signature acquisition and signal processing capabilities of all the sensors and the data combining capabilities of the fusion algorithm, and potentially lets each sensor operate at a higher false alarm probability. The latter, in turn, increases the detection probability of each sensor, especially when target signatures are suppressed.

An example of the improved object discrimination capabilities provided by combining active and passive MMW sensor data is illustrated in Figure 2.10. The abscissa shows the active radar signature and the ordinate, the passive radiometer

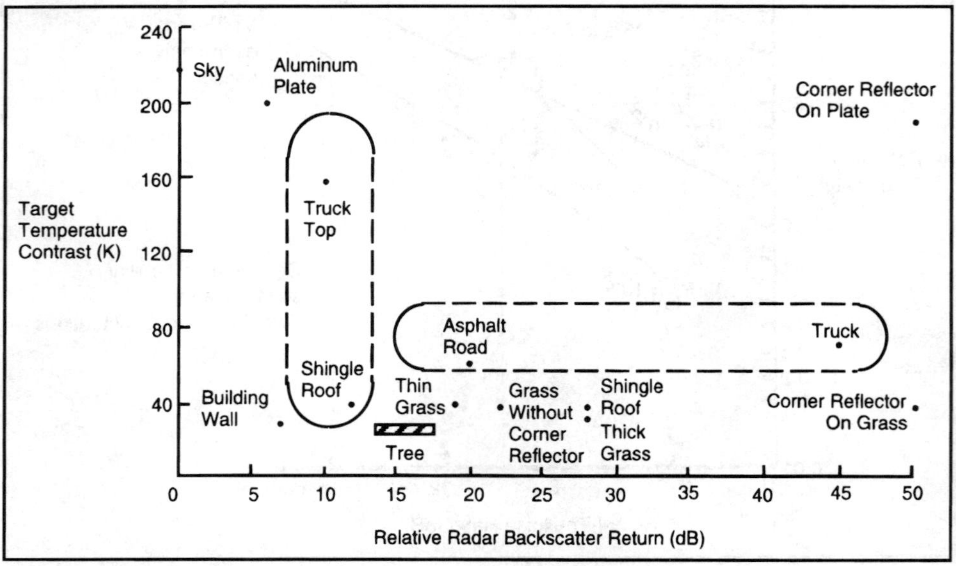

Figure 2.10 Target discrimination using MMW radar and radiometer data [25].

signature. Examination of the truck top and shingle roof signatures on the left of the figure (enclosed by dashed lines) shows that it is difficult to discriminate between them using only radar data, because they have similar radar cross sections and, hence, relative radar backscatter returns. However, if a radiometer is added to the sensor mix, the difference in the two objects' signatures stands out as shown on the vertical target temperature contrast scale. By contrast, if only a radiometer is available, it is difficult to distinguish between an asphalt road and a truck, as shown in the dashed region on the right of the figure. However, the radar now adds the discriminating signatures, making object differentiation possible [25].

Multiple sensors also act synergistically when operating in high-clutter environments and in adverse weather. A mission using a sensor, such as a MMW radar, that may be hampered by the high clutter of snow is aided in detecting targets by a passive sensor that is not similarly affected. This synergy is illustrated later in Chapter 6. On the other hand, a mission employing an IR sensor that may be impaired by dust, clouds, or fog is aided by the MMW radar in detecting targets under these conditions.

An active sensor can enhance the performance of passive high resolution IR imagers by providing range data that allow the absolute size of objects in the imagery to be calculated. Without the range information, the passive IR sensor would only be able to provide information about the relative size of the imaged objects. Knowledge of absolute object size, however, generally allows automatic target recognition systems to function at higher detection probabilities and lower false alarm rates as illustrated in Figure 2.11 [26]. The bottom curve shows the detection and false alarm parameters that are achieved when only a passive imaging IR (IIR) sensor is used to detect ground-based military targets. Range data from an independent, spatially registered laser radar allow higher detection probability with fewer false alarms as shown in the next curve to the left. The combination of the passive IR and range data improves the performance further. Range and Doppler data from the laser radar have the potential to enhance performance still further when moving targets are present and their speed is such that they can be easily differentiated from stationary background objects and clutter.

Another example of the synergy of multiple sensors is the information they provide about the location of a potential target's vulnerable area. A passive MMW radiometer supplies data to compute the centroid of the object that can potentially be used to select a target aimpoint for a precision guided munition. A high-resolution FLIR can provide data to locate the boundary of an object and a region of warmer temperatures within that area. With suitable knowledge about the targets, the warmer region can be inferred to belong to the area over the engine, which is an ideal aimpoint for a munition. This imagery, as well as the passive MMW centroid data, allow the aimpoint to be located within the boundary of the object and avoids simple hot-spot detection that can declare a "false aimpoint" (e.g., from tracking hot exhaust gases) located outside the physical area of the target.

Benefits from multiple sensor systems also accrue from their ability to defeat countermeasures deployed to make a sensor ineffective either by jamming,

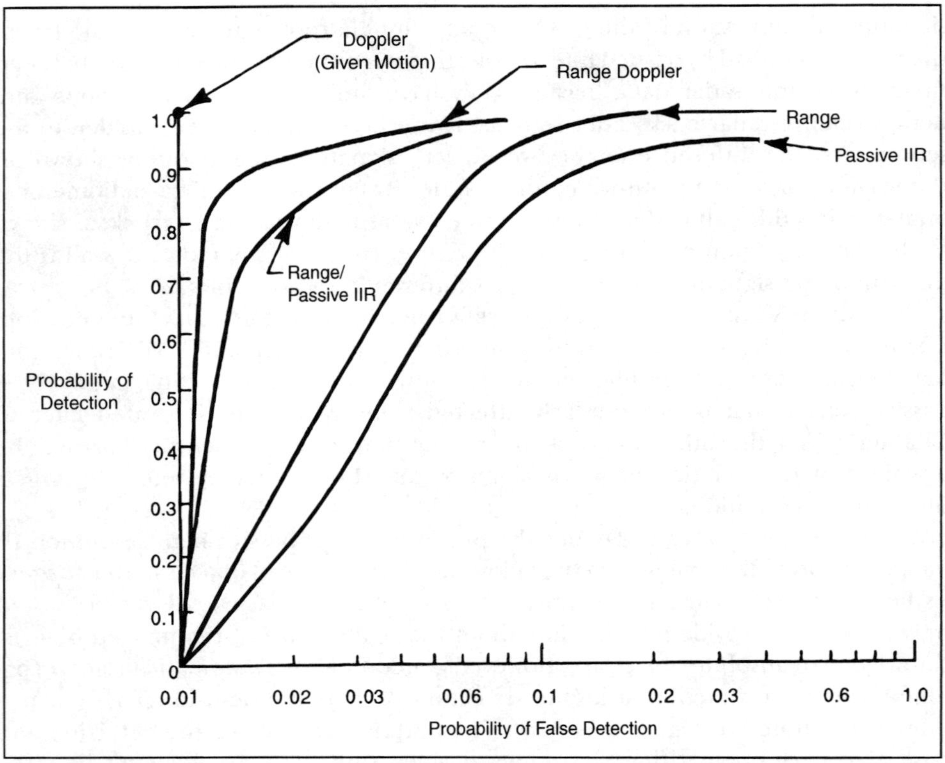

Figure 2.11 Detection probability and false alarm dependency on number and types of sensors used in an automatic target recognition system. (*Source:* [26]. Reprinted with permission.)

reducing, or eliminating a signature generation mechanism at select frequencies, or by mimicking target signatures that pull the sensor off the true target track. Multiple sensors either completely or partially defeat these countermeasures by exploiting target signature phenomena that are not countered or by driving up the cost of the countermeasure by requiring it to be more complex to replicate target signatures over a wide spectral band in the active and passive signature domains. An example of such a complex countermeasure is a tank decoy that utilizes camouflage paint to make visible images look authentic, heaters to simulate the IR signature of the engine, and crude transmitters to produce radar signals that emulate a tank signature. In this case the approximately $23,000 cost of the countermeasure is small in comparison to the greater than one-million cost for a real tank [27].

The robustness of a combined MMW and IR sensor is advantageous where smoke countermeasures may impair the operation of the IR sensor, but leave the MMW sensor unaffected. Conversely, chaff and corner reflectors that may degrade the operation of the MMW sensor are not effective against the IR sensor.

A two-color IR sensor (one that detects energy emitted in two distinct wavelength bands such as the near and mid IR) is also effective in differentiating real targets from countermeasures (such as flares), burning objects (such as fuel or targets previously engaged by munitions), and sunglint (the mimicking of a target signature by reflection of sunlight into the sensor's aperture) [28]. The false targets usually contain emissions in both bands, whereas the real target emits energy in the mid-band region (3 to 5 μm) only as shown in Figure 2.12.

2.4 RESOLUTION VERSUS WAVELENGTH

Electromagnetic sensor resolution increases as the wavelength decreases or, conversely, as the frequency increases as shown in Figure 2.13. In this illustration, the aperture and effective range of the sensors are held constant at 8 inches (20 cm) and 5 km, respectively, as the operating frequency increases from microwave through visible. As frequency increases, atmospheric constituents and obscurants impede the performance of the sensor to a greater degree. Some active MMW sensors also become more difficult and expensive to design and build as the frequency increases because high output power sources and low noise figure mixers and amplifiers become more scarce, generally because of lower demand for components at these frequencies.

2.5 EFFECTS OF OPERATING FREQUENCY ON AVAILABILITY AND PERFORMANCE OF MMW SENSOR HARDWARE COMPONENTS

A summary of the major influences of operating frequency on MMW sensor design parameters and components is shown in Table 2.5. The first three are aperture

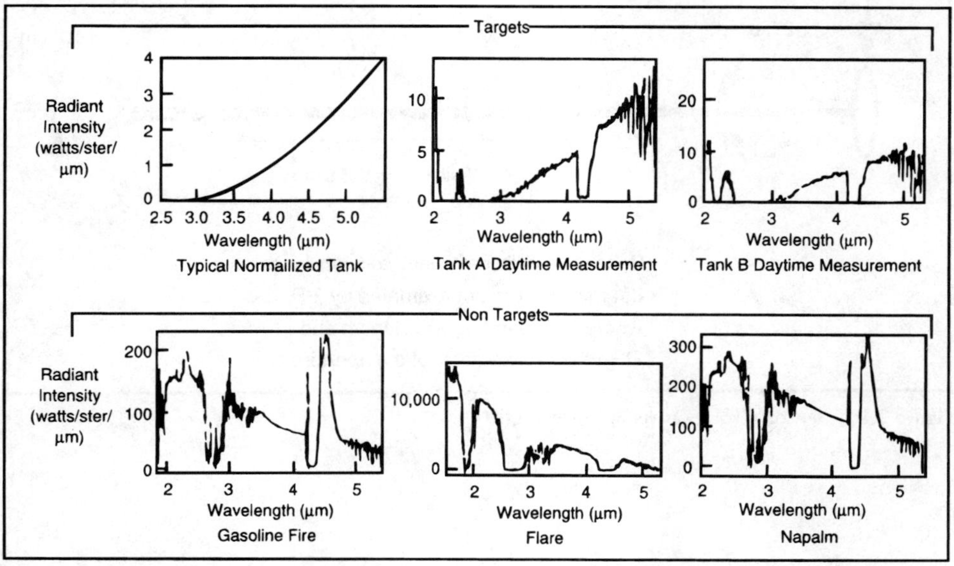

Figure 2.12 Target discrimination capability of two-color IR sensor [28].

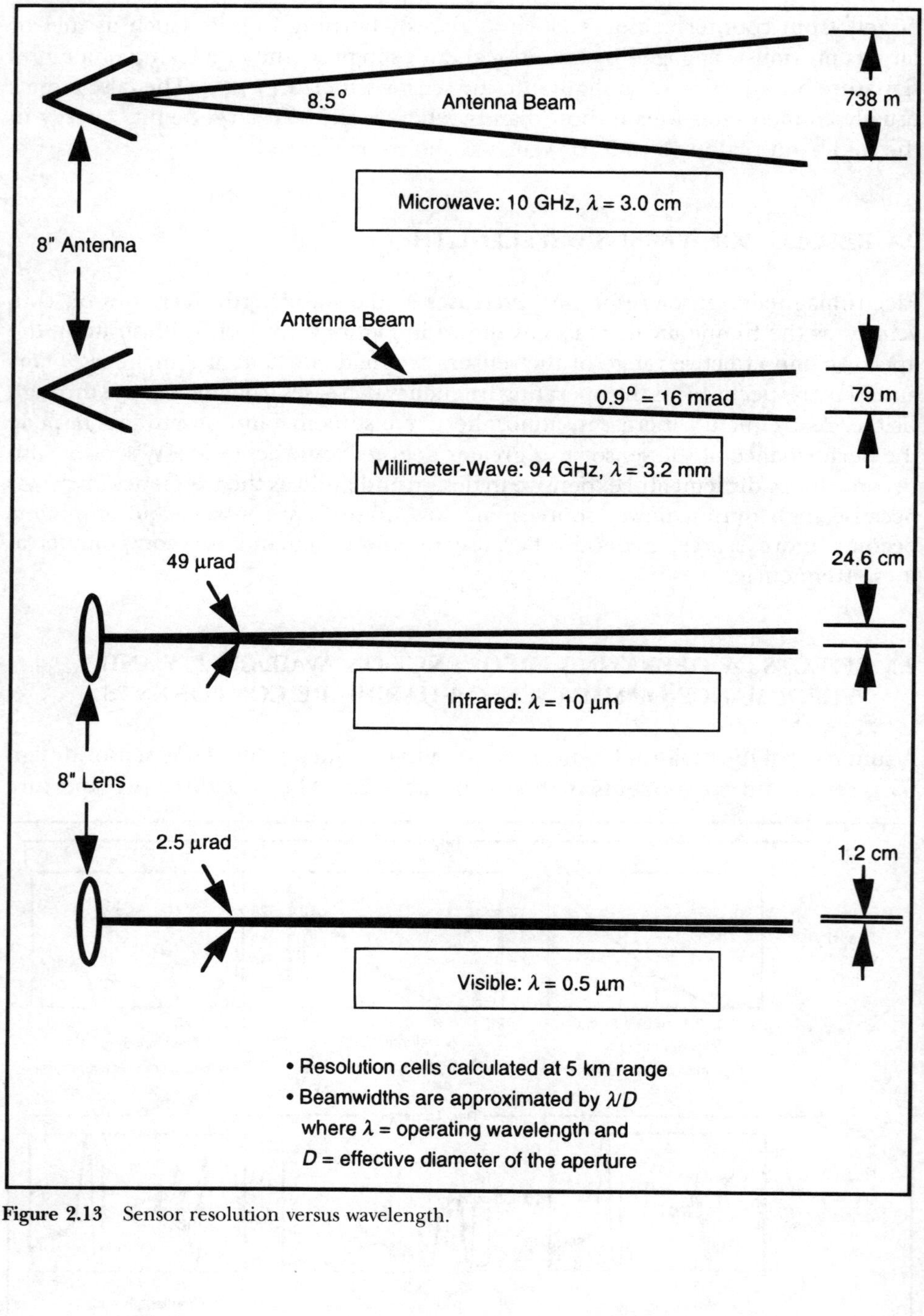

Figure 2.13 Sensor resolution versus wavelength.

Table 2.5
Influence of MMW Frequency on Sensor Design Parameters

Parameter	*Effect of Higher Frequency*
Aperture	Higher gain
Pointing accuracy	Smaller error (standard deviation)
Clutter cell size	Smaller
Attenuation in air	Generally higher
Attenuation in rain and fog	Higher
Power available	Generally less
Component size	Smaller
Receiver noise figure	Generally higher
Integrated components in production	Less likely

related. At a fixed aperture, a higher operating frequency reduces the antenna beamwidth and increases resolution. Increased resolution increases pointing accuracy and reduces clutter cell size, but may adversely affect the ability to search large areas within an acceptable time. The search time is a function of the inverse relation between sensor resolution and field of view (higher resolution, smaller field of view), or equivalently, the direct relationship between resolution and scan rate (higher resolution implies higher scan rate to search a given area in the same allotted time). Atmospheric attenuation in clear weather and in rain and fog also increases as frequency increases. More detailed descriptions of the effects of the atmosphere and obscurants are provided in Chapter 3.

Average power outputs from gallium arsenide (GaAs) Impact Avalanche and Transit Time (IMPATT) diodes operating at 10 GHz and silicon (Si) IMPATT diodes operating at 100 GHz have grown at approximately 3 dB/decade for the past five to ten years [29]. The increase in power available from solid state millimeter-wave power sources is illustrated by the data in Figure 2.14 for IMPATT and Gunn diodes [30]. Indium phosphide (InP) IMPATTs are capable of higher frequency operation than GaAs IMPATTs due to their higher saturation velocity and smaller diffusion coefficient. The InP IMPATTs produce higher output power because of their higher operating voltage made possible by a higher breakdown voltage. The InP diodes also have a higher conversion efficiency by approximately 50 percent because of their higher peak-to-valley ratio. Gunn diode output power is about an order of magnitude less than that from IMPATTs. However, the amplitude modulation noise produced by Gunn diodes is about 10 to 30 dB less than from IMPATTs. Therefore, Gunn diodes are the component of choice in local oscillators and other applications requiring low noise. In amplifier applications, noise from InP Gunn diodes is approximately 5 dB less than that from GaAs Gunn diodes.

Monolithic microwave integrated circuit (MMIC) Gunn diode voltage-controlled oscillators (VCOs) have been developed to provide high volume, low cost power sources for the precision guided munition market. One such design is given

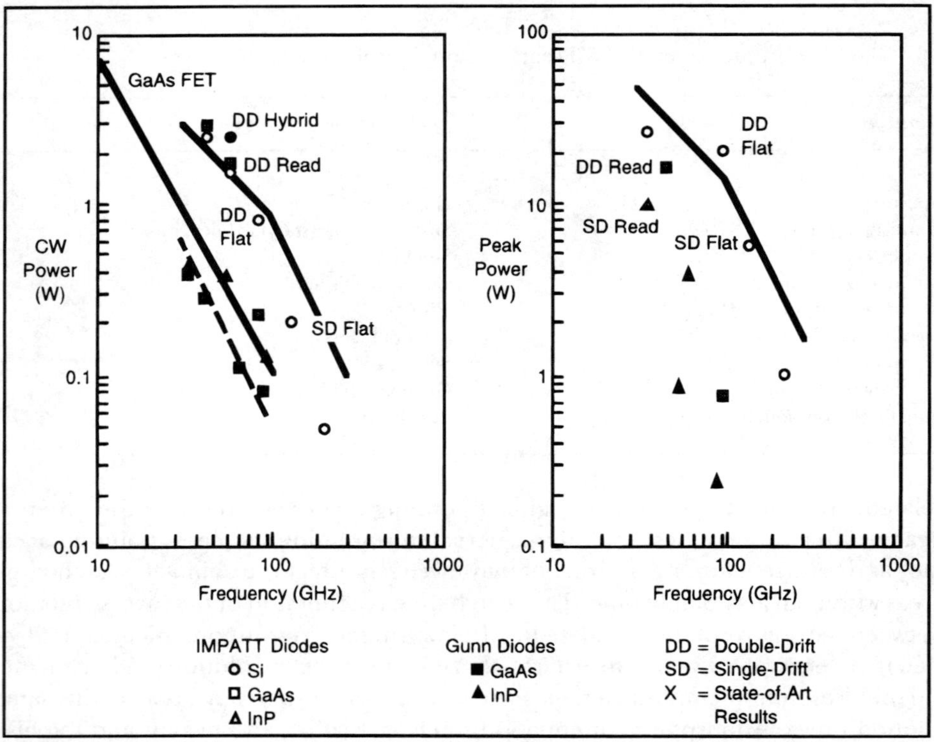

Figure 2.14 Millimeter-wave solid state IMPATT and Gunn diode power sources. (*Source:* [30]. Reprinted with permission.)

in Figure 2.15 [31]. Here the combined output power from fourteen Gunn diodes (located on an arc surrounding the power combiner element) produces a higher power device. The voltage applied to a varactor diode, grown along with the Gunn diodes to the left of center in Figure 2.15(a), controls the frequency at which the power is output and thus enables the MMIC VCO to produce a frequency-modulated

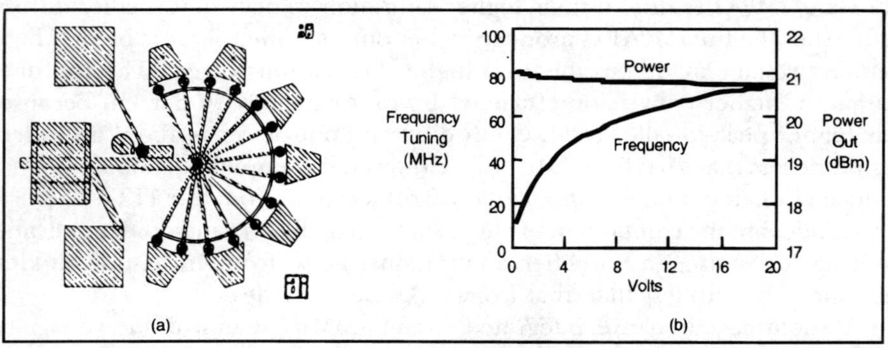

Figure 2.15 (a) MMIC Gunn diode voltage-controlled oscillator circuit with (b) its output power and tuning bandwidth. (*Source:* [31]. © 1988 IEEE. Reprinted with permission.)

continuous wave (FMCW) signal that is useful in range finding and target detection applications. The output power is fairly constant over small deviations from the optimum center frequency of operation as shown in Figure 2.15(b). The decrease of power with increasing varactor tuning bandwidth and center frequency is illustrated in Figure 2.16. The decrease of power with increasing center frequency is partially due to the particular design of the VCO which was optimized for 32 GHz operation. By modifying the size or location of features in the integrated circuit layout, the output power can be maximized at other center frequencies of operation.

MMIC VCOs incorporating field-effect transistors (FETs) as the power source have also been designed. They operate over larger tuning bandwidths and with greater efficiency than Gunn diode VCOs, but generally have lower output power when used without power amplifiers. Buffer amplifiers may also be added to the MMIC design to isolate the VCO from antenna loading effects.

An implementation of an MMIC FET VCO with power amplifiers is shown in Figure 2.17 [32]. The curves in the upper left illustrate the influence of temperature on the output frequency and power. Output power is relatively constant with respect to temperature. The center frequency of operation, however, is sensitive to temperature. When the varactor diode voltage is used to control frequency, as shown by the curves in the lower left, the desired result is achieved as the output power remains fairly constant as the voltage or output frequency changes. The transmitter circuit and MMIC chip layout for the VCO and amplifiers are shown in the upper and lower right of the figure, respectively.

Receiver noise figures at 35 and 95 GHz are roughly equal, although they are dependent on the technology used to manufacture the mixer diodes, e.g., microwave integrated circuit (MIC) technology that incorporates discrete components versus MMIC. Mixers are commonly used to convert the input frequency signal at the radio frequency (RF) port into a lower frequency signal at the output or intermediate frequency (IF) port. The operation of mixers is discussed further in Chapter 4. Noise figures are larger at still higher frequencies. Since higher frequency technologies generally appear in fewer applications, there are currently less high frequency components available in integrated circuit designs.

An example of a two-inch wafer containing over 150 potential 35 GHz MMIC mixer chips is shown in Figure 2.18 [33]. Three-inch wafers increase the yield by somewhat less than the ratio of the wafer areas because of the increased number of incomplete circuits found along the edge of the larger wafer. The mixer contains two Schottky-barrier diodes arranged in a 180-degree rat-race hybrid ring as shown in the lower part of the figure. For effective mixer operation, it is desirable for the isolation between the local oscillator (LO) port and the RF port to be large. This is achieved in the MMIC mixer where the isolation is greater than 20 dB over a 20 percent bandwidth. The noise figure is less than 5 dB at room temperature as illustrated in Figure 2.19 for 35 GHz MMIC mixers from two manufacturers [33,34].

Programs such as the U.S. government-sponsored Microwave/Millimeter-Wave Monolithic Integrated Circuit (MIMIC) program have spurred the development

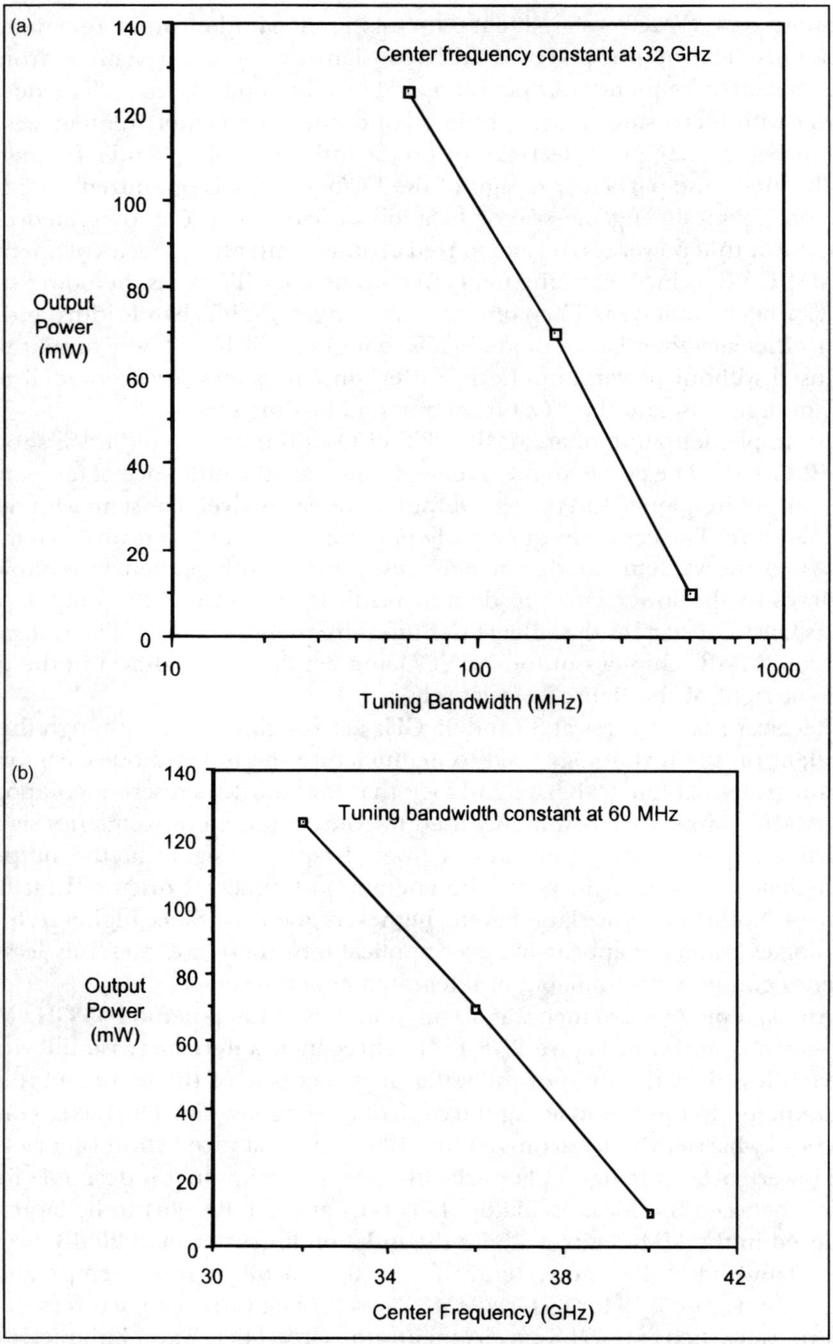

Figure 2.16 MMIC Gunn VCO output power as a function of tuning bandwidth and center frequency: (a) plotted as semi-log and (b) plotted as linear [31].

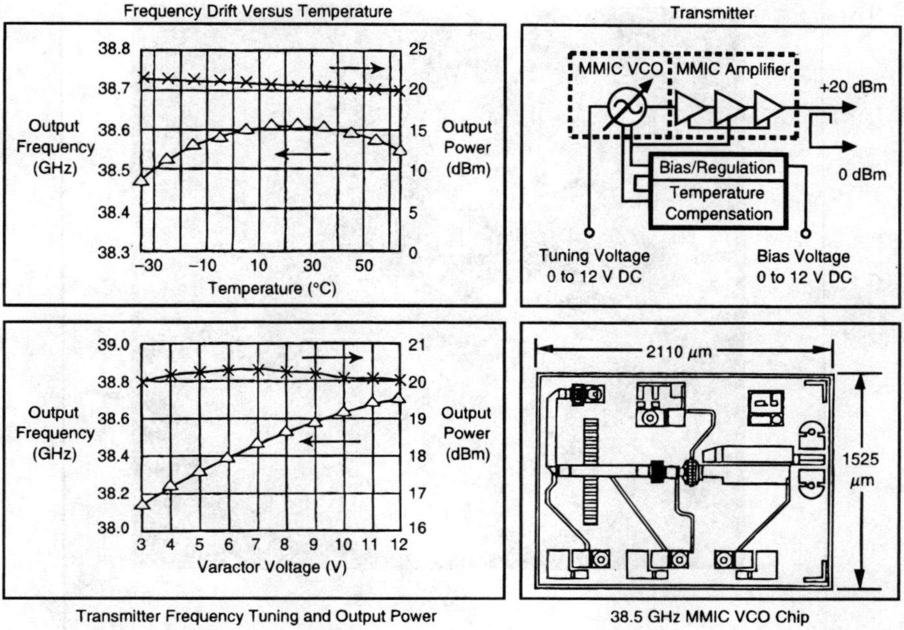

Figure 2.17 MMIC FET VCO and amplifier circuit design showing output power and tuning bandwidth versus temperature and tuning voltage. (*Source:* [32]. Reprinted with permission.)

of GaAs integrated circuitry in military and civilian applications at microwave and MMW frequencies. The Phase 1 projects focused on the development or improvement of GaAs material manufacturing, computer-aided design of the circuits used on the chips, and chip packaging and testing [35,36]. Phase 2 sought further improvements to the performance and yield of the devices as summarized in Table 2.6 [37].

Smaller MIMIC Phase 3 contracts were awarded to various companies and universities to further MMIC development through projects such as on-wafer chip characterization, advancing multichip ceramic packaging, improving models for predicting process yield, automatic testing of chip parameters, improving the processes used to grow and dope GaAs substrates, and advancing the computer-aided design models used to manufacture chips and optimize their layout on the wafers [38].

Other efforts, such as the Manufacturing Methods and Techniques (MMT) programs, have focused on increasing the yield of higher output power MMIC devices. Early results from a 56 and 94 GHz InP Gunn diode MMT program are displayed in Figure 2.20. The goals of this project were to maximize the yield of the diodes at maximum power and efficiency [39]. The output powers are consistent with those shown in Figure 2.14.

2.6 IR THERMAL AND QUANTUM DETECTORS

Most infrared detectors used in electro-optical applications are either thermal detectors or quantum detectors. Thermal detectors have larger time constants (i.e.,

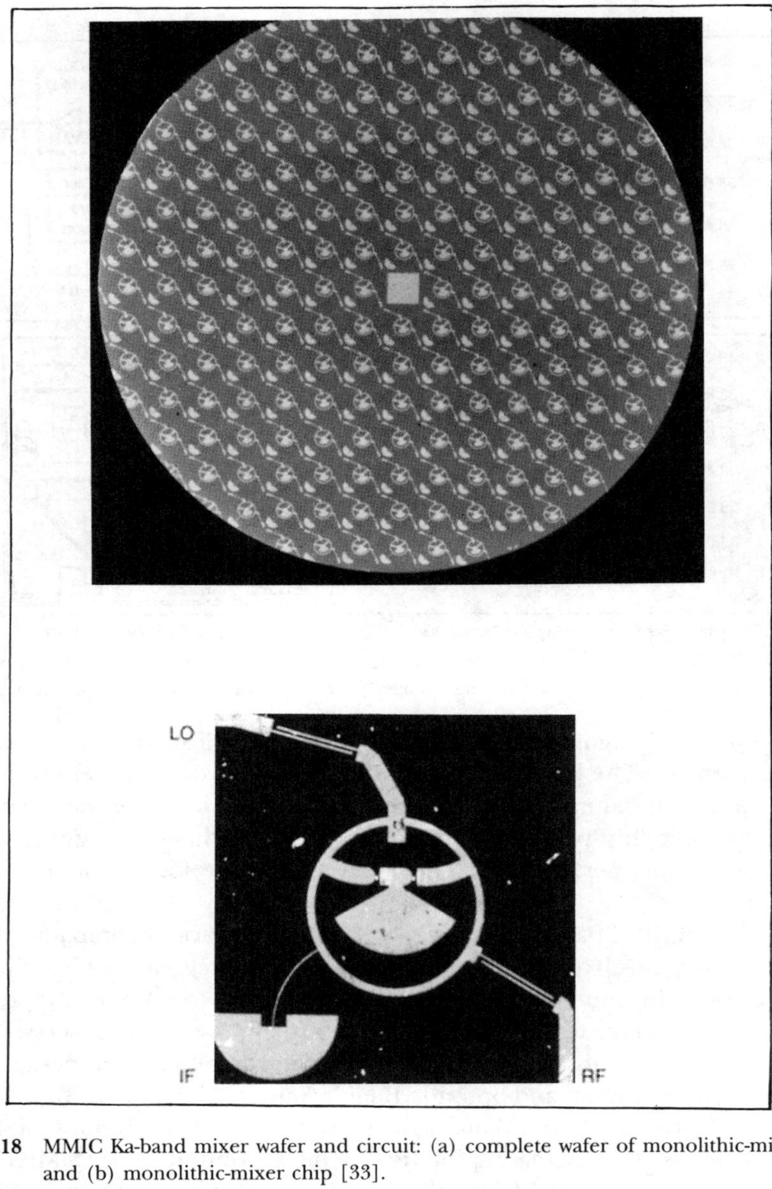

Figure 2.18 MMIC Ka-band mixer wafer and circuit: (a) complete wafer of monolithic-mixer circuit and (b) monolithic-mixer chip [33].

a slower response) than quantum detectors, but are sensitive to optical energy over a broader wavelength band. Thermal detectors include thermistors, thermocouples, thermopiles, bolometers, and pyroelectric devices.

Thermistors are composed of a temperature-sensitive resistor made of metallic oxides or semiconductors. They have a high temperature coefficient of resistance that minimizes heating caused by current flow in the detector.

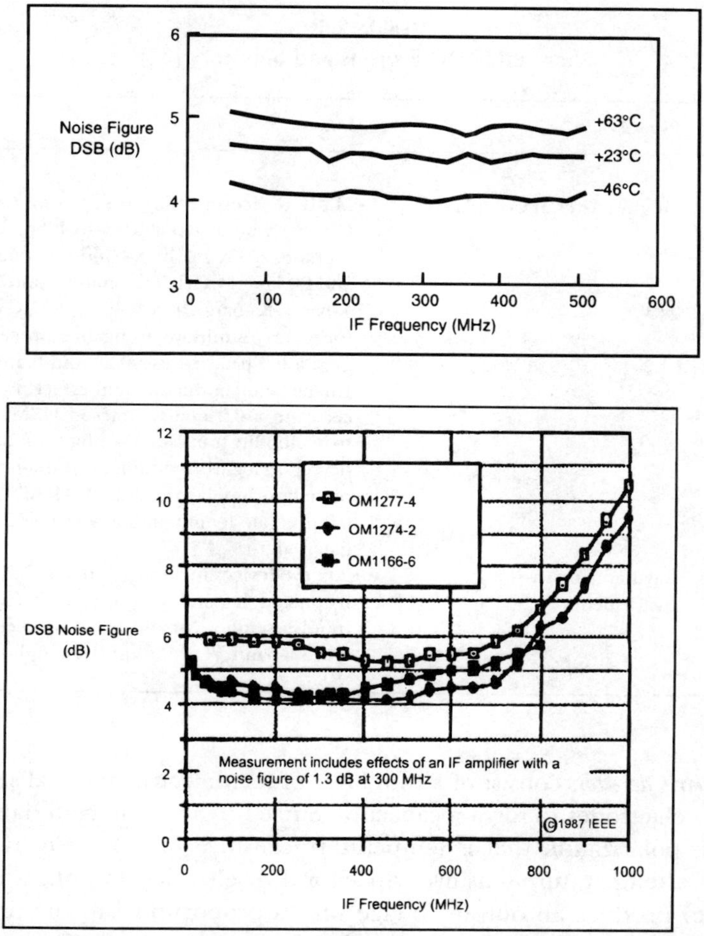

Figure 2.19 35 GHz MMIC mixer noise figures as a function of IF bandwidth and operating temperature from different manufacturers. (*Source:* [33,34]. Reprinted with permission.)

Thermocouples are composed of two junctions of dissimilar metals where each junction operates at a different temperature. Metals and semiconductors may be used to manufacture thermocouples, with semiconductors making the more sensitive junction. When thermocouples are used as optical radiation detectors, one junction is exposed to the radiation and the other is shielded from it. Since the output of a thermocouple is small, the output is increased by connecting several in series forming a *thermopile.*

Bolometers are constructed from temperature-sensitive resistors (thermistors or metallic detectors). Their output is measured as the voltage across a load resistor in series with the thermistor and a bias voltage.

Table 2.6
Phase 2 MIMIC Projects and Sponsors [37]

Sponsor	*Team Leader*	*Projects*
Air Force	Hughes Aircraft Company	• MMIC-device insertion on airbone radar systems. Circuits produced include a six-bit phase shifter for a phased-array radar, X-band low noise amplifier, and 20 and 44 GHz transmitters and receivers.
Army	TRW	• X-band, Ka-band, and W-band MMIC components for smart munitions using five process technologies: a 0.5 μm process; 0.25 and 0.5 μm MESFET (metal semiconductor field effect transistor) processes in a 0.2 quarter-micron HEMT (high electron mobility transistor) configuration; a successive detection logarithmic amplier, built-in HBT (heterojunction bipolar transistor), HEMT process with 0.2 μm gate length; and a power FET with 0.5 μm gate length.
Navy	Raytheon-Texas Instruments	• 26 new devices including wideband amplifiers and mixers, high gain-high power TWT (traveling wave tube) assembly, X-band transmit/receive module, and Ka-band VCO and buffer amplifier assembly.

Pyroelectric detectors consist of a thin layer of ferroelectric material sandwiched between two electrodes to form a capacitor. Since ferroelectric materials exhibit a spontaneous polarization that is temperature sensitive, the pyroelectric detector functions as a temperature-sensitive capacitor. Pyroelectrics do not require a bias voltage. They produce an output voltage that is proportional to the time rate of change of the temperature and not to the temperature itself. The response of pyroelectrics can be faster than from thermistors, thermocouples, thermopiles, and bolometers.

Quantum detectors include photoemissive, photoconductive, and photovoltaic materials. Since photon energy is inversely proportional to wavelength, the minimum photon energy required to produce a charge carrier determines the maximum wavelength to which a quantum detector will respond. Response to minimum wavelengths (higher energy photons) is limited by the ability of the detector material to absorb or stop the photons. Absorption efficiency of most materials decreases as the photon energy increases. The maximum and minimum constraints on the wavelengths absorbed by a quantum detector restrict quantum detector operation to a relatively narrow spectral region.

Photoemissive devices have a high surface work function, requiring absorbed photons to possess sufficient energy to produce free electrons that escape from the material. Thus photon detection consists of converting the photons into emitted

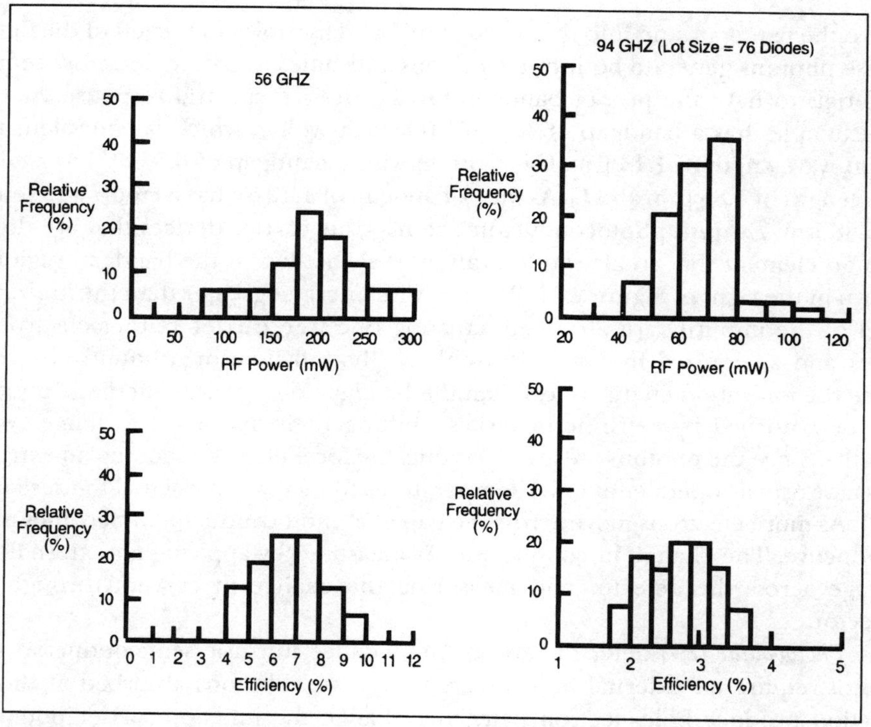

Figure 2.20 InP Gunn diode MM&T yield. (*Source:* [39]. Reprinted with permission.)

electrons. The photoemitter is typically enclosed in a vacuum tube and responds to relatively high energy photons at visible and ultraviolet wavelengths.

Photoconductive (PC) detectors are high impedance semiconductors whose conductance is increased when absorbed photons produce free charge carriers in the material. The electron need not escape from the PC semiconductor altogether when a photon is absorbed. Photoconductive quantum detectors can be intrinsic or extrinsic devices as illustrated in Figure 2.21. *Intrinsic* PC semiconductors absorb photons with enough energy to cause an electron to jump from the valence band,

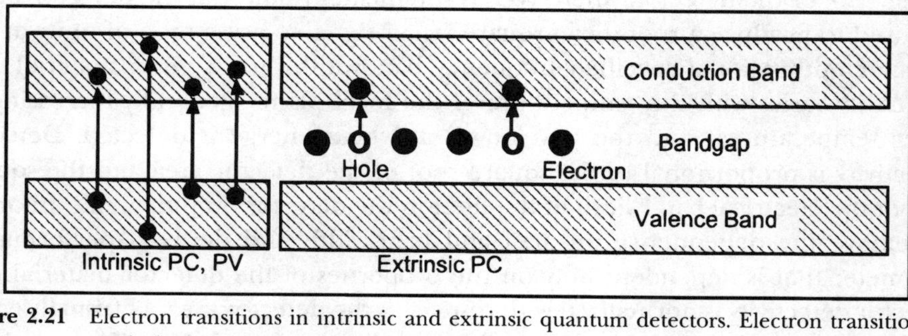

Figure 2.21 Electron transitions in intrinsic and extrinsic quantum detectors. Electron transitions in extrinsic detectors require less energy, allowing them to operate at longer wavelengths.

across the bandgap, and into the conduction band as shown in the left of the figure. These photons generate both free electrons and holes. Intrinsic detectors require materials to have the proper bandgap for the desired spectral response. Silicon, for example, has a bandgap of 1.11 eV (electron volts), which is equivalent to a cutoff wavelength of 1.13 μm. Germanium with a bandgap of 0.66 eV has a cutoff wavelength of 1.88 μm, and GaAs with a bandgap of 1.42 eV has a cutoff wavelength of 0.88 μm. *Extrinsic* photoconductors are majority carrier devices that are doped with an element that creates an ionization level that lies in the bandgap region as shown in the right of Figure 2.21. Photons with energies greater than the ionization level of the impurity are absorbed, creating one free carrier (either electron or hole) and an ionized impurity site (either a bound hole or a bound electron). Since the ionization energy is less than the bandgap energy, less energetic photons can be absorbed by extrinsic materials, shifting their operation to longer wavelengths. Since the photons are less energetic, the focal plane containing an extrinsic detector usually operates at lower temperatures to increase the signal-to-noise ratio [40]. As more electrons move across the gap of a photoconductor, it becomes more conductive. The change in conductance is measured by applying an external bias voltage across the detector and measuring the change in current through the detector.

Photovoltaic (PV) detectors are low impedance intrinsic semiconductors that do not require an external bias voltage to operate. Photons absorbed at the *p-n* junction produce hole-electron pairs that change the junction-barrier potential. This changes the open circuit voltage of the detector and provides a measure of the incident energy. As with a photoconductor, the spectral sensitivity of a photovoltaic detector is determined by the energy levels in the semiconductor material.

2.6.1 Detector Detectivity

One of the primary parameters used to characterize detector materials is their detectivity D_d. The detectivity is defined as the reciprocal of the noise equivalent power *NEP* of the detector, where *NEP* is the incident power at the detector input required to produce a root mean square (rms) signal-to-noise ratio of unity at the detector output for a specified frequency bandwidth. For a given material, the detectivity varies with the temperature of the focal plane (generally increasing at lower temperatures) and the wavelength at which energy is detected. Detector detectivity is proportional to the square root of the detector area and the square root of the electrical bandwidth of the filter in the detector output circuit. In order to remove the dependence on area and bandwidth and derive a performance parameter that is dependent only on the properties of the detector material, the detector detectivity is normalized with respect to the detector area A_D (usually equal to 1 cm^2) and electrical bandwidth ΔF (usually equal to 1 Hz). The resulting parameter, specific detector detectivity D_d^*, is a function of wavelength and focal plane temperature and is defined as

$$D_d^* = \frac{(A_D hr\,\Delta F)^{1/2}}{NEP}\,\frac{\text{cm Hz}^{1/2}}{\text{W}} \qquad (2.1)$$

When thermal radiation from the background is greater than that from another signal source or from internally generated noise, the detector input is dominated by the background radiation and the detector is said to be operating as a background-limited infrared photodetector (BLIP). In BLIP performance, the background photon source is the dominant source of noise. Equations that describe the *NEP* under these conditions are discussed in Chapter 7.

The operating parameters of thermal detectors, and intrinsic and extrinsic quantum detectors are given in Tables 2.7 through 2.9, respectively [41]. The time constant, operating mode (photoconductive or photovoltaic) of the quantum detectors, focal plane temperature, wavelength at which the peak value of D_d^* occurs, usable wavelength region, and D_d^* are listed. The thermal detectors are slower to respond to changes in the temperature of objects in their field of view because of their longer time constant. They are, however, sensitive to photon emission over a large wavelength region and operate at room temperatures, in contrast to the quantum detectors. The specific detectivities of a number of quantum detectors are plotted in Figures 2.22 and 2.23 as a function of wavelength. Figure 2.22 contains data on photoconductive materials such as the lead salts, lead selenide (PbSe) and lead sulfide (PbS), and the tertiary compound mercury-cadmium-telluride (HgCdTe). The D_d^* value is dependent on the field of view and background temperature used in the measurement as indicated in the upper right portion of the figure. The incident photon flux on the detector is large for the 2π steradian field of view and the 295K background temperature specified for these D_d^* data. The actual D_d^* value for a detector enclosed in a well designed sensor can be significantly

Table 2.7

Operating Parameters of Thermocouple, Thermopile, Bolometer, and Pyroelectric Thermal Detectors

Detector Type	*Focal Plane Temperature (K)*	*Time Constant (ms)*	*Usable Wavelength Region (μm)*	D_d^* *at 500 K (cm Hz$^{1/2}$ W^{-1})*
Thermocouples	300	10 to 100	0.8 to 20	$\leq 1.5 \times 10^9$
Thermopiles	300	0.5 to 400	0.8 to 35	$<1.5 \times 10^9$
Bolometers	300	1 to 100	0.8 to 40	$<1.5 \times 10^9$
Pyroelectric	300	0.1 to 100	0.6 to 35	$\leq 8 \times 10^8$

Source: Stahl, K.J., "IR-Detectors: State-of-the-Art, Future Trends," *Photonics Spectra*, Sept. 1989, pp. 95–98.

Table 2.8
Operating Parameters of Intrinsic Quantum Detectors [41]

Detector Type	*Operating Mode*	*Focal Plane Temperature (K)*	*Time Constant (μs)*	*Peak Wavelength (μm)*	*Usable Wavelength Region (μm)*	D^*_λ *at 500 K (cm Hz$^{1/2}$ W^{-1})*
Si	PV	300	0.01 to 0.5	0.9	0.6 to 1.1	$\approx 10^7$
Si	Avalanche	300	$\approx 10^{-5}$	0.9	0.6 to 1.1	$>2 \times 10^7$
Ge	PV	300	$\approx 10^{-5}$	1.5	0.9 to 1.9	5×10^9
PbS	PC	300	>100	2.4	1.1 to 3.5	10^{11}
PbS	PC	195	500	2.8		$\approx 5 \times 10^{11}$
InAs	PV	77	0.5	3.1	1.8 to 3.8	2×10^{10}
GaAs		295	1	0.8	0.4 to 1.1	$\approx 10^{14}$
PtSi	Schottky barrier	<90	$\approx 10^3$	0.9	0.8 to 5.0	2×10^9
PbSe	PC	300	1 to 5	3.9	1 to 5.0	$\approx 10^{10}$
PbSe	PC	195	10 to 40	4.8	2 to 5.0	3.5×10^{10}
InSb	PV	77	0.1 to 2	4.9	2 to 5.5	7×10^{10}
HgCdTe	PV	195	≈ 2	4.5	1 to 5.5	4×10^{10}
HgCdTe	PC	77	0.1 to 2	10.5	8 to 14	2.5×10^{10}
HgCdTe	PV	77	0.5 to 2	10.5	8 to 11.5	2×10^{10}
PbSnTe	PV	77	≈ 0.5	11	8 to 11.5	2×10^{10}

Table 2.9
Operating Parameters of Extrinsic Quantum Detectors

Detector Type	*Focal Plane Temperature (K)*	*Time Constant (ns)*	*Peak Wavelength (μm)*	*Usable Wavelength Region (μm)*	D^*_λ *at 500 K (cm Hz$^{1/2}$ W^{-1})*
Si:Zn	70 to 110	10 to 50	2.4	0.9 to 2.5	8×10^9
Si:Te	<120	50 to 100	3.5	0.9 to 4.2	2×10^{10}
Si:In	≤50	≈100	4.6	0.9 to 7	2×10^{10}
Si:Mg	<50	≈5	11.5	0.9 to 12	5×10^{10}
Si:Ga	<30	≈50	15	0.9 to 17	2.5×10^{10}
Si:Bi	<30	1 to 10	17.5	0.9 to 22	2.5×10^{10}
Si:As	<30	≈100	24	0.9 to 28	1×10^{10}
Ge:Cu	4.2	1 to 10	21	1.5 to 27	1×10^{10}
Ge:Zn	4.2	0.5 to 5	37	1.5 to 45	2.5×10^{10}
Ge:Be	4.2	0.1 to 10	50	1.5 to 65	BLIP
Ge:Cd	4.2	1 to 50	90	1.5 to 100	BLIP
Ge:In (stressed)	4.2	1 to 10	110	1.5 to 240	BLIP

Source: Stahl, K. J., "IR-Detectors: State-of-the-Art, Future Trends," *Photonics Spectra,* Sept. 1989, pp. 95–98.

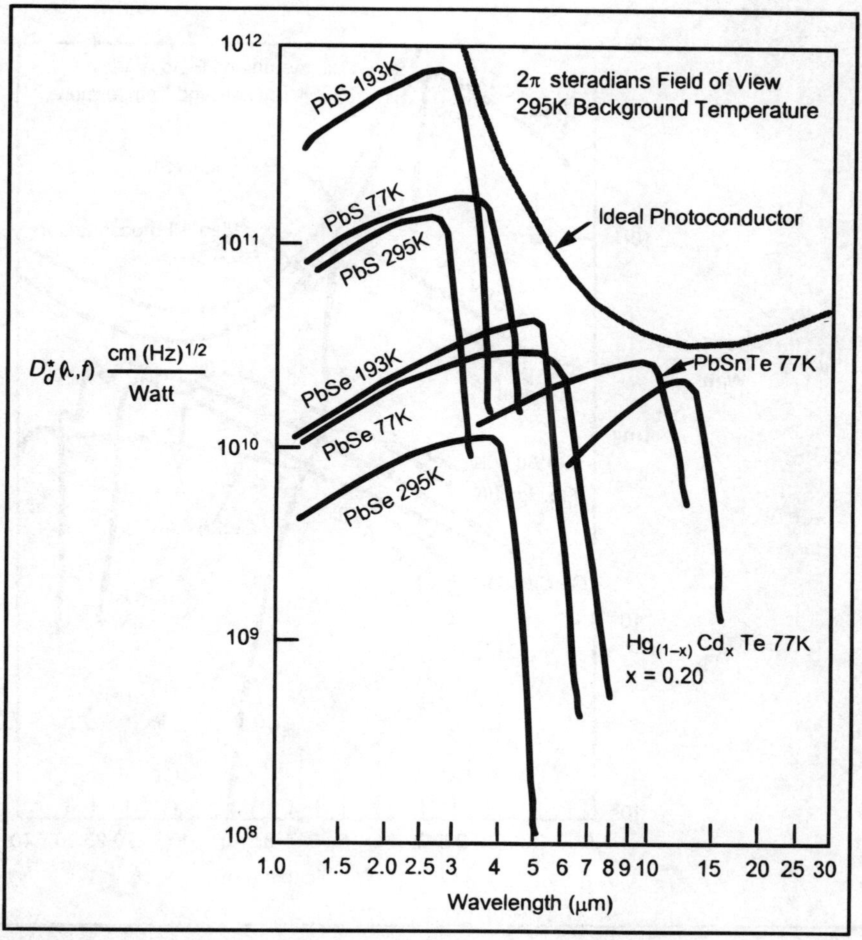

Figure 2.22 Specific detector detectivity of photoconductive quantum detector materials. (*Source:* Hughes Santa Barbara Research Center, Goleta, CA. Reprinted with permission.)

greater than that shown in Figures 2.22 and 2.23 if the detector is enclosed in a cold shield to limit photons absorbed from background radiation [42,43]. However, maximizing D_d^* with the use of a cold shield may degrade other performance parameters of the sensor, for example its ability to search larger areas in a given time because of a reduction in the instantaneous field of view. Figure 2.23 contains D_d^* data for photoconductive, photovoltaic, and extrinsic detector materials. The specific detector detectivities for an ideal photoconductive and photovoltaic quantum detector are also illustrated for the same background temperature and field of view. The theoretical maximum realizable detectivity for a photovoltaic detector is 40 percent greater than for a photoconductive detector. Figure 2.24 illustrates the decrease of D_d^* that generally occurs with increasing focal plane temperature and shows the temperature threshold effect exhibited by many quantum detectors [43].

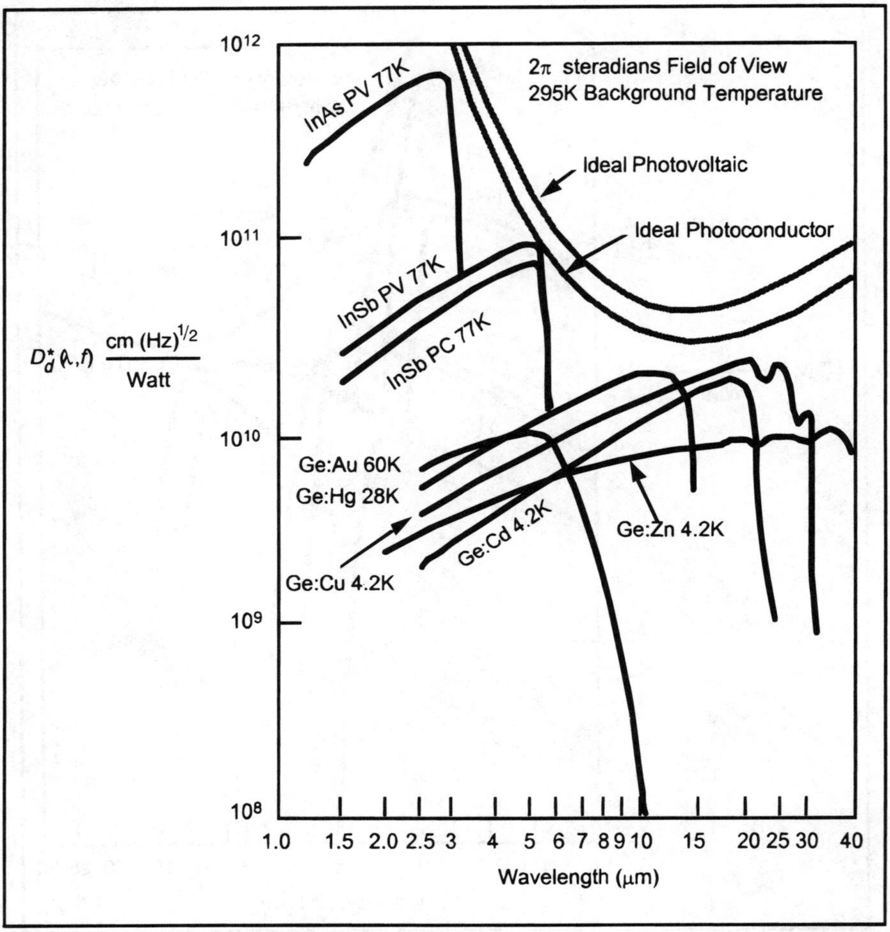

Figure 2.23 Specific detector detectivity of photoconductive and photovoltaic quantum detector materials. (*Source:* Hughes Santa Barbara Research Center, Goleta, CA. Reprinted with permission.)

This produces the almost instantaneous drop in D_d^* that appears when the temperature of the detector rises above the threshold value. Tradeoffs between D_d^* and time constant (that decreases at warmer temperatures) can also be made in selecting the focal plane operating temperature.

2.6.2 Detector Material Trade for Dual Band Focal Plane Arrays

Suppose that there is an automatic target recognition or missile launch and trajectory detection requirement for a staring IR sensor to operate in the far- and mid-infrared wavelength bands with low power dissipation, uniformity of detector element response, and low-risk production. The materials under consideration at this

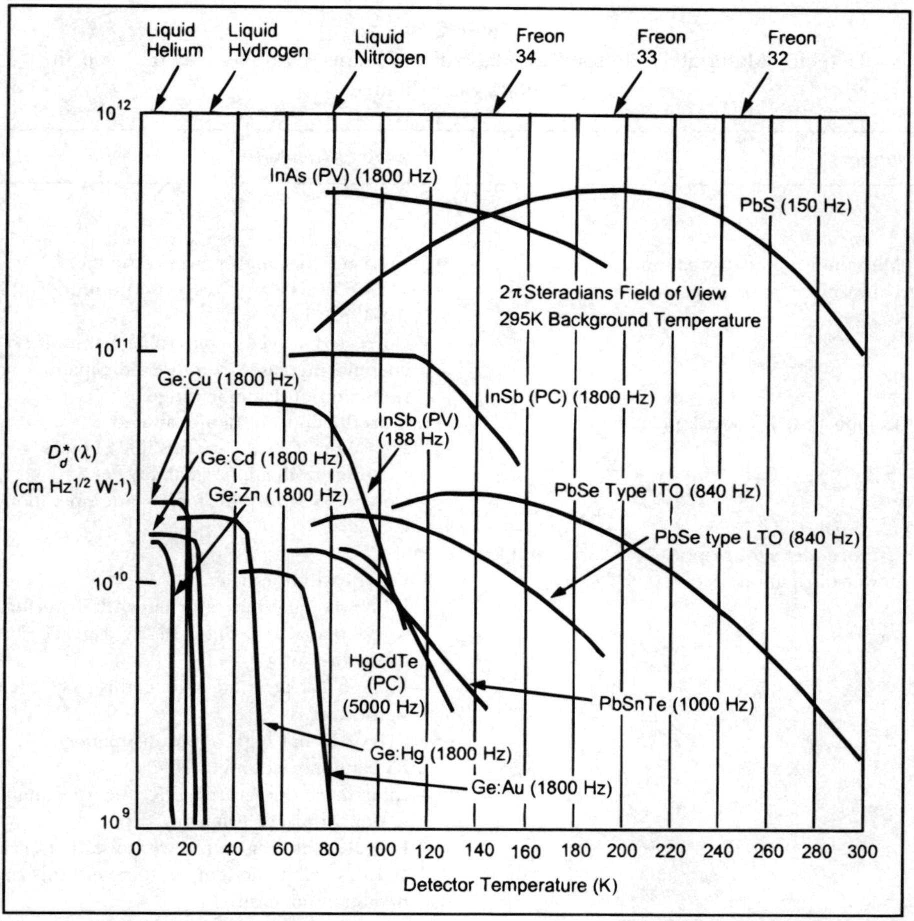

Figure 2.24 Specific detector detectivity as a function of focal plane temperature. (*Source:* [43]. Reprinted by permission of John Wiley & Sons, Inc.)

time may include PC HgCdTe, PV HgCdTe, IrSi (iridium silicide), and extrinsic silicon (Si:x) for the far-infrared band, and PtSi (platinum silicide), PC HgCdTe, and InSb (indium antimonide) for the mid-infrared band. All of these materials have some drawback associated with their selection as shown in Table 2.10. Among these are higher power dissipation than desired, nonuniform response over the array, production limitations, and noise and sensititivity issues. Other materials will likely be considered for the same application as they are developed and mature.

2.6.3 Quantum Detectors Used in Defense Applications

The dominant types of IR quantum detector materials found in defense applications are shown in Figure 2.25. Mercury-cadmium-telluride is the most widely used,

Table 2.10
Detector Material Trade for Dual-Band Focal Plane Arrays as May be Used in Military Surveillance

Requirement	*Materials and Properties*
8–12 μm band	
• Minimum power dissipation	• PC HgCdTe has high power dissipation
• Uniform detector response over array	• PV HgCdTe has large response nonuniformities in 10–12μm band —Corrected with large bias voltage, but with concurrent increase in power dissipation and reduction of dynamic range
• Demonstrated, low-risk production	• IrSi production not demonstrated • PC Si:Ga or PC Si:As are possible choices —Existing production capability —Coolers exist for 10K focal plane operation
3.5 μm band	
• Uniform detector response over array and low power dissipation	• PtSi —Low quantum efficiency —With low quantum efficiency, PtSi performs worse than InSb, Si:In, or PV HgCdTe by a factor of 6 or 7 —In high background clutter, this may not be significant • PC HgCdTe has high power dissipation • InSb band gap not adjustable —InSb must operate at 10K due to common dewer with 8–12 μm array —Requires filtering to reduce noise that occurs at 15K focal plane temperature with this large band gap material
Dual band	• As technology matures, PV HgCdTe may be a potential material for use in both bands

followed by extrinsic silicon, InSb, PtSi, and other materials that include the lead salts [44]. Detectors used in specific defense programs are listed in Table 2.11 along with the imaging technique (scanning or staring), wavelength band, and focal plane temperature [45].

2.6.4 Limits on IR Sensor Design and Performance

Passive IR sensors provide high resolution in a relatively compact package and, since they detect energy that is naturally emitted by their surroundings, their operation is covert. Furthermore, they can provide spectral coverage flexibility through detector material selection and lens filter coatings. However, they are more subject to degradation from atmospheric effects, such as fog, smoke, and air

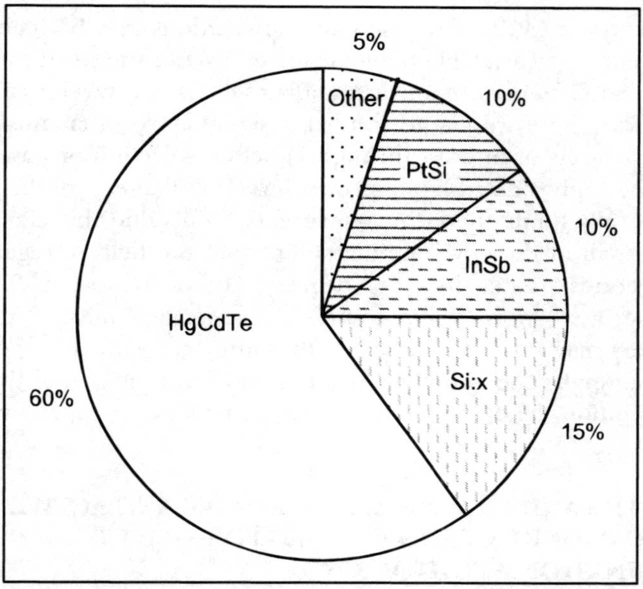

Figure 2.25 IR detector material usage in defense applications. The category "Other" includes PbSe, PbS, superconductor, and super lattice detectors [44].

Table 2.11
Defense Applications of Focal Plane Array Detectors

Program	*Strategic/ Tactical*	*Scan/ Stare*	*Application*	*Material*	*Wave-length Band*	*Focal Plane Temp. (K)*
Abrams M-1	Tactical	Scan	Tank	HgCdTe	LWIR	77
AGB	Tactical	Scan	Missile launch	HgCdTe	LWIR	77
AGM-130	Tactical	Scan	Missile seeker	HgCdTe	LWIR	77
AH-64B Apache	Tactical	Scan	Helicopter	HgCdTe	LWIR	77
B-1B	Tactical	Scan	Aircraft	HgCdTe	LWIR	77
Boost Surv & Tkg	Strategic	Both	Space surv	HgCdTe,InSb	MWIR	120
Bradley M-2	Tactical	Scan	Vehicle	HgCdTe	LWIR	77
Chaparral (MIM-72)	Tactical	Stare	Missile seeker	HgCdTe	LWIR	77
F-14D	Tactical	Stare	Aircraft	HgCdTe	LWIR	77
F-22	Tactical	Scan	Aircraft	HgCdTe	LWIR	77
Hellfire	Tactical	Scan	Missile seeker	HgCdTe	LWIR	77
High Endo Interceptor	Tactical	Scan	Missile seeker	HgCdTe,InSb	MWIR	77
Javelin	Tactical	Stare	Missile seeker	HgCdTe	LWIR	77
LANTIRN	Tactical	Scan	Aircraft	HgCdTe	LWIR	77
LHX	Tactical	Scan	Aircraft	HgCdTe	LWIR	77
Maverick	Tactical	Stare	Missile seeker	HgCdTe	LWIR	77
Sidewinder (AIM-9)	Tactical	Stare	Missile seeker	PtSi,InSb,HgCdTe	MWIR	77
Stinger	Tactical	Stare	Missile seeker	PtSi,InSb,HgCdTe	MWIR	77
TOW 2B	Tactical	Scan	Missile launch	HgCdTe	LWIR	77

turbulence, than are MMW sensors. Some applications may be hampered by the properties of currently available detector materials and the need to provide cryogenic cooling. Since the entire universe radiates energy at wavelengths detected by passive IR sensors, there are a large number of signal sources that must be differentiated from the objects of interest through judicious selection of wavelength bands and by applying sophisticated signal processing. The diameter of the sensor design envelope generally limits the clear aperture D_0 to a value that lies between one third and one half of the envelope diameter when the field of regard or angular search dimension is greater than ± 45 degrees. The restriction on D_0 subsequently limits the sensor resolution. When mounted on supersonic missiles, the shock wave boundary layers may produce emissions that interfere with the operation of the sensor. Finally, weight and volume allocation may limit the size of the sensor and, hence, its resolution, field of regard, and signal processing capability.

2.7 STRENGTHS AND WEAKNESSES OF PASSIVE AND ACTIVE ELECTROMAGNETIC SENSORS USED FOR OBJECT DISCRIMINATION AND TRACKING

Resolution, weather, day/night operation capability, clutter, and countermeasures influence the choice of particular electromagnetic sensors for object discrimination and track estimation, as shown in Table 2.12. As frequency is increased, resolution improves and designs are more compact, but degradation by the atmosphere and man-made obscurants increases. Furthermore, the ability to rapidly search large areas can degrade because of the decreased field of view of the sensor. Active sensors provide range and velocity data, while passive sensors provide stealth operation and angle tracks that can be associated with targets.

2.8 SUMMARY

The attributes of active and passive sensors in the microwave, millimeter-wave, and infrared portions of the electromagnetic spectrum have been enumerated to illustrate the advantages they bring to a high-performance, multisensor suite in defense and civilian applications. Several sensor and data fusion architectures have been discussed to show the ways in which multiple sensor data may be combined to enhance target detection, classification, and tracking. These techniques include fusing the raw sensor data in a centralized fusion algorithm, processing the data fully in each sensor and then combining the results using an autonomous fusion algorithm, or using combinations of these approaches. Data may also be fused at the pixel, feature, or decision level depending upon the resolution and registration of the sensors. The trades in selecting MMW and IR sensor operating frequencies have been discussed to show the implications on resolution, hardware availability and specifications, and operating frequency compatibility with the expected

Table 2.12
Operating Characteristics of Electromagnetic Sensors

Sensor	*Advantages*	*Disadvantages*
• Microwave/ millimeter-wave radar	• All weather operation • Lower frequencies penetrate foliage • Large search area • Day/night operation • Range and image data • Velocity data with coherent system	• Moderate resolution • Not covert • Simpler radar designs exhibit more susceptibility to corner reflector decoys and active jammers
• Microwave/ millimeter-wave radiometer	• Covert imagery, detection, and tracking • All weather operation • Lower frequencies penetrate foliage • Large search area • Day/night operation	• Somewhat less resolution than radar for same aperture • Large bandwidth increases susceptibility to jamming • Range data, in theory, by performing a maneuver
• Infrared imager (FLIR)	• Fine spatial and spectral resolution imagery • Covert • Day/night operation	• Affected by rain, fog, haze, dust, smoke • Poor foliage and cloud penetration • Requires cooled focal plane • Large search areas require scan mechanism or large detector array • Range data by performing a maneuver
• Infrared tracker (IRST)	• Hot-spot detection and tracking • Covert target tracking • Compact • Day/night operation	• Same disadvantages as infrared imager
• Laser radar	• Fine spatial and spectral resolution imagery • Range and reflectance data • Velocity and track data • Can be compact • Day/night operation • Countermeasure immune	• Affected by rain, fog, haze, dust, smoke • Poor foliage penetration • Most effective when cued by another sensor to search a relatively small area
• Visible imager	• Best-resolution imager • Covert • Technology well understood	• Day operation only • Affected by clouds, rain, fog, haze, dust, smoke and any other atmospheric obscurants • No foliage penetration • No range data

signatures from the objects of interest and the background in which the sensors operate. For civilian applications, the longer wavelength microwave and millimeter-wave sensors provide cloud penetration and data used in weather forecasting, pollution and earth resource management, and land use monitoring. Multispectral IR imagery provides information that includes land cover and geological features, cloud cover, river expansion from floods, and changes in the ocean ecosystem. The relatively good performance of the active mode microwave and MMW sensors in inclement weather and various countermeasures can be used to complement an IR sensor to provide reliable target detection, range, and tracking for military applications. The higher resolution IR sensors provide imagery used to classify potential military targets and improve the selection of missile impact aimpoints.

References

[1] Goldberg, J. H., "The (Secret) Shape of Things to Come," *National Defense*, July/Aug. 1989, pp. 18–23.

[2] Gagliano, J. A., "Millimeter Wave Radiometric Measurements for Airborne and Ground Based Applications," *SPIE Vol. 791 Millimeter Wave Technology IV and Radio Frequency Power Sources*, 1987, pp. 99–103.

[3] Willis, N. J., *Bistatic Radar*, Artech House, Norwood, MA, 1991.

[4] Aboutalib, A. O., and Luu, T. K., "An efficient target extraction technique for laser radar imagery," *SPIE Vol. 1096: Digital Signal Processing, Association, and Tracking of Point Source, Small, and Cluster Targets*, 1989.

[5] Staelin, D. H., and Rosenkranz, P. W., (editors), *High Resolution Passive Microwave Satellites, Applications Review Panel Report*, Final Report, Research Laboratory of Electronics, Massachusetts Institute of Technology, 1978.

[6] Weissman, D. E., Ed., Special Issue on the SEASAT-1 Sensors, *IEEE J. Oceanic Eng., OE-5*, 1980, pp. 71–181.

[7] Evans, D. L., and Smith, M. O., "Separation of Vegetation and Rock Signatures in Thematic Mapper and Polarimetric SAR Images," *Remote Sensing of Environment 37*, 1991, pp. 63–75.

[8] Salu, Y., and Tilton, J., "Classification of Multispectral Image Data by the Binary Diamond Neural Network and by Nonparametric, Pixel-by-Pixel Methods," *IEEE Trans. on Geos. and Remote Sensing*, Vol. 31, No. 3, May 1993, pp. 606–617.

[9] "Remote Sensing," *IEEE Spectrum*, March 1995, pp. 24–31.

[10] "Multispectral Satellite Sensors," Technote 95-02, *IR Electro-Optics Spectral Reflections*, The Infrared Information Analysis Center, ERIM, April 1995, Vol. 25, No. 2.

[11] Krishnamurthy, S., Sitharama, S., Holyer, R. J., and Lybanon, M., "Histogram-Based Morphological Edge Detector,"*IEEE Trans. on Geos. and Remote Sensing*, Vol. 32, No. 4, July 1994, pp. 759–767.

[12] Yhann, S. R., and Simpson, J. J., "Application of Neural Networks to AVHRR Cloud Segmentation," *IEEE Trans. on Geos. and Remote Sensing*, Vol. 33, No. 3, May 1995, pp. 590–604.

[13] Waltz, E., and Llinas, J., *Multisensor Data Fusion*, Artech House, Norwood, MA, 1990.

[14] Klein, L. A., *Sensor and Data Fusion Concepts and Applications*, Vol. TT 14, SPIE Press, Bellingham, WA, 1993.

[15] Blackman, S. S., "Multiple Sensor Tracking and Data Fusion," Chapter 7 in S. A. Hovanessian, *Introduction to Sensor Systems*, Artech House, Norwood, MA, 1988.

[16] Comparato, V. G., "Fusion-The Key to Tactical Mission Success," *SPIE Vol. 931: Sensor Fusion*, 1988, pp. 2–7.

[17] Blackman, S. S., "Association and Fusion of Multiple Sensor Data," Chapter 6 in *Multitarget-Multisensor Tracking: Advanced Applications,* Editor: Y. Bar-Shalom, Artech House, Norwood, MA, 1990.

[18] Data Fusion Subpanel, *Data Fusion Lexicon,* Joint Directors of Laboratories, Technical Panel for C^3, October 1991.

[19] Blackman, S. S., "Theoretical Approaches to Data Association and Fusion," *SPIE Vol. 931: Sensor Fusion,* 1988, pp. 50–55.

[20] Robinson, G. S., and Aboutalib, A. O., "Trade-off Analysis of Multisensor Fusion Levels," *Proc. of The 2nd National Symposium On Sensors And Sensor Fusion* Vol. II, GACIAC PR 89-01, IIT Research Institute, Chicago, 1990, pp. 21–34.

[21] Klein, L. A., and Kassis, S. Y., "Processing Requirements for Multisensor Low-Cost Brilliant Munitions," *IEEE Trans. on Aerospace and Electronic Systems,* Vol. AES-29, No. 4, Oct. 1993, pp. 1084–1094.

[22] Luo, R. C., and Kay, M. G., "Multisensor Integration and Fusion: Issues and Approaches," *SPIE Vol. 931: Sensor Fusion,* 1988, pp. 42–49.

[23] Castalez, P. F., and Dana, J. L., "Neural Networks in Data Fusion Applications," *Proc. of The 2nd National Symposium On Sensors And Sensor Fusion,* Vol. II, GACIAC PR 89-01, IIT Research Institute, Chicago, 1990, pp. 105–114.

[24] Klein, L. A., "A Boolean Algebra Approach to Multiple Sensor Voting Fusion," *IEEE Trans. on Aerospace and Electronic Systems,* Vol. AES-29, April 1993, pp. 1–11.

[25] Foiani, D. L., and Pearce, R. H., "Combined Radar and Radiometer at Millimeter Wavelengths," *Symposium on Submillimeter Waves,* Polytechnic Institute of Brooklyn, March 1970.

[26] Delashmit, W. H., "Present Status and Future Needs for Automatic Target Recognition," *SPIE Vol. 1075: Digital Image Processing Applications,* 1989.

[27] "Decoys: Tanks but No Tanks," *Time,* Feb. 4, 1991, p 25.

[28] Hame, T. G., "Precision Guided Munition Seeker Technology," *AGARD Conference Proceedings No. 320,* AGARD CP-320, May 1982.

[29] Parker, R. K., and Abrams, R. H., Jr., "Radio Frequency Vacuum Electronics: A Resurgent Technology for Tomorrow," *SPIE Vol. 791: Millimeter-Wave Technology IV and Radio Frequency Power Sources,* 1987, pp. 2–12.

[30] Ma, Y. E., "Millimeter-Wave Active Solid-State Devices," *SPIE Vol. 544: Millimeter-Wave Technology III,* 1985, pp. 95–102.

[31] Goldwasser, R., et al., *Proc. IEEE 1988 Microwave and Millimeter-Wave Monolithic Circuits Symposium,* June 1988, pp. 55–58.

[32] Raffaelli, L., and Stewart, E., "A Smart Monolithic Transmitter for 38 GHz PCN Applications," *Microwave Journal,* Oct. 1992, pp. 24–30.

[33] Goldwasser, R., Donoghue, D., Crossley, I., Nash, S., and Fingerman, C., "Millimeter-Wave Monolithic IC in Production," *Journal of Monolithic Technology,* July 1988, pp. 16–19.

[34] Chu, A., Chu, L., Sloat, D., Theobald, M., Teunas, J., Litchfield, T., and Moroney, W., "Low Cost Millimeter Wave Monolithic Receivers," *Proc. IEEE 1987 Circuits Microwave and Millimeter-Wave Monolithic Symposium,* June 1987, pp. 63–65.

[35] Public Affairs News Release, Office of the Assistant Secretary of Defense, May 20, 1988.

[36] Stiglitz, M. R., and Resnick, L., "MMIC Phase 1 Awards: Shocks and Aftershocks," *Microwave Journal,* Sept. 1988, pp. 30–40.

[37] Stiglitz, M. R., and Blanchard, C., "Phase II of the DoD MIMIC Program: A Status Report," *Microwave Journal,* Jan. 1992, pp. 90–100.

[38] *Aerospace Daily,* June 20, 1989, pp. 468–469.

[39] Fank, B., et al., "InP Gunn Diode Sources," *SPIE Vol. 544: Millimeter-Wave Technology III,* 1985, pp. 22–28.

[40] Kruse, P. W., McGlauchlin, L. D., and McQuistan, R. B., *Elements of Infrared Technology: Generation, Transmission, and Detection,* John Wiley and Sons, New York, 1962.

[41] Stahl, K. J., "IR-Detectors: State-of-the-Art, Future Trends," *Photonics Spectra,* Sept. 1989, pp. 95–98.

[42] Limperis, T., and Mudar, J., "Detectors," Chapter 11 in *The Infrared Handbook, Revised Edition*, 3rd Printing, W. J. Wolfe and G. J. Zissis, Editors, Infrared Information and Analysis Center, ERIM, 1989.
[43] Pinson, L. J., *Electro-Optics*, John Wiley and Sons, 1985.
[44] Rotolante, R. A., "Military IR-Detector Market Remains Stable," *Laser Focus World*, Feb. 1991, p. 23.
[45] Sullivan, P. D. (Ed.), *The Design of Electro-Optic Sensors for Military Applications (Draft)*, AIAA, NY, to be published.

CHAPTER 3

Effects of the Atmosphere and Obscurants on Millimeter-Wave and Infrared Sensors

Atmospheric attenuation is produced by two phenomena—absorption and scattering. Absorption is dependent on the frequency of operation and the gases and pollutants that are present. Scattering is dependent on the size, shape, and dielectric constant of the scattering objects and the wavelength of the incident energy. Atmospheric constituents such as oxygen, water vapor, and carbon dioxide play a dominant role in determining MMW and IR attenuation. The internal energy states of these molecules define frequencies at which the molecules absorb energy, thus creating frequency bands of high attenuation. These regions of the electromagnetic spectrum may be used to broadcast short range communications that are intended to be difficult to intercept, and to gather information used for weather forecasting and cloud top location. Relatively low absorption exists at still other portions of the electromagnetic spectrum called windows. Sensors that operate at these frequencies can propagate energy over greater distances for long range target detection and for earth resource monitoring. Weather-related obscurants such as rain, snow, and fog add to the absorption and scattering experienced under clear weather conditions and further limit sensor performance. Models that adequately predict atmospheric absorption and scattering in the MMW and IR spectra may be used when measured data are not readily available at specific frequencies or atmospheric conditions. Both measured and modeled MMW and IR absorption and scattering data are presented and discussed in this chapter.

3.1 INFLUENCE OF WAVELENGTH ON ATMOSPHERIC ATTENUATION

Figure 3.1 compares the atmospheric attenuation of higher resolution IR and visible sensors to millimeter-wave sensors. The oscillatory curve with the peaks and valleys

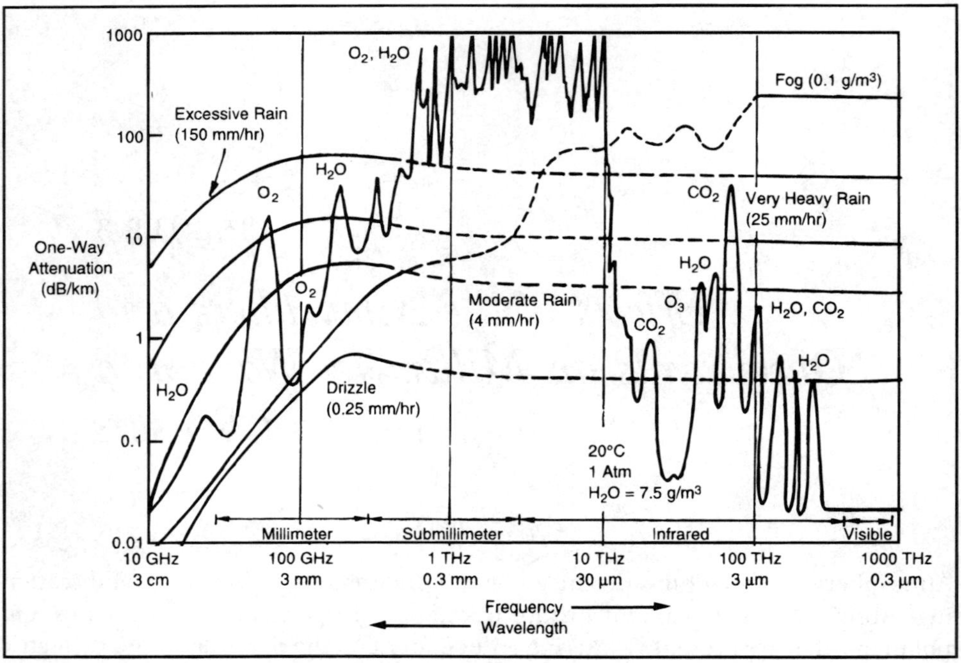

Figure 3.1 Atmospheric attenuation spectrum from 0.3 μm to 3 cm.

in attenuation corresponds to 1 atmosphere of pressure at a temperature of 20°C and water density of 7.5 g/m^3. Window frequencies in the MMW spectrum, denoted by absorption minima, occur at approximately 35, 94, 140, 225, and 350 GHz. Sensors that detect and classify targets at long ranges are typically designed for operation at these windows. Peak absorption occurs in the microwave and MMW spectra at approximately 22, 60, 118, 183, and 320 GHz. Absorption at 60 and 118 GHz is due to oxygen, while absorption at the other frequencies is due to water vapor. The absorption frequency bands find use in sensors flown in weather satellites and applications that require secure communications.

The infrared absorption spectra are due to molecular rotations and vibrations that occur in atmospheric molecules. The near-IR wavelength band extending from 0.77 to 1.5 μm is constrained at the upper end by water vapor absorption. The mid-IR wavelength band from 3 to 5 μm is bounded on the lower and upper ends by water vapor absorption. An absorption peak in the middle of the band is due to carbon dioxide. The far-IR band or thermal IR extends from approximately 8 to 12 μm and beyond. The lower wavelength is restricted by water vapor and the upper by a combination of water vapor and carbon dioxide.

In moderate and heavier rain, MMW frequencies of 97 GHz and above are generally subject to similar attenuation effects as the near IR as the rain rate curves of 4, 25, and 150 mm/hr show. At frequencies below 100 GHz, drizzle (0.25 mm/hr) produces less attenuation on MMW energy than on IR. The final curve in Figure 3.1 illustrates that a fog with 0.1 g/m^3 liquid water content is a greater attenuator of IR energy than MMW energy.

3.2 ABSORPTION AND ATTENUATION OF MMW ENERGY BY ATMOSPHERIC CONSTITUENTS

The major atmospheric effect on the performance of MMW sensors is attenuation due to absorption. Figure 3.2 expands the MMW spectrum to show the window frequencies corresponding to relative absorption minima and the frequencies at which maximum absorption occurs. In addition to absorption, MMW propagation through the atmosphere is affected by forward and backward scattering, atmospheric ducting, and atmospheric turbulence. The magnitude of the latter are dependent on frequency, polarization, and operating range. Ducting in the ionosphere is usually associated with transmission over long ranges. Since active MMW sensors are generally designed for short to moderate range operation that does

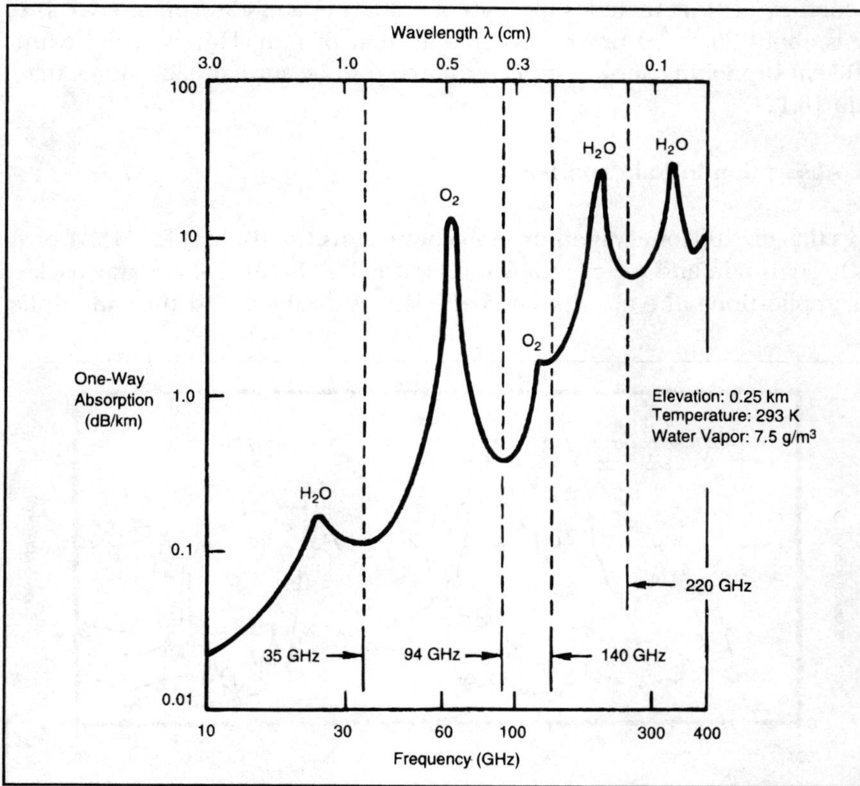

Figure 3.2 Regions of minimum and maximum atmospheric absorption in the MMW spectrum. Attenuation values are based on measured data [5,6].

not exceed several thousand feet, ducting is not experienced with most of their applications [1,2].

3.2.1 Attenuation by Snow

Figure 3.3 contains attenuation measurements made during the Snow-One-A data collection experiments on January 31, 1982, under the sponsorship of the U.S. Army Cold Regions Research and Engineering Laboratories (CRREL) at Camp Ethan Allen in Jericho, VT. The correlation of attenuation (from Ballistic Research Laboratory measurements) to snow rate (from U. S. Air Force Geophysics Laboratory measurements) and center frequency is shown for 95, 140, and 217 GHz [3,4]. The higher frequency energy is subject to the greatest attenuation, which increases as snow rate increases. The separation of attenuation into absorption and scattering components was not provided in this measurement.

At frequencies below 60 GHz, the attenuation caused by frozen particles, such as snow or ice crystals, is small and is often neglected since attenuation by liquid raindrops dominates [5]. For the same precipitation rate, the extinction or attenuation coefficient (that includes the sum of scattering and absorption effects) of dry snow is about 20 to 50 times smaller than that of rain. However, the extinction coefficient of melting snow has been observed to be substantially larger than that of rain [6,12].

3.2.2 Absorption in Rain and Fog

Curves that give the one-way absorption coefficients (in dB/km) for MMW propagation through rain and fog are shown in Figure 3.4 [7–10]. For two-way backscatter radar applications, the absorption coefficient is doubled and then multiplied by

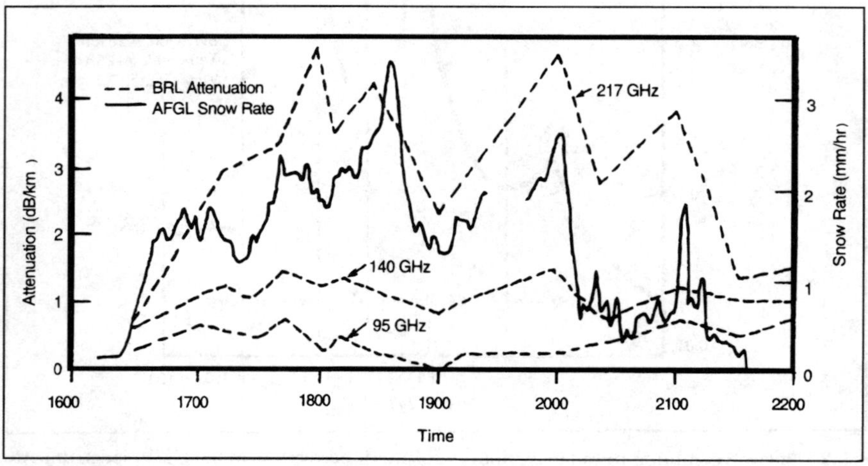

Figure 3.3 Attenuation in snow at 95, 140, and 217 Ghz. (*Source:* [3]. Reprinted with permission.)

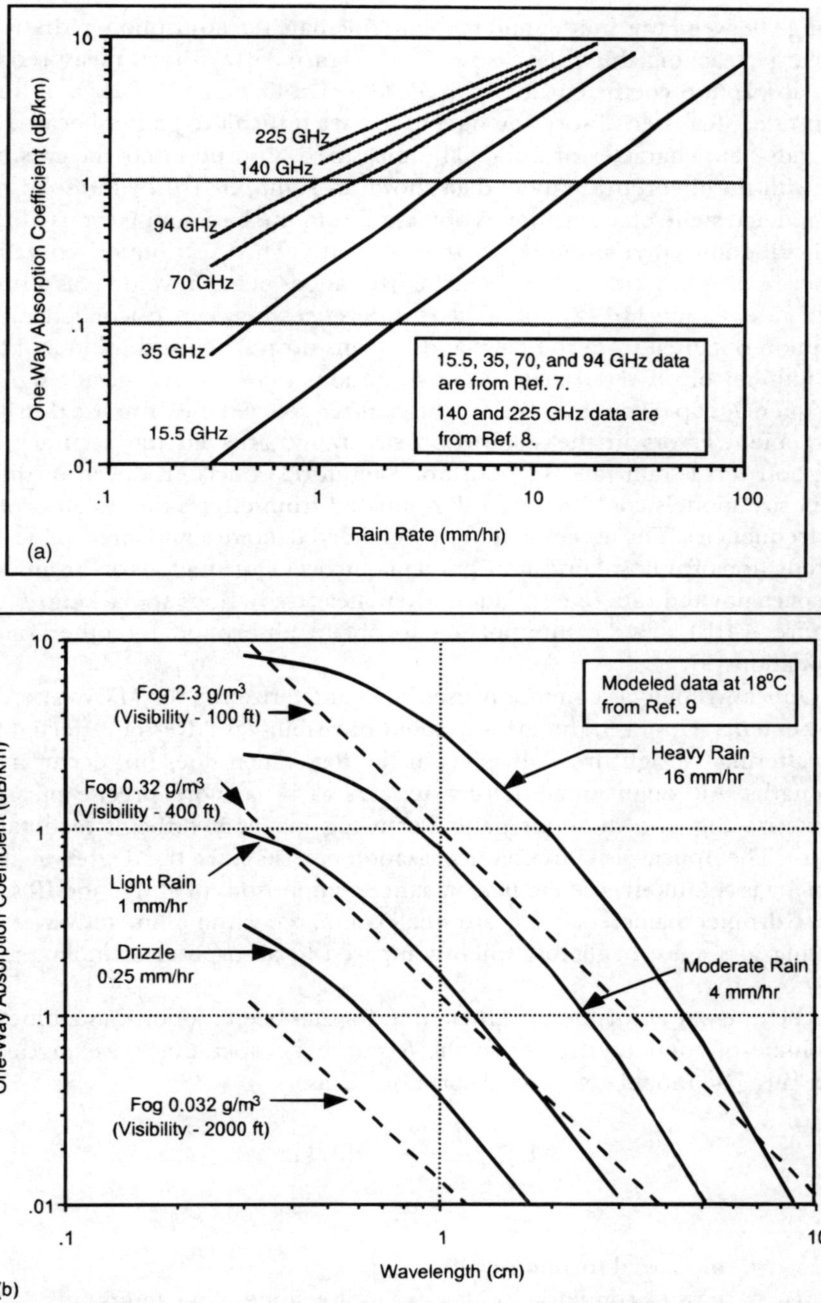

Figure 3.4 Absorption coefficient in rain and fog as a function of MMW frequency and rain rate: (a) measured data and (b) modeled data. (*Source:* [9] Reprinted with permission.)

the range between transmitter and target to calculate the absorption in dB by which the energy reaching the sensor is reduced. Figure 3.4(a) shows measured values of the absorption coefficient for 15.5, 35, 70, 94, 140, and 225 GHz as a function of rain rate. Measured absorption data in fog are difficult to gather because of the nonsteady-state character of a fog. The measured absorption coefficients in rain agree with the theoretical model data shown in Figure 3.4(b) by the solid curves. The modeled value of absorption is obtained from the Laws and Parsons [11] drop size distribution corresponding to the rain rate. This distribution contains the number of droplets with radii in specific size ranges per unit volume as a function of rain rate. Crane [10,12] found that differences between calculated values of absorption obtained from the Laws and Parsons drop size distribution and from a large number of observed drop size distributions were not statistically significant for frequencies up through 50 GHz. At higher frequencies, the drop size distribution measurement errors in the small drop size range affected the accuracy of the absorption versus rain rate relationship. Therefore, effects produced by different droplet size models could not be differentiated from effects due to absorption at these frequencies. The agreement of the modeled data with measured values allows the prediction of atmospheric absorption in rain over large regions of the millimeter-wave spectrum and rain rate variation when measured values are lacking. The data in Figure 3.4(b) may be interpolated to obtain absorption for other values of precipitation [9].

Optical visibility is commonly used to characterize fog for MMW attenuation measurements. However, the measurement of visibility is adversely affected by the Mie scattering[1] of light from droplets in the fog, which does not occur at MMW wavelengths. Although water density appears to be a more precise measure to characterize fog, the transient nature of a fog makes it difficult to obtain this measure. The optical visibility characterization persists since the degree of attenuation in fog is commonly used in performance comparisons of MMW and IR systems. Because droplet diameters in fog are small compared with millimeter wavelengths, scattering losses are negligible when compared to absorption of millimeter-wave energy by a fog.

The one-way absorption coefficient in fog has been modeled as a function of the volume of condensed water in the fog and the operating wavelength of the sensor [9]. The model gives the absorption K_α as

$$K_\alpha = \frac{0.438 M_W}{\lambda^2} \text{ dB/km} \tag{3.1}$$

where

K_α = one-way absorption coefficient;
M_W = mass of condensed water per unit volume of air in gm/m^3;
λ = sensor wavelength of operation in cm.

1 Mie scattering theory gives the general solution for the scattering of electromagnetic waves by a dielectric sphere of arbitrary radius. Rayleigh scattering, a limiting case of Mie scattering, applies when the wavelength is much larger than the scatterer's diameter [6].

Equation 3.1 is accurate within 5 percent when $0.5 \text{ cm} \leq \lambda \leq 10 \text{ cm}$ and when the droplets are extremely small with diameters of the order of 0.001 to 0.005 cm. A value of $M_W = 1 \text{ g/m}^3$ represents about the maximum water content of most fogs, with the possible exception of heavy sea fogs. In most fogs, M_W is much less than 1. The FASCODE-1 weather model developed by the U.S. Air Force Geophysics Laboratory simulates two heavy fogs with liquid water contents of 0.37 and 0.19 g/m^3 and two moderate fogs with liquid water contents of 0.06 and 0.02 g/m^3 [13,30,31]. For both types of simulated fog, the condensed water mass is less than 1. The modeled absorption data for fog, shown in Figure 3.4(b) by the dashed lines, are plotted from (3.1).

Ryde and Ryde, as reported by Goldstein, have given an empirical relation between an average $\overline{M_W}$ and optical visibility in fog, namely [9]

$$\overline{M_W} = 1660 V_i^{-1.43} \tag{3.2}$$

where V_i is the optical visibility in feet and $\overline{M_W}$ is such that in 95 percent of the cases, M_W lies between $0.5M_W$ and $2M_W$. Such a relation may be useful when more precise values of M_W are not available.

Fogs found over land are of two types, advective fog (formed by cool air passing over a colder surface) typical of coastal regions, and radiative fog (formed by radiative cooling of the Earth's surface below its dew point level) found in inland regions. Advective fogs contain a greater number of large water drops and generally higher liquid water content than radiative fogs [14]. When the size of a particle in fog, cloud, rain, dust, etc. is comparable to the wavelength of the incident energy, the phase of the wave is not uniform over the particle. These phase differences give rise to the observed scattering of energy. Therefore, when the ratio of particle size to wavelength approaches unity, the energy attenuation increases. Thus, attenuation of shorter wavelengths (higher frequencies) can be greater in advective fogs because of the greater number of large particles and also because of the larger liquid water content of the fog. Calculations made by Richard [15] show that there can be a difference of 8 dB/km in attenuation at 140 GHz between advective and radiation fogs at 0.1 km visibility. Earlier measurements by Richer [16] at the Ballistic Research Laboratories found a maximum one-way attenuation of 23 dB/km at 140 GHz during a 30-second time period, that returned to a lower value of 15 dB/km during the following 30-second interval. The change in attenuation was not accompanied by an appreciable change in visibility. The measured 8-dB variation was attributed to an increase in fog density beyond the limits of human visibility or to the condensation of fog into rain along the propagation path.

3.2.3 Attenuation by Smoke and Dust

The attenuation produced by various types of smokes and dust on 35, 94, and 140-GHz transmissions was measured by the U.S. Army during Smoke Week II in November 1978. Seven levels of obscurants were tested from among hexachloroethane, white and red phosphorous packaged in several ways to achieve desired burning rates and smoke concentrations, fog oil, and dust dispersed by detonating high explosives. Of these, only the high-explosive dust affected the transmission of the radars, and then only produced a maximum attenuation at 140 GHz of 0.4 dB (corresponding to a transmissivity of 91 percent) over the 800 m range. At 35 and 94 GHz, the attenuation ranged from 0 to 0.3 dB over the five dust trials. At 140 GHz, the attenuation ranged from 0 to 0.4 dB over the five trials. The variation in measured attenuation was attributed to variations in the size of the dust particles falling through the line of sight of the radars during the measurement period [17].

3.3 BACKSCATTER OF MMW ENERGY BY RAIN

Although absorption is the major atmospheric effect on MMW propagation, rain does scatter transmitted energy back toward the receiver, thereby interfering with or masking the return from targets. Rain backscatter coefficient (in m^2/m^3) data are shown in Figure 3.5 for a linearly polarized transmitter and receiver. The

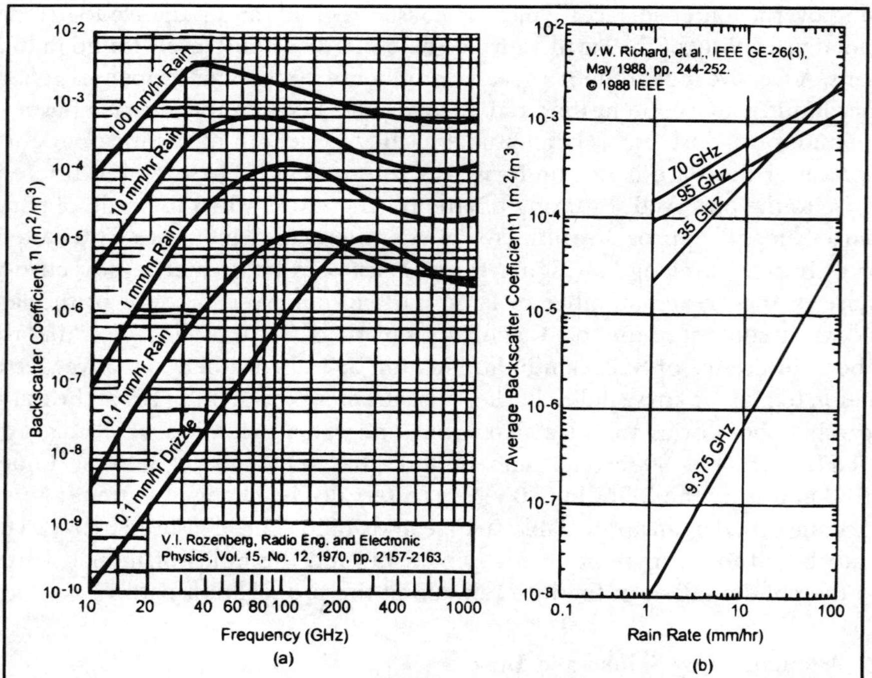

Figure 3.5 Rain backscatter coefficient as a function of frequency and rain rate: (a) theoretical results at 18°C based on Marshall-Palmer drop size distribution and (b) measured results (V-V pol.). (*Source:* [18,19]. Reprinted with permission.)

backscatter coefficients η on the left of the figure are theoretical results based on a Marshall-Palmer drop size distribution [18], while those on the right are measured values for 9 through 70 GHz [19]. The theoretical results by Rozenberg classify rain as precipitation in the form of water drops with diameters in the 0.5 to 7 mm range. Drizzle is classified as precipitation not exceeding 0.25 mm/hr consisting of small droplets with diameters less than 0.5 mm. In the drizzle model used to produce the curve in Figure 3.5, the minimum diameter of the drops was 0.1 mm and the maximum diameter was 0.5 mm. The Marshall-Palmer and Laws and Parsons distributions for the number of drops of a given size are nearly equivalent for drop size diameters greater than 1.0 to 1.5 mm. For backscatter applications where larger drop sizes dominate, the exponential Marshall-Palmer distribution is used [10]. Once again, the theory adequately models the measured values. If backscatter is large and inhibits radar operation at the selected frequency of operation, a potential solution is to use circular polarization. Figure 3.6 shows that this approach can reduce the backscatter by 20 dB at 9.375 GHz and by 18 dB at 95 GHz [19].

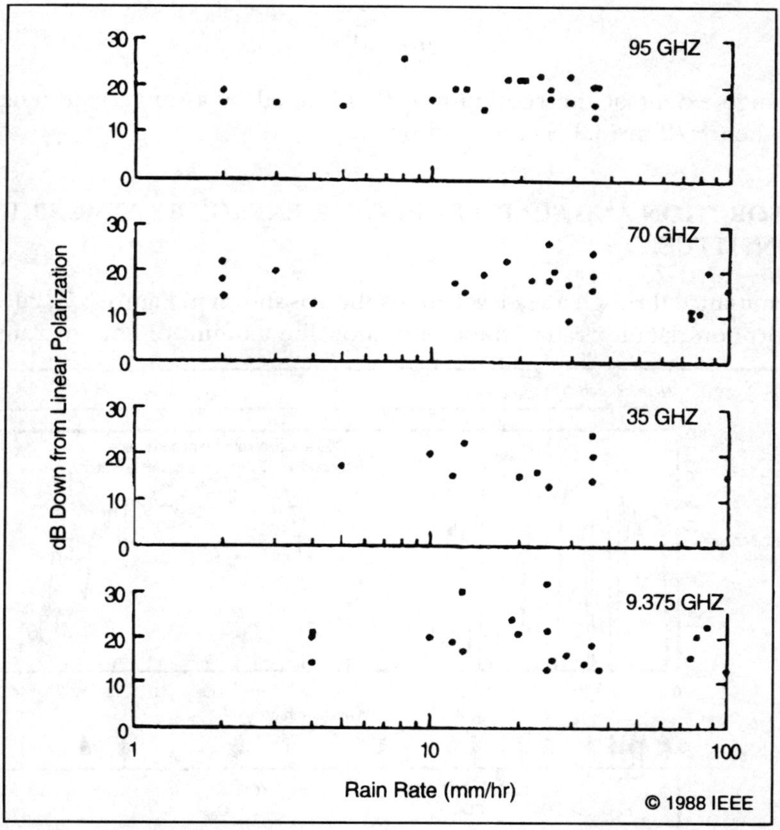

Figure 3.6 Rain backscatter coefficient reduction by circular polarization. *Source:* [19]. Reprinted with permission.)

Since backscatter is a volumetric effect, the backscatter coefficient η is multiplied by the volume of the radar resolution cell V to obtain the equivalent radar cross section of the rain. Rain can thus act as a"pseudotarget"and scatter energy toward the radar receiver that competes with the energy scattered from the real target.

If the resolution cell volume V of the radar is given by

$$V = \pi/4(R\theta_a)(R\theta_e)(c\tau/2) \tag{3.3}$$

where

R = range from the radar to the rain resolution cell in meters;
θ_a, θ_e = two-way 3dB azimuth and elevation antenna beamwidths, respectively, in radians;
τ = width of the transmitted pulse in sec;
c = speed of light in m/sec.

then the radar cross section σ of the rain cell in m^2 is given by

$$\sigma = \eta V \tag{3.4}$$

If the range extent of the resolution cell is limited by a range gate length L in meters, then $c\tau/2$ in (3.3) is replaced by L.

3.4 ABSORPTION AND SCATTERING OF IR ENERGY BY ATMOSPHERIC CONSTITUENTS

IR transmittance through a sea-level atmosphere is shown in Figure 3.7 [20]. Unlike the absorption data for radar, these data show the amount of energy transmitted.

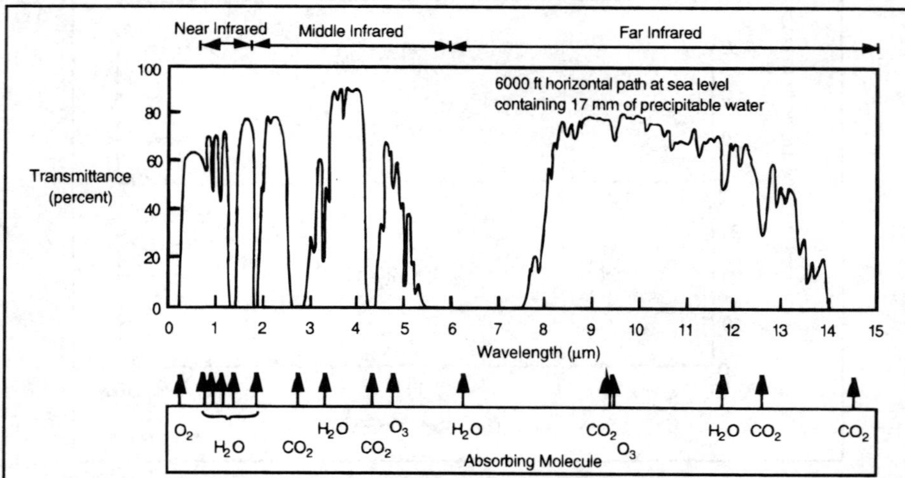

Figure 3.7 IR transmittance of the atmosphere. *Source:* [20]. Reprinted with permission.)

The principle permanent atmospheric constituents contributing to the absorption of energy at IR wavelengths are carbon dioxide, nitrogen, and methane. Variable constituents that absorb IR energy include water and ozone. In addition to absorption, IR energy is scattered from molecules and aerosols in the atmosphere. Wavelengths less than 2 μm experience negligible molecular scattering, while scattering from aerosols is a function of the radius of the scatterer divided by the wavelength. Aerosol-type scatterers include rain, dust, fog, and smoke.

Atmospheric transmittance $\tau_a(\lambda)$ is modeled by the Lambert-Beer Law as [21,22]

$$\tau_a(\lambda) = \exp[-\gamma(\lambda)R] \tag{3.5}$$

where

- γ = extinction coefficient or power attenuation coefficient in nepers (Np)/km;
- R = range or path length in km.

Multiplying the attenuation coefficient in dB/km by 0.23 converts it into Np/km.

The extinction coefficient $\gamma(\lambda)$ is the sum of the absorption and scattering coefficients $K_\alpha(\lambda)$ and $K_\sigma(\lambda)$, respectively, and can be written as

$$\gamma(\lambda) = K_\alpha(\lambda) + K_\sigma(\lambda) \tag{3.6}$$

Absorption and scattering coefficients, in turn, are sums of molecular and aerosol components denoted by the subscripts *m* and *a*, respectively, such that

$$K_\alpha(\lambda) = K_{\alpha m}(\lambda) + K_{\alpha a}(\lambda) \tag{3.7}$$

and

$$K_\sigma(\lambda) = K_{\sigma m}(\lambda) + K_{\sigma a}(\lambda) \tag{3.8}$$

The extinction coefficient is a complex function of wavelength as may be inferred from Figure 3.7. An expression for the average value of the transmission $\overline{\tau_a}$ over a wavelength band λ_1 to λ_2 is given by

$$\overline{\tau_a} = \frac{1}{\lambda_2 - \lambda_1}\int_{\lambda_1}^{\lambda_2} \exp[-\gamma(\lambda)R]d\lambda \tag{3.9}$$

Values for $\overline{\tau_a}$ are generally obtained from computer-hosted programs such as LOWTRAN [23–25], which covers a spectral range of 0 to 50,000 cm^{-1} (0.2 μm to infinity) with a resolution of 20 cm^{-1} (full width at half maximum). It models radiance from single and multiple scattering and path geometries from space-viewing ground-based sensors, air-to-air scenarios, surface point-to-point paths, and earth-viewing airborne sensors.

In LOWTRAN 7 (the most recent LOWTRAN code), the user specifies the geographical atmospheric model (from one of six contained in the LOWTRAN program or from user-generated input), the altitude- and seasonal-dependent aerosol profiles, and the extinction coefficients. The six program-defined atmospheric models are tropical, midlatitude summer, midlatitude winter, subarctic summer, subarctic winter, and 1976 U.S. standard. Each defines the temperature, pressure, density, and atmospheric gases mixing ratio as a function of altitude. Table 3.1 shows the rural, urban, maritime, tropospheric, and fog aerosol models that are available for the atmospheric boundary layers [23,24]. Figures 3.8 through 3.10 illustrate the extinction and absorption coefficients in Np/km as derived from aerosol models. The extinction coefficients have been normalized to unity at 0.55 μm. The clear atmosphere condition referenced in these plots is based on a visibility range of 23 km. Seasonal models for the aerosol profiles of the troposphere

Table 3.1
LOWTRAN Aerosol Models for the Atmospheric Boundary Layer

Aerosol Model	*Representative Region*	*Constituent*
Rural (0 to 2 km altitude)	Continental areas not directly influenced by urban/industrial aerosol sources	Atmospheric gases and surface dust particles
Urban (0 to 2 km altitude)	Modifies rural background by adding aerosols from combustion products and industrial sources	20%/80% mixture of carbonaceous aerosols to rural type aerosols, respectively
Maritime (0 to 2 km altitude)	Aerosols of oceanic origin	Sea salt particles
Tropospheric (2 to 10 km altitude)	Troposphere with extremely clear conditions and uniform aerosol properties	Rural model constituents without large particles
Fog 1 (0 to 2 km altitude)	Advection fog	Default visibility = 0.2 km
Fog 2 (0 to 2 km altitude)	Radiation fog	Default visibility = 0.5 km

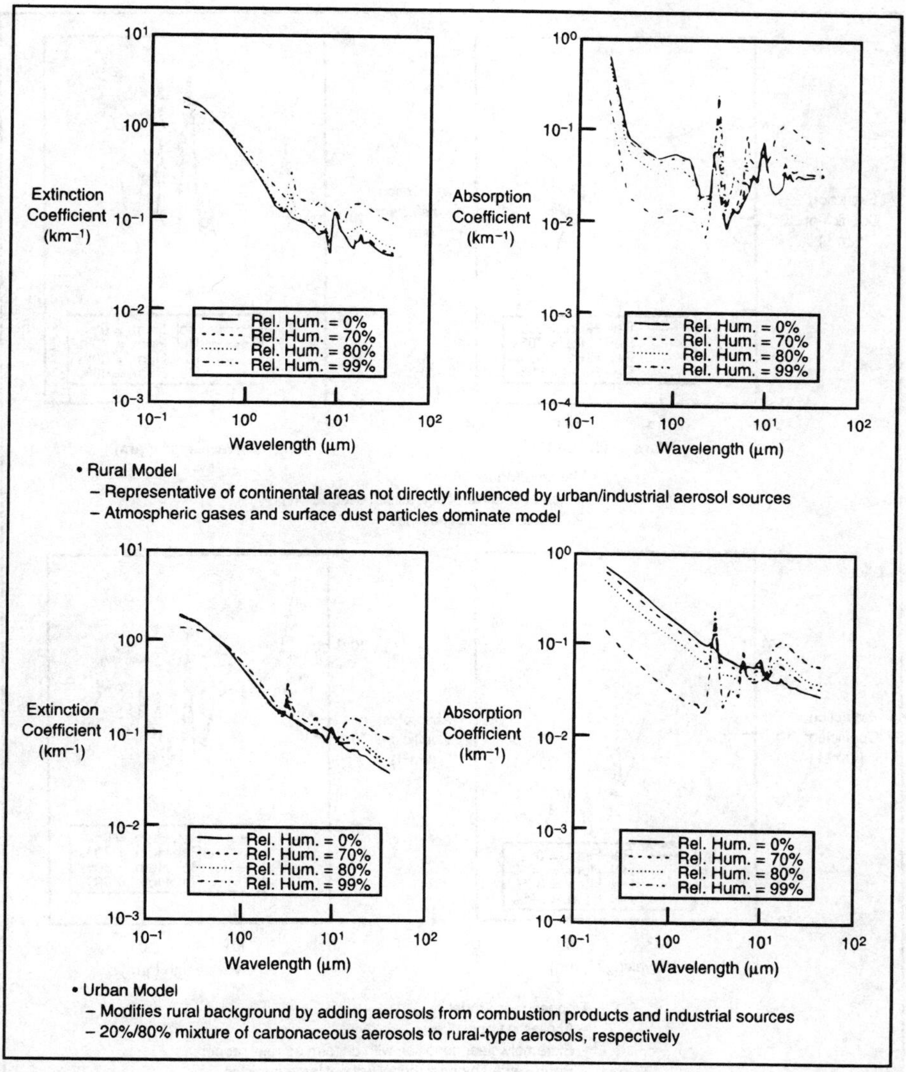

Figure 3.8 Extinction and absorption coefficients in urban and rural aerosols as derived from LOWTRAN 5 model [23].

(2–10km) and stratosphere (10–30 km), and transition profiles for the stratosphere up to 100 km are also available. Rain rate, cloud models, wind speed, and other atmospheric parameters can be varied to tailor the model to the conditions under which the transmission is desired.

In addition to LOWTRAN, there are other models available to assess the effects of weather on sensor systems. These include FASCODE and MODTRAN that are supported by the U.S. Air Force Geophysics Laboratory at Hanscom Air

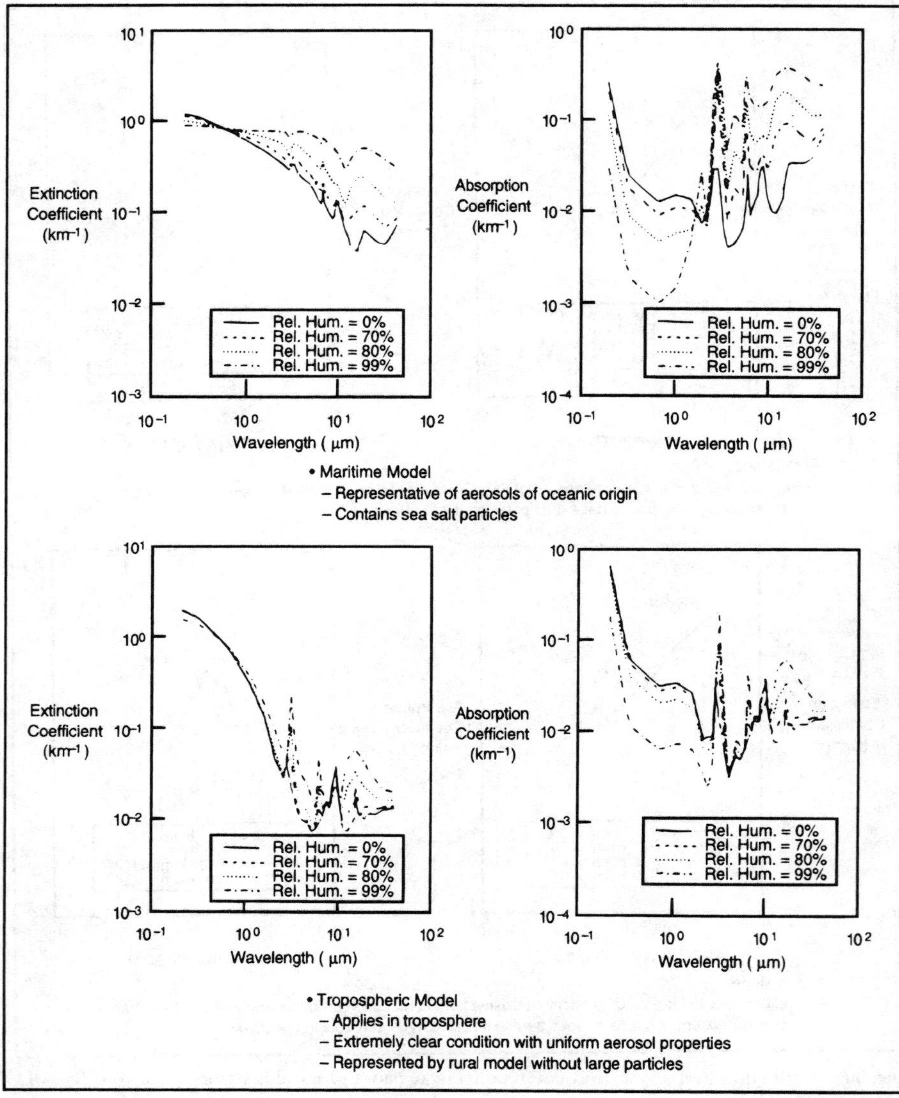

Figure 3.9 Extinction and absorption coefficients in maritime and tropospheric aerosols as derived from LOWTRAN 5 model [23].

Force Base, Bedford, Massachusetts 01731 [26–31]. FASCODE models very high altitude (<70 km) and very narrow spectral bands that are applicable to laser-line resolution. It is useful for extinction dominated by molecular absorption, improving upon the resolution offered by LOWTRAN in this region. MODTRAN was written for moderate resolution calculations that do not require FASCODE and provides

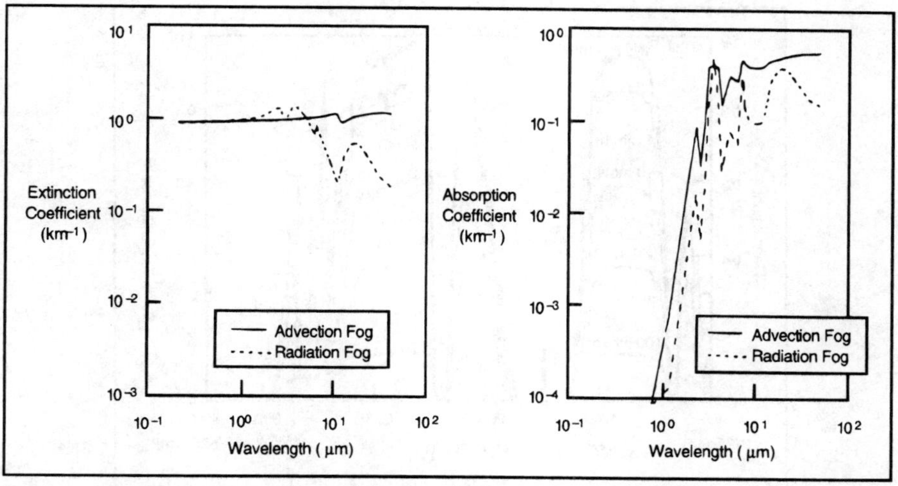

Figure 3.10 Extinction and absorption coefficients in advection and radiation fog as derived from LOWTRAN 5 aerosol model [23].

2 cm^{-1} (full width at half maximum resolution [27]. Both programs are available from the Geophysical Laboratory via e-mail.[2]

3.4.1 Attenuation by Rain

Rain attenuates target-to-background contrast in IR imagery in two ways, first by introducing an attenuation loss over the signal path to the receiver, and second by cooling the target [32]. A set of atmospheric transmission curves produced by LOWTRAN 6 for several values of rain rate is shown in Figure 3.11. Here wavenumber is defined as the reciprocal of wavelength, measurement path is 300 meters, surface and dew point temperatures are both equal to 10°C, and the meteorological visibility range is 23 km in the absence of rain.

3.4.2 Transmittance Through Snow

Transmittance through snow, as measured in the visible and IR spectral bands during the Snow-One-A data collection experiments, is shown in Figure 3.12. Transmittance values, which increase upwards, are listed on the left of the graph and snow rate, which increases downward, on the right. The spectral wavelength bands in μm are shown in the lower right portion of the figure. Two general conclusions are apparent from the data. First, the shorter wavelengths have higher values of transmittance, and second, the transmittance decreases as the snow rate increases. There appears to be a crossover in the transmittance of the visible and near IR bands with snow

2 The contact personnel are Gail Anderson: GAnderson@plh.af.mil and Jim Chetwynd: Chetwynd@plh.af.mil.

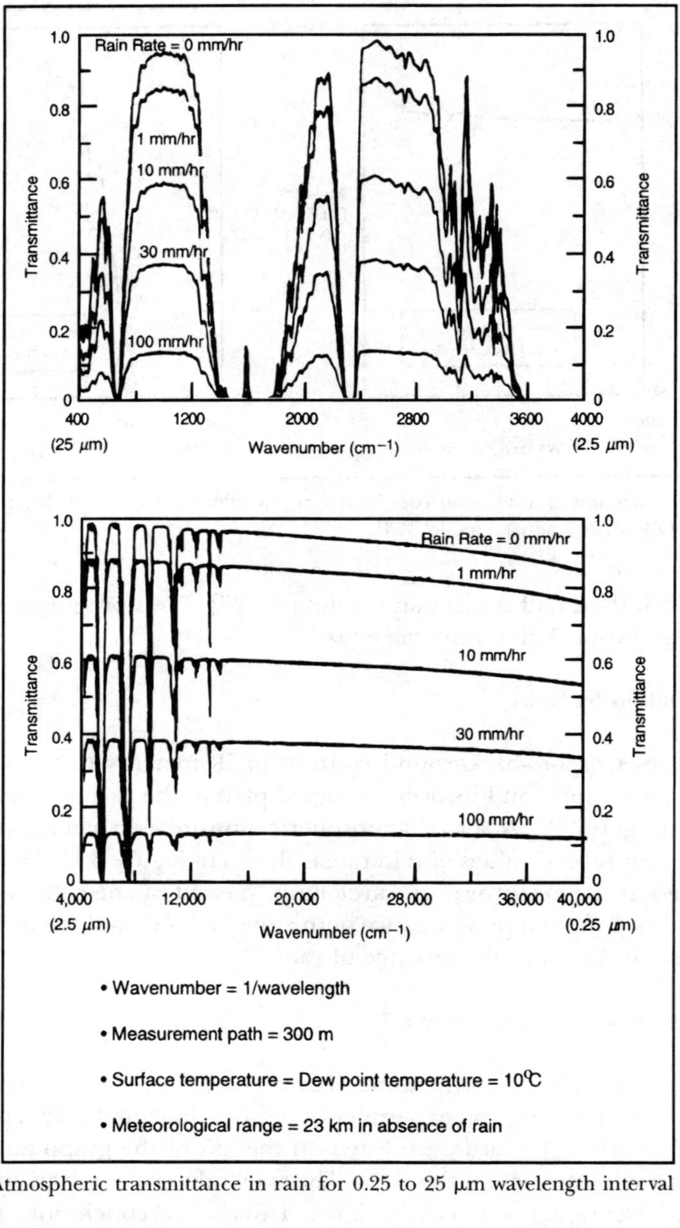

Figure 3.11 Atmospheric transmittance in rain for 0.25 to 25 μm wavelength interval [24].

rate, the near IR having the highest transmittance before approximately 1830 hours at lower values of snow rate and the visible having the highest transmittance after that time when the snow rate decreased [3,4]. A possible interpretation for the crossover of transmittance with wavelength is the occurrence of fog during the snowfall before 1830 hours, causing the visible transmittance to be less than infrared [33].

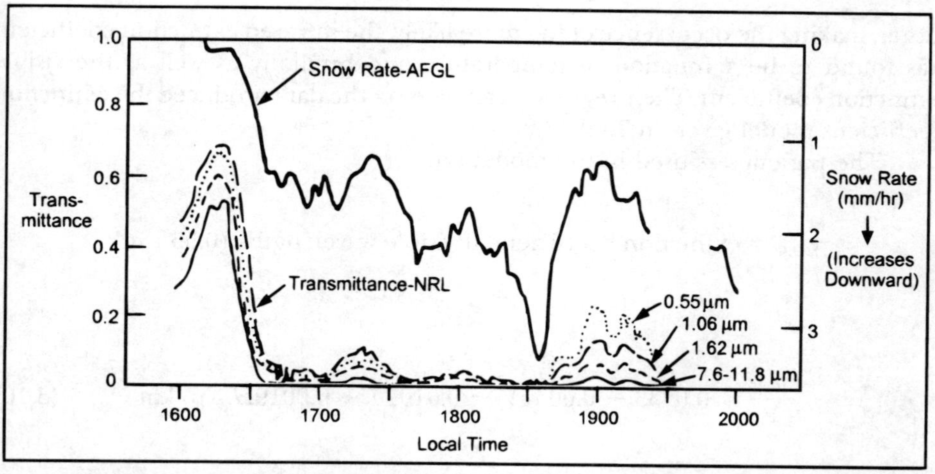

Figure 3.12 Transmittance in snow through the visible and IR spectral bands over a path length of 320 meters. (*Source:* [3]. Reprinted with permission.)

Using data from the Snow-One measurements conducted between January 5, 1981, and February 20, 1981, at Camp Ethan Allen, Seagraves and Ebersole [33] found that the extinction coefficient at mid- and far-infrared wavelengths could be expressed as a function of only the visible extinction coefficient when the relative humidity was less than or equal to 94 percent. When the relative humidity was

Table 3.2
Extinction Coefficient Model for Snow [33]

Applicable Wavelength	*Applicable Humidity*	*Extinction Coefficient Model*
3.0 μm	≤94 percent	$\gamma_{3.0} = 1.21\gamma_{0.55}$ Np/km
3.0 μm	>94 percent	$\gamma_{3.0} = \gamma_{0.55}(-0.107T - 0.101H - 0.042V_i + 10.74)$ Np/km
10.4 μm	≤94 percent	$\gamma_{10.4} = 1.18\gamma_{0.55}$ Np/km
10.4 μm	>94 percent	$\gamma_{10.4} = \gamma_{0.55}(-0.182T - 0.223H - 0.426V_i + 25.35)$ Np/km

larger, making the occurrence of fog more likely, the infrared extinction coefficient was found to be a function of temperature and humidity as well as the visible extinction coefficient. Their regression analysis on the data produced the extinction coefficient model given in Table 3.2.

The parameters used in the model are:

$$\gamma_{0.55} = \text{extinction coefficient at visible wavelengths } (0.55\ \mu\text{m})$$

$$\frac{\gamma_{0.55}}{V_c} = 0.0233 - 0.0031V_c - 0.0101T + 0.0019H\ \text{Np/km} \tag{3.10}$$

$$V_c = \text{volume concentration of snow in } 10^{-8}\ \text{m}^3/\text{m}^3$$

$$= R/v \tag{3.11}$$

with

R = equivalent liquid water precipitation rate;

v = particle settling velocity;

T = surface temperature in °C;

H = surface relative humidity in percent.

and

$$V_i = \text{visibility in km}$$

$$= \frac{3.0}{\gamma_{0.55}} \tag{3.12}$$

The model was derived from data with visibility, temperature, and humidity parameters having the ranges of 1.2 km $\leq V_i \leq$ 7.5 km, -11.9°C $\leq T \leq$ 2.0°C, and 68% $\leq H \leq$ 100%, respectively. It should be used with caution elsewhere. The model gives the largest errors in transmittance as compared to measured data when the relative humidity is between 90 and 95 percent, probably because the presence of fog is most in doubt in this region.

3.4.3 Extinction Coefficients for Natural and Man-made Obscurants

Ranges for the extinction coefficients of atmospheric obscurants are shown in Table 3.3 for the visible, IR, and MMW spectral bands [34]. Table 3.4 presents qualitative summary information about the susceptibility of visible and infrared sensors to battlefield smoke, MMW chaff, and inclement weather as a function of

Table 3.3
Extinction Coefficients of Atmospheric Obscurants

Approximate Ranges of Extinction Coefficients (Np/km)					
	Spectral Region				
Atmospheric Obscurant	*Visible 0.4 to 0.7 μm*	*Mid IR 3 to 5 μm*	*Far IR 8 to 12 μm*	*MMW (35 GHz) 8.6 mm*	*MMW (95 GHz) 3.2 mm*
Gases	Very low: ≅0.02	Low/med: 0.25 to 0.73	Very low/ med: 0.03 to 0.8	Very low: 0.02 to 0.06	Very low/ low: 0.03 to 0.2
Haze	Low/med: 0.2 to 2.0	Very low/ med: 0.02 to 1.0	Very low/ low: 0.02 to 0.4	Very low: ≅0.001	Very low ≅0.001
Fog	High: 2.0 to 20	Med/high: 1.0 to 20	Med/high: 0.4 to 20	Very low/ low: 0.001 to 0.1	Very low/ low: 0.01 to 0.4
Rain	Low/med: 0.3 to 1.6	Low/med: 0.3 to 1.6	Low/med: 0.3 to 1.6	Very low/ med: 0.05 to 1.0	Low/med: 0.3 to 2.0
Snow	Med/high: 2.0 to 12	Med/high: 2.0 to 12	Med/high: 2.0 to 12	Very low/ med: 0.004 to 1.0	Very low/ med: 0.03 to 1.0
Dust	Low/high: 0.2 to 4.0	Low/high: 0.2 to 4.0	Low/high: 0.2 to 4.0	Very low: 0.0005 to 0.005	Very low: 0.0005 to 0.005

Extinction Coefficient	*Descriptive Term*	*Visual Range*
<0.1 Np/km	Very low	>30 km, very clear
0.1 to 0.5 Np/km	Low	6 to 30 km, clear to hazy
0.5 to 2 Np/km	Medium	2 to 6 km, hazy
>2 Np/km	High	<2 km, foggy

Source: Smoke and Natural Aerosol Parameters (SNAP) Manual, Joint Technical Coördinating Group for Munitions Effectiveness, Smoke and Aerosol Working Group, Report 61, JTCG/ME-85-2, April 26, 1985.

Table 3.4
Qualitative Susceptibility of Visible and Infrared Sensors to Battlefield Smoke, MMW Chaff, and Inclement Weather

	← *Smoke in Current Inventory* →			← *Developmental* →		*Inclement Weather*
Type of Obsurant	*Oil Smoke*	*Hexachlorethane*	*White Phosphorous*	*IR Smokes*	*MMW Chaff*	*Fog, Rain, Snow*
Application	*Generators Engine Exhaust*	*Pots Artillery*	*Projectiles Grenades*	*Generators Grenades*	*Generators Projectiles*	
Visible	●	●	●	●	○	●
Near IR	●	●	●	●	○	●
Mid IR	○	○	◓	●	○	◓
Far IR	○	○	◓	●	○	◓

Note: ○ Poses minimal problem; ◓ poses moderate problem; ● poses severe problem.
Source: U.S. Army RD&A Bulletin PB 70-88-1 (Test), *Smoke/Obscurants Technology,* Jan.–Feb. 1988.

the spectral band [35]. The MMW chaff was included in this table by the U.S. Army to show an example of the effect of countermeasures designed for sensors operating at other than IR wavelengths on infrared sensors. As expected, the effects of the smokes and inclement weather are greatest at the shorter wavelengths.

3.5 EOSAEL SENSOR PERFORMANCE MODEL

One of the more comprehensive models for analyzing a variety of physical processes that affect the performance of MMW and IR sensors, as well as those that operate in the visible, ultraviolet, and on 53 laser lines, is EOSAEL (Electro-Optical Systems Atmospheric Effects Library) [36,37]. It addresses the phenomena listed below, which describe various aspects of electromagnetic energy propagation and scenarios of interest in defense applications.

- Spectral transmission and contrast transmission;
- Multiple scattering;
- Sensor performance;
- Transport and diffusion;
- Turbulence effects on imaging;
- High energy laser propagation;
- Radiative transfer;
- Thermal contrast;

Table 3.5
EOSAEL Modules and Their Functions

Category	*Module*	*Valid Range*	*Function*
Gases	LOWTRAN	0.25 to 28.5 μm	Calculates atmospheric transmittance, radiance, and contrast due to specific molecules at 20 inverse cm spectral resolution on a linear wave-number scale
	LZTRAN	Visible to far IR (0.5 to 11.0 μm)	Calculates transmission through atmospheric gases at specific laser frequencies for slant or horizontal paths
	NMMW	10 to 1000 GHz (0.3 to 30.0 mm)	Calculates transmission, backscatter, and refractivity due to gaseous absorption, fog, rain, and snow
Natural Aerosols	XSCALE	0.2 to 12.5 μm	Calculates fog and haze transmission for horizontal or slant paths, and rain and snow transmission for horizontal paths
	CLIMAT	Not applicable	Provides values of meteorological parameters for select European, Mideastern, Korean, Alaskan, Scandinavian, Central American, Indian, S.E. Asian, South American, and Mexican locales
	CLTRAN	0.2 to 2.0 μm, 3.0 to 5.0 μm, 8.0 to 12.0 μm	Calculates slant path transmission through six cloud types
	COPTER	0.4 to 0.7 μm, 3.0 to 5.0 μm, 8.0 to 12.0 μm, 0.3 to 30.0 mm	Calculates effects of looses snow or dust lofted by helicopter downwash
Battlefield Aerosols	COMBIC	0.4 to 1.2 μm, 3.0 to 5.0 μm, 8.0 to 12.0 μm, and 94 GHz (3.2 mm)	Calculates size, path length, concentration, and transmission through various smokes and artillery or vehicular dirt and dust particles
	SABRE	Not applicable	Calculates effects of terrain on wind flow and smoke screens
	KWIK	Not applicable	Provides a placement and number of smoke munitions necessary to produce a screen that reduces the probability of target detection to a given level
	GRNADE	0.4 to 1.2 μm, 3.0 to 5.0 μm, 8.0 to 12.0 μm, and 94 GHz (3.2 mm)	Models obscuration produced by tube-launched grenades used in self-screening applications
	FITTE	0.4 to 12.0 μm	Calculates dimensions of and transmittance through plumes from burning vegetation and vehicles
	MPLUME	Not applicable	Calculates performance degradation of target designation systems by missile smoke plumes

Table 3.5
(Continued)

Category	*Module*	*Valid Range*	*Function*
Radiative Transfer	FCLOUD	Any wavelength included in the phase function file PFNDAT*	Calculates beam transmittance, path radiance, and contrast transmittance through a homogeneous ellipsoidal cloud
	OVRCST	Any wavelength	Calculates beam transmittance, path radiance, and contrast transmittance along an arbitrary line of sight under an overcast sky
	MSCAT	Any wavelength included in the phase function file PFNDAT	Calculates multiple scattering of a laser beam into a detector from an ellipsoidal aerosol cloud
	ASCAT	Any wavelength included in the phase function file PFNDAT	Calculates effects of multiple scattering for transmissometer and lidar configurations
	ILUMA	Photopic	Prediction of natural illumination under realistic atmospheric conditions
	FASCAT	0.55 and 1.06 μm	Determination of path radiance and contrast effects
	GSCAT	Visible	Determines path radiance effects
	LASS	Visible	Determines the effectiveness of smoke screens deployed against large fixed and semifixed installations
	REFRAC	≈0.4 to ≈20.0 μm	Calculates the amount of curvature a ray of light experiences as it passes over a complex terrain surface
Laser Propagation	IMTURB	<14 μm	Predicts the effects of laser propagation through optical turbulence in the weak fluctuation regime
	NOVAE	<14 μm	Calculates linear and nonlinear effects on high-energy laser beam propagation from clear air, smokes, and aerosols
Target Acquisition and System Performance	TARGAC	Visible to mid-IR	Evaluates the combined atmospheric and system effects to determine the range for target detection and classification
	RADAR	0.1 to 350 GHz	Calculates millimeter-wave system performance

* PFNDAT contains phase functions, extinction and scattering coefficients, and the single-scattering albedo for 38 natural and man-made aerosols at 16 wavelengths ranging from 0.55 to 12.0 μm.

- Generation of battlefield obscurants;
- Climatology for 47 nonoverlapping climatic regions.

The EOSAEL software consists of 26 modules that are divided into six categories: atmospheric gases, natural aerosols, battlefield aerosols, radiative transfer, laser propagation, and target acquisition and system performance. The modules are more engineering oriented than based on first principals. The EOSAEL development philosophy was to include modules that give reasonably accurate results, while minimizing computer time, for conditions that may be expected on a battlefield. The modules can be run under the control of an executive routine or executed in a stand-alone mode. Versions are available in FORTRAN 77 that allow the program to run on a variety of computers. The 26 EOSAEL modules and their functions are listed in Table 3.5. Three modules of particular interest are the previously discussed LOWTRAN and NMMW (both found in the gases category) and TARGAC. NMMW models the effects of atmospheric gases on MMW sensors. TARGAC, in the target acquisition and system performance category, is built into the FLIR target detection and classification model discussed in Chapter 7. Government agencies and organizations under contract to the U.S. government can obtain copies of EOSAEL from the U.S. Army Atmospheric Sciences Laboratory at White Sands Missile Range, New Mexico.

3.6 SUMMARY

Measured data and models were presented for calculating atmospheric absorption and backscatter of MMW and IR energy in clear weather, rain, fog, snow, and other obscurants. Attenuation of MMW and IR energy may be modeled using an extinction coefficient that contains terms to account for absorption and scattering. The modeled data generally agree with measured data and can, therefore, be used to predict sensor performance when actual absorption and backscatter measurements are not available. Some of the models are relatively simple and address only one phenomenon while others, such as EOSAEL, model more complex problems and scenarios.

References

[1] Bohlander, R. A., McMillan, R. W., Patterson, E. M., Clifford, S. F., Hill, R. J., Priestley, J. T. and Schoenfeld, W. P., "Fluctuations in Millimeter-Wave Signals Propagated Through Inclement Weather," *IEEE Trans. on Geos. and Remote Sensing*, Vol. 26, No. 3, May 1988, pp. 343–354.

[2] Wallace, H. Bruce, "Millimeter-Wave Propagation Measurements at the Ballistic Research Laboratory," *IEEE Trans. on Geos. and Remote Sensing*, Vol. 26, No. 3, May 1988, pp. 253–258.

[3] Plank, V. G., Berthel, R. O., and Main, B. A., "Snow Characterization Measurements and E/O Correlation Obtained During Snow-One-A and Snow-One-B," *SPIE Vol. 414*, 1983, pp. 97–102.

[4] Snow-One-A Data Report, Aitken, G. W., Ed., Special Report 82-8, AD-B068569L, U.S. Army Cold Regions Research and Engineering Laboratory/CRREL-RG, Hanover, NH, May 1982.

[5] Crane, R. K., "Prediction of the Effects of Rain on Satellite Communication Systems," *Proc. IEEE*, Vol. 65, 1977, pp. 456–474.

[6] Ulaby, F. T., Moore, R. K., and Fung, A. K., *Microwave Remote Sensing Active and Passive, Vol. 1: Microwave Remote Sensing Fundamentals and Radiometry*, Artech House, Norwood, MA, 1981.

[7] Atlas, D., "Advances in Radar Meteorology," in *Advances in Geophysics*, Vol. 10, Academic Press, NY, 1964, pp. 317–478.

[8] Nemarich, J., Wellman, R. J., and Lacombe, J., "Backscatter and attenuation by falling snow and rain at 96, 140, and 225 GHz," *IEEE Trans. on Geos. and Remote Sensing*, GE-26(3), 319–329 (1988).

[9] Goldstein, H., "Attenuation by Condensed Water," Chapter 8 in *Propagation of Short Radio Waves*, Editor: Kerr, D. E., McGraw-Hill, NY, 1951.

[10] Crane, R. K., *Microwave Scattering Parameters for New England Rain*, Technical Report 426, Lincoln Laboratory, MIT, Oct. 3, 1966.

[11] Bogush, Jr., A. J., *Radar and the Atmosphere*, Artech House, Norwood, MA, 1989.

[12] Crane, R. K., "Prediction of Attenuation by Rain," *IEEE Trans. on Comm.*, Vol. COM-28, No. 9, Sept. 1980, pp. 1717–1733.

[13] Falcone, Jr., V. J., and Abreu, L. W., "Atmospheric Attenuation of Millimeter and Submillimeter Waves," *IEEE EASCON-79 Conference Record*, Vol. 1, pp. 36–41 and *Millimeter Wave Radar*, Johnston, S. L., Editor, Artech House, Norwood, MA, 1980.

[14] Falcone, Jr., V. J., Abreu, L. W., and Shettle, E. P., "Atmospheric attenuation of millimeter and submillimeter waves: models and computer code," AFGL-TR-79-0253 (AD A084485), Air Force Geophysics Laboratory, Hanscom AFB, MA, October 1979.

[15] Richard, V. W., Kammerer, J. E., and Reitz, R. G., "140 GHz Attenuation and Optical Visibility Measurements of Fog, Rain, and Snow," ARBRL-MR-2800, 1977.

[16] Richer, K. A., "Environmental Effects on Radar and Radiometric Systems at Millimeter Wavelengths," A72-15610, *Proc. Symposium on Submillimeter Waves*, Polytechnic Institute of Brooklyn, March 31–April 2, 1970, pp. 533–543.

[17] Knox, J. E., "Millimeter Wave Propagation in Smoke," *IEEE EASCON-79 Conference Record*, Vol. 2, pp. 357–361 and *Millimeter Wave Radar*, Johnston, Stephen L., Editor, Artech House, Norwood, MA, 1980.

[18] Rozenberg, V. I., "Radar characteristics of rain in submillimeter range," *Radio Eng. and Electronic Physics*, Vol. 15, No. 12, 1970, pp. 2157–2163.

[19] Richard, V. W., Kammerer, J. E., and Wallace, H. B., "Backscatter Measurements at Millimeter Wavelengths," *IEEE Trans. on Geos. and Remote Sensing*, GE-26(3), 1988, pp. 244–252.

[20] Hudson, R. D., *Infrared System Engineering*, John Wiley & Sons, NY, 1969.

[21] Weichel, H., *Laser Beam Propagation in the Atmosphere*, Vol. TT3, SPIE Optical Engineering Press, Bellingham, WA, 1990.

[22] LaRocca, A. J., "Methods of Calculating Atmospheric Transmittance and Radiance in the Infrared," *Proc. of the IEEE*, Vol. 63, No. 1, Jan. 1975, pp. 75–94.

[23] Kneizys, F. X., Shettle, E. P., Gallery, W. O., Chetwynd, Jr., J. H., Abreu, L. W., Selby, J. E. A., Fenn, R. W., and McClatchey, R. A., *Atmospheric Transmittance/Radiance: Computer Code: LOWTRAN 5*, AFGL-TR-80-0067, AFGL, Hanscom AFB, MA 01731, 1980.

[24] Kneizys, F. X., Shettle, E. P., Galler W. O, Chetwynd, Jr., J. H., Abreu, L. W., Selby, J. E. A., Clough, S. A., and Fenn, R. W., *Atmospheric Transmittance/Radiance: Computer Code LOWTRAN 6*, AFGL-TR-83-0187, AFGL, Hanscom AFB, MA 01731, 1983.

[25] Kneizys, F. X., Shettle, E. P., Abreu, L. W., Chetwynd, Jr., J. H., Anderson, G. P., Gallery, W. O., Selby, J. E. A., and Clough, S. A., *Users Guide to LOWTRAN 7*, AFGL-TR-88-0177, AFGL, Hanscom AFB, MA 01731, 1988.

[26] Fenn, R. W., Clough, S. A., Gallery, W. O., Good, R. E., Kneizys, F. X., Mill, J. D., Rothman, L. S., Shettle, E. P., and Volz, F. E., "Optical and Infrared Properties of the Atmosphere," Chapter 18 in *Handbook of Geophysics and the Space Environment*, Jursa, A. S., Ed., AFGL-TR-85-0315 (AD A167000), USAF Geophysics Laboratory, Hanscom AFB, MA, 1985.

[27] Berk, A., Bernstein, L. S., and Robertson, D. C., *MODTRAN: A Moderate Resolution Model for LOWTRAN 7*, GL-TR-89-0122, USAF Geophysics Laboratory, Hascom AFB, MA, 1989.

[28] Rothman, L. S., et al., "The HITRAN Molecular Database: Editions of 1991 and 1992," *J. Quant. Spectrosc. Radiat. Transfer*, Vol. 48, 1992, p. 469.

[29] Anderson, G. P., Chetwynd, J. H., Theriault, J. M., Acharya, P., Berk, A, Robertson, D. C., Kneizys, F. X., Hoke, M. L., Abreu, L. W., And Shettle, E. P., "MODTRAN 2: Suitability for Remote Sensing," *SPIE Vol. 1968: Atmospheric Propagation and Remote Sensing II*, 1993, pp. 514–525.

[30] Chetwynd, J. H., Wang, J., and Anderson, G. P., "FASCODE: An Update and Applications in Atmospheric Remote Sensing," *SPIE Vol. 2266*, 1994, pp. 613–623.

[31] Wang, J., Anderson, G. P., Revercomb, H. E., and Knutseon, R. O., et al., "Validation of FASCODE 3 and MODTRAN 3: Comparison of Model Calculations with Ground-Based and Airborne Interferometer Observations Under Clear-Sky Conditions," *Applied Optics*, Vol. 35, No. 30, Oct. 1996, pp. 6028–6040.

[32] Watkins, W. R., Kantrowitz, F. T., and Crow, S. B., "Optical, Infrared, and Millimeter Wave Propagation Engineering," *Proc. SPIE 926*, 1988, pp. 69–84.

[33] Seagraves, M. A. and Ebersole, J. F., "Visible and infrared transmission through snow," *Optical Engineering*, Jan.–Feb. 1983, Vol. 22, No. 1, pp. 90–93.

[34] *Smoke and Natural Aerosol Parameters (SNAP) Manual*, Joint Technical Coordinating Group for Munitions Effectiveness, Smoke and Aerosol Working Group, Report 61, JTCG/ME-85-2, April 26, 1985.

[35] U.S. Army RD&A Bulletin PB 70-88-1 (Test), *Smoke/Obscurants Technology*, Jan.–Feb. 1988.

[36] Shirkey, R. C., "Effects of atmospheric and man-made obscurants on visual contrast," *SPIE Vol. 305 Atmospheric Effects on Systems Performance*, 1981, pp. 37–44.

[37] Shirkey, R. C., Duncan, L. D., and Niles, F. E., *EOSAEL 87, Vol. 1, Executive Summary*, Report TR-0221-1, U.S. Army Laboratory Command, Atmospheric Sciences Laboratory, White Sands Missile Range, NM 88002-5501, Oct. 1987.

CHAPTER 4

Millimeter-Wave Radar Sensors

The ability of radars to quickly scan large angular sectors makes them particularly suitable for detecting and warning of the approach of potentially dangerous objects. Radar frequencies penetrate the atmosphere under conditions unmatched by visible and infrared sensors. Although radars were originally developed in the 1920s, the major impetus that accelerated their maturity and propelled them into prominence was the air defense requirement of the Second World War. The first military radars operated at long wavelengths with pulse waveforms that were used to detect air and sea targets. Typical of the problems that had to be overcome were increasing the isolation between the transmitted and received waveforms and developing power sources and low-noise receivers that could detect distant targets and measure their range. The isolation problems can now be addressed through devices such as circulators, isolators, separate receive and transmit antennas, and antennas that incorporate high-transmit/receive isolation techniques, such as cross-polarized designs. Modern vacuum tube and solid state devices provide power over a frequency range whose upper limit borders on the infrared spectrum. Low-noise parametric and solid state integrated circuit amplifiers are now used to enhance the detection of weak signals.

Waveform selection and signal-processing technique are two of the more important radar design issues that are controlled primarily by the application. Radars designed to detect high-value targets, such as enemy aircraft, at long range generally incorporate more sophisticated signal processing and are more complex and expensive than those designed solely to detect lower value ground combat vehicles. Packaging volume and signal-processing time constraints in missiles and other fire-and-forget weapons also limit waveform selection and, to some extent, the signal-processing algorithms that can be applied. In an effort to increase the effectiveness of the lower cost radar designs against lower value targets, they are often combined with other sensors that exploit alternative signature-generation phenomena. In this way, their effectiveness is increased in a multiple target environment where inclement weather and countermeasures are also present.

The types of radars described in this chapter are monostatic pulse, frequency-modulated continuous wave (FMCW), Doppler, and synthetic aperture radars that have the size, power, cost, and performance attributes that make them suitable for fire-and-forget munitions and surveillance applications. The general operating principles presented apply to microwave and millimeter-wave (MMW) radars. The ability of polarimetric radars to provide target features for use in classification algorithms is also discussed. The operation of mixers and their characteristics are described, as are the derivation and interpretation of single-sideband and double-sideband noise figures. Since size is a major constraint in munitions applications, many of the design parameters and background and target signature data presented later in the chapter are characteristic of MMW frequencies. An exception is the microwave synthetic aperture radar that finds use in surveillance applications. Target detection theory for optimizing radar performance in ground clutter is contrasted with techniques suitable for detection of airborne targets where white Gaussian noise limits performance. Equations that give the angle-tracking accuracy of a radar are presented to show its dependence on the antenna's two-way 3-dB beamwidth and target scatterer fluctuations. Tree and snow high-clutter backgrounds are described in terms of their scattering coefficients, which are functions of viewing angle, frequency, polarization, and other factors such as surface and bulk water content. Several time- and frequency-domain target detection and classification algorithms are discussed, including parametric and nonparametric classification algorithms. A detailed discussion of bistatic radars, those that function with spatially separated transmitters and receivers, is given by Willis [1]. The operating principles of laser radars are described in Chapter 8.

4.1 RADAR CONFIGURATIONS

Radars detect that portion of the transmitted signal that is scattered from a target in the direction of the receiver. A large variety of waveforms can be transmitted from the radar, including pulsed, continuous wave, frequency- and polarization-agile, and compressed. The particular application dictates the waveform selection. For example, pulsed waveforms are typically used for longer range applications on air-to-air missiles where larger amounts of energy and high resolution are needed to detect a target. Modulated continuous waveforms find use in shorter range applications such as radar altimeters and missile fuzing. Unmodulated continuous wave Doppler radars are used to measure target speed. Frequency- and polarization-agile waveforms find applications in target classification and identification where information is extracted from as many of the different sized, shaped, and oriented scatterers on the target as the waveform agility allows. Compressed waveforms are used to obtain finer resolution of the target than might otherwise be obtainable from the time duration of the uncompressed transmitted pulse.

A major portion of radar design consists of selecting the transmitted signal waveform, transmitted and received signal polarizations, and signal-processing technique from among the options listed in Table 4.1. Polarization refers to the orientation of the electric field vector [2]. In parallel-polarized radars, the transmitted and received signal polarizations are the same. In cross-polarized designs, the transmitted and received signal polarizations are orthogonal to one another. In polarimetric radars, the transmitted signal is capable of switching polarizations from pulse burst to pulse burst, while a multiple channel receiver simultaneously processes several received signal polarizations. Time-domain and frequency-domain processing techniques are used to extract as much of the target information as possible from the received waveform. Factors that influence the choice of the signal-processing approach include sensor resolution, scan rate, field of view, processing time and processor volume constraints, degree of desired target discrimination (i.e., detection, orientation, classification, or identification), target and clutter density, artifacts such as glint, atmospheric and weather effects, and countermeasures.

4.2 MIXER OPERATION

A key component of most radars is a mixer. A mixer is a nonlinear device composed of one or more diodes used to convert a signal from one frequency to another,

Table 4.1
Radar Configuration Elements

Transmitted Signal Waveform	*Transmitted Signal Polarization*	*Received Signal Polarization*	*Signal-Processing Techniques*
• Pulsed single frequency • Pulsed frequency agility • Pulse compression • FMCW • Spread spectrum • Coherent	• Horizontal • Vertical • RHC* • LHC† • Linear • Elliptical • Combinations of above	• Horizontal • Vertical • RHC* • LHC† • Phase angle between pairs of the above • Linear • Elliptical • Combinations of above	• Averaging • Statistical measures • Frequency transform • Doppler beam sharpening • Synthetic aperture • Autocorrelation

* RHC = right hand circular.
† LHC = left hand circular.

where information extraction and signal processing are usually simplified. Referring to Figure 4.1, the nonlinear relation that expresses the output signal e_o (referred to as the intermediate frequency or IF signal) from a mixer in terms of the input e_i can be written as

$$e_o = a_1 e_i + a_2 e_i^2 + a_3 e_i^3 + \cdots + a_n e_i^n \tag{4.1}$$

Ideally, the input signal contains two terms, one from the local oscillator and one from the received signal, usually from an antenna, as shown in Figure 4.2. The input is written as

$$e_i = E_L \sin \omega_L t + E_S \sin(\omega_S t + \psi) \tag{4.2}$$

where

ω_L = local oscillator (LO) frequency;
ω_S = received signal frequency, sometimes called the radio frequency (RF) signal;
ψ = arbitrary phase angle.

Upon substituting (4.2) into (4.1), the mixer output becomes

$$\begin{aligned} e_o = {} & a_1[E_L \sin \omega_L t + E_S \sin(\omega_S t + \psi)] \\ & + a_2[E_L^2 \sin^2 \omega_L t + 2E_L E_S \sin \omega_L t \sin(\omega_S t + \psi) \\ & + E_S^2 \sin^2(\omega_S t + \psi)] + \cdots \end{aligned} \tag{4.3}$$

The middle term with coefficient a_2 is the one of interest as it contains the

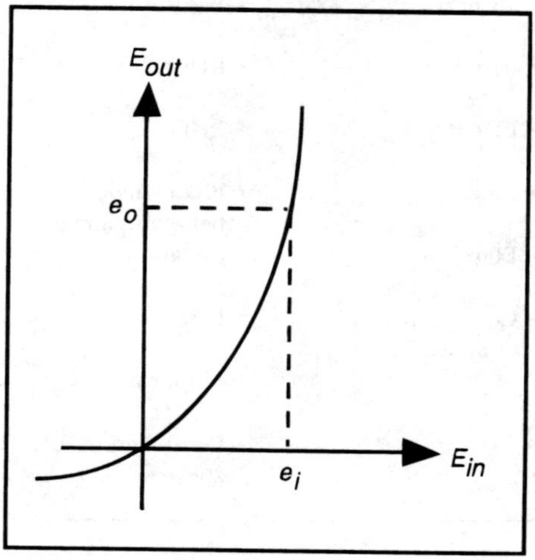

Figure 4.1 Nonlinear mixer diode operation.

desired signal information translated to the sum and difference frequencies formed by the LO and received signal. The other terms, including the squared sinusoids with coefficient a_2, are filtered out with low-pass and high-pass filters. The remaining desired terms are

$$2E_LE_S \sin \omega_L t \sin (\omega_S t + \psi) = E_LE_S\{\cos[(\omega_L - \omega_S)t - \psi] - \cos[(\omega_L + \omega_S)t + \psi]\} \quad (4.4)$$

The term with the $(\omega_L - \omega_S)$ difference frequency is the down-converted signal and the term with $(\omega_L + \omega_S)$ is the up-converted signal. In Figure 4.2, the down-converted term is kept as the desired output.

As the number of diodes used in a mixer design is increased, the performance of the mixer with respect to high input-to-output isolation and reduced spurious frequencies[1] is increased [3,4]. However, cost and volume occupied by the mixer also increase.

The single balanced mixer shown in Figure 4.3 is a good compromise for low-cost radar applications that require a mixer. It uses two diodes to cancel direct current (dc) components at the intermediate frequency, achieves a low-voltage standing wave ratio (VSWR) if the diodes are well-matched, and finally produces high local oscillator-to-input (RF) isolation [3]. All of these attributes enhance mixer performance by reducing the spurious response to harmonics of the LO and RF signals.

4.3 NOISE FIGURE AND RECEIVER SIDEBANDS

The *noise figure F* of a linear two-port device provides a measure of the degradation in signal-to-noise ratio between the input and output ports of the device and is equal to

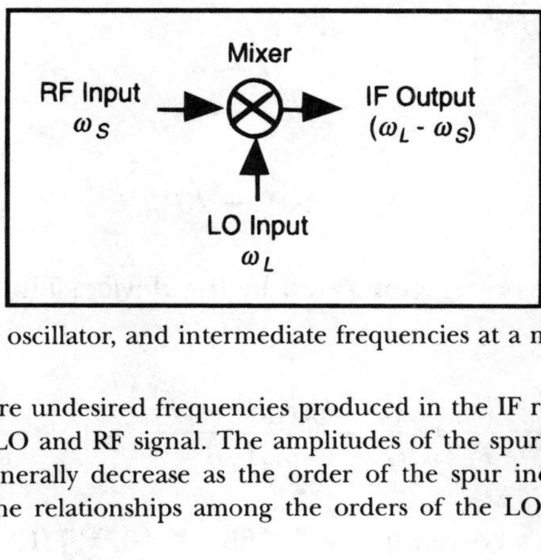

Figure 4.2 RF input, local oscillator, and intermediate frequencies at a mixer.

1. Spurious frequencies are undesired frequencies produced in the IF region from combinations of the harmonics of the LO and RF signal. The amplitudes of the spurious frequencies depend on the LO power and generally decrease as the order of the spur increases. Spur charts can be constructed to show the relationships among the orders of the LO and RF signals and the IF mixer output.

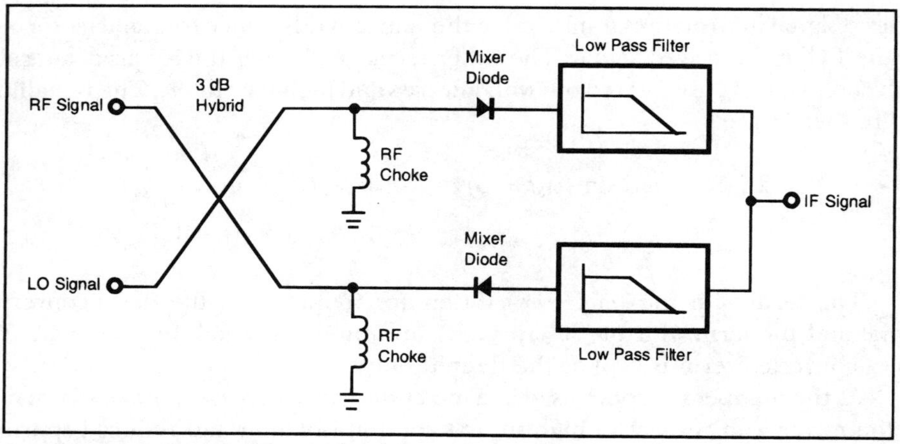

Figure 4.3 Single balanced mixer.

$$F = \frac{P_{si}/P_{ni}}{P_{so}/P_{no}} \tag{4.5}$$

where

P_{si} = available input signal power in W;
P_{ni} = available input noise power in W;
P_{so} = available output signal power in W;
P_{no} = available output noise power in W.

Noise figure is defined for a specific input noise power, namely the power provided by a resistor matched to the input port of the device at a temperature $T_O = 290\text{K}$. The average power gain G of the device is defined over the noise bandwidth B such that

$$P_{so} = GP_{si} \tag{4.6}$$

and

$$P_{no} = GP_{ni} + \Delta P_{no} \tag{4.7}$$

where ΔP_{no} is the noise power generated by the device. Thus, the noise figure can be rewritten as

$$F = \frac{P_{si}}{P_{so}} \cdot \frac{P_{no}}{P_{ni}} = \frac{1}{G} \cdot \frac{Gk_BT_OB + \Delta P_{no}}{k_BT_OB} = 1 + \frac{\Delta P_{no}}{Gk_BT_OB} \tag{4.8}$$

where k_B = Boltzmann's constant = 1.380662×10^{-23} J/K.

In practice, the noise figure of a real device is always larger than unity. Only for an ideal noise-free device where $\Delta P_{no} = 0$, does $F = 1$. The output noise power given by (4.7) may be written in terms of F as

$$P_{no} = Gk_B T_O B + Gk_B T_O B(F - 1) = FGk_B T_O B \tag{4.9}$$

Thus, the output noise power from a noisy device is equivalent to that from a noise-free device having an input noise temperature of $T_O + (F - 1)T_O = FT_O$. When noise figure is used to describe noise performance, it is necessary to specify the value of T_O at which F is measured. This temperature for T_O has been standardized at 290K.

An *equivalent input noise temperature* T_E for the device may also be defined. If ΔP_{no} is the noise power generated with the input of the device connected to a noise-free termination, then T_E is the temperature of a thermal resistor that, if placed at the input of an equivalent noise-free device, would produce the same noise power ΔP_{no} at the output terminals of the device. Thus,

$$\Delta P_{no} = Gk_B T_E B \tag{4.10}$$

The relation between equivalent input noise temperature and noise figure is found by equating the term representing ΔP_{no} term in (4.9) to (4.10). The result is

$$T_E = (F - 1)T_O \tag{4.11}$$

The number of *sidebands* in mixers and RF amplifiers used in MMW radars and radiometers affects the signal power reception and noise power of the device. Mixers and amplifiers can be designed for double-sideband or single-sideband operation. Examples of double-sideband receivers are a superheterodyne receiver that does not use image rejection mixers and a superheterodyne receiver that does not contain an RF amplifier or an RF filter in front of the mixer to limit the mixer input to only one of the two RF bands [3,5]. Double-sideband receivers have two RF spectral bands centered at

$$\omega_S = \omega_L \pm \omega_{IF} \tag{4.12}$$

that produce the same IF output frequencies as shown in Figure 4.4. If one of these is the desired RF signal, the other is referred to as the image frequency because the pair are mirror images with respect to the LO frequency. Each band contains similar signal information, thus making the double-sideband signal power twice that of a single-sideband receiver [6]. For this reason, double-sideband receivers are used in radiometers and radio telescopes to detect wideband signals.

Superheterodyne receivers containing RF to IF translation of a single frequency band of width ΔF are examples of single-sideband receivers. Single-sideband

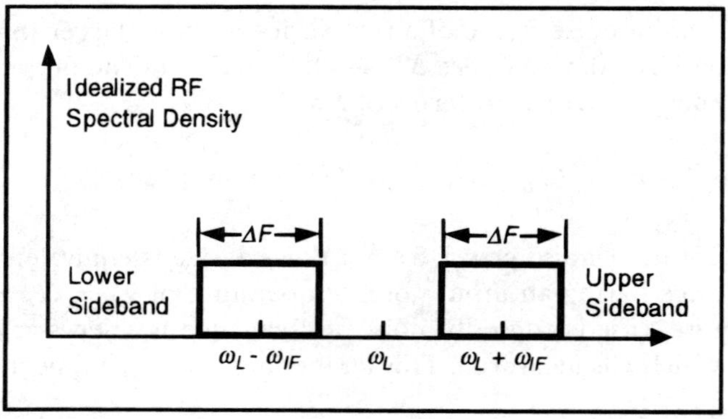

Figure 4.4 Receiver sidebands.

receivers find use in radars to limit interference from transmitters that have emissions in the rejected RF band, in radio astronomy applications where the signal spectrum is narrow, and in radio communications to maximize the transmitted power over a narrow spectral band.

The double-sideband receiver noise figure F_{DSB} is related to the single sideband receiver noise figure F_{SSB} by

$$F_{DSB} = \frac{1}{2}(F_{SSB} + 1) \tag{4.13}$$

as illustrated by the solid curve in Figure 4.5. Equation (4.13) is derived from (4.8) by writing F_{DSB} using a total RF bandwidth of $2B$, and F_{SSB} using a total RF bandwidth equal to B. The desired result is obtained upon solving for ΔP_{no} in the equation for F_{SSB} and substituting it into the equation for F_{DSB}. The factor of unity in (4.13) becomes especially important when the receiver noise figure is small, as indicated

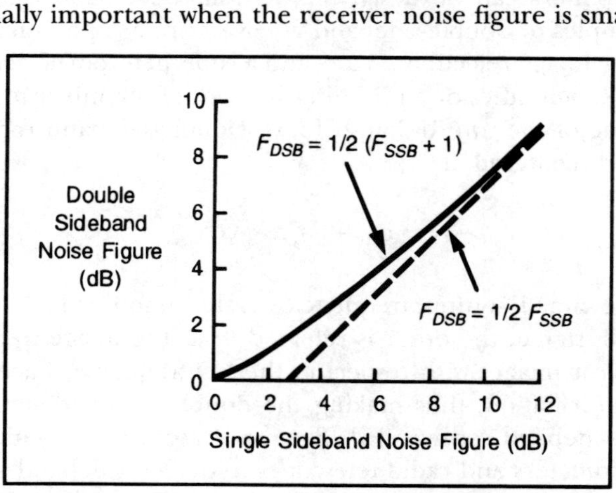

Figure 4.5 Relation between single- and double-sideband noise figures. The single- and double-sideband noise figures have been converted into decibels from the linear values calculated using (4.13).

by the gap between the solid and dashed curves, and its omission may lead to erroneous predictions of receiver performance.

The relation between the corresponding noise temperatures is found by substituting T_E from (4.11) into (4.13) to obtain

$$T_{DSB} = \frac{1}{2} T_{SSB} \tag{4.14}$$

The values T_{DSB} and T_{SSB} represent the equivalent input double-sideband and single-sideband noise temperatures, respectively. Thus, it is the noise temperatures in the double-sideband and single-sideband receivers that are related by the factor of one-half. In constrained volume and cost applications, single-sideband receiver noise figures are generally larger than double-sideband noise figures at MMW frequencies, although the reverse situation can occur when low-noise RF amplifiers are used.

4.4 PULSE RADAR

Target detection by pulse radars allows the reception of weak echoes from long-range targets by not allowing strong echoes from close-range clutter to occur at the same time instant, or the leakage signal from the transmitter to be received at the same time. The time delay of the echo from the long-range target also allows the range from radar transmitter to target to be measured.

A noncoherent pulse radar, illustrated in Figure 4.6, provides radar-to-target range R by measuring the time delay Δt between the transmitted and received signal. Thus,

$$R = \frac{c\,\Delta t}{2} \tag{4.15}$$

where c is the speed of light.

The MMW pulse transmitter outputs a series of pulses at time t through the transmit/receive (TR) switch and the antenna. In this example, the same antenna

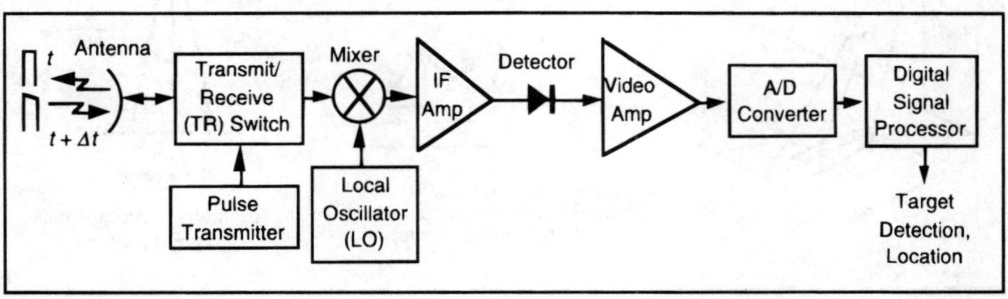

Figure 4.6 Pulse radar.

is used to both transmit and receive energy. At some time Δt later, depending on the range to the target, a portion of the transmitted energy is received back at the antenna. It passes through the TR switch into the mixer, where it is nonlinearly combined with the LO signal to produce the IF signal. After amplification and filtering, the IF signal is noncoherently detected using a detector that typically optimizes the signal-to-noise ratio or signal-to-clutter ratio, depending on the radar application and dominant source of signal interference. After further amplification and filtering, the video signal is digitized in an analog-to-digital (A/D) converter and input to a signal processor. Here, algorithms tailored to the resolution, polarization, and relative motion of the radar and targets are used to declare valid targets or objects of interest and compute their range relative to the radar location.

Pulse radar resolution for a *side-looking* geometry is described in Figure 4.7. Resolution r_a in the along-track direction is governed by the antenna dimension L

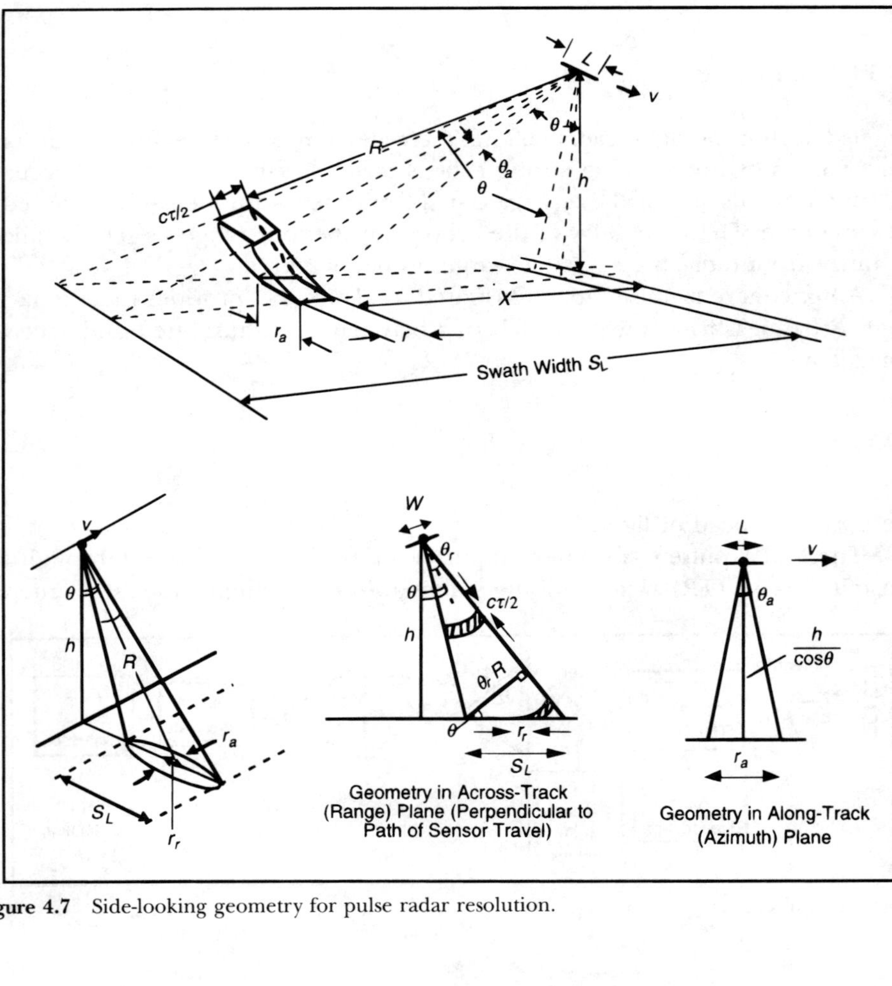

Figure 4.7 Side-looking geometry for pulse radar resolution.

through its effect on the two-way, half-power (3 dB) antenna beamwidth θ_a. The relation between these parameters is given by

$$r_a = 2R\sin(\theta_a/2) = 2h\sin(\theta_a/2)/\cos\theta \tag{4.16}$$

where

θ_a = along-track two-way 3-dB beamwidth of antenna;
θ = angle of incidence of the radar measured from nadir;
h = height of the radar above the ground.

For $\theta_a << 1$,

$$\sin(\theta_a/2) = \frac{\lambda}{2L} \text{ radians} \tag{4.17}$$

and

$$r_a = h\lambda/(L\cos\theta) \tag{4.18}$$

In the across-track (range) direction, resolution r_r is determined by the pulse length τ as

$$r_r = c\tau/(2\sin\theta) \tag{4.19}$$

The swath length S_L, representing the total range dimension scanned by the radar, is a function of the antenna dimension W in the across-track direction. Swath S_L is given by

$$S_L \approx h\lambda/(W\cos^2\theta) \tag{4.20}$$

when $\theta_r << 1$ (i.e., $\theta_r \approx \lambda/W$).

If the pulse radar is *forward-looking*, then the expressions for the along-track and cross-track resolutions are reversed. As shown in Figure 4.8, the swath for the forward-looking configuration is given by

$$S_L = h[\tan(\theta + \theta_a/2) - \tan(\theta - \theta_a/2)] = h\left[\tan\left(\theta + \frac{\lambda}{2L}\right) - \tan\left(\theta - \frac{\lambda}{2L}\right)\right] \tag{4.21}$$

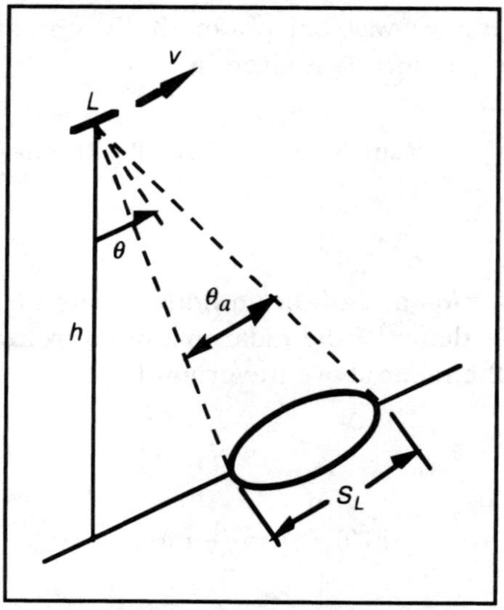

Figure 4.8 Swath geometry for forward-looking pulse radar.

4.5 FMCW RADAR

Frequency-modulated continuous wave (FMCW) radars are used to measure ranges to targets at closer distances than pulse radars can provide. Separate but collocated transmit and receive antennas are illustrated in the block diagram of Figure 4.9, although the same antenna can be employed for both functions when a circulator is used to provide isolation between the transmit and receive signals. A voltage-controlled oscillator (VCO), driven by a triangular modulation electronic circuit, is used as the transmitter. The VCO also acts as the LO by providing a portion of

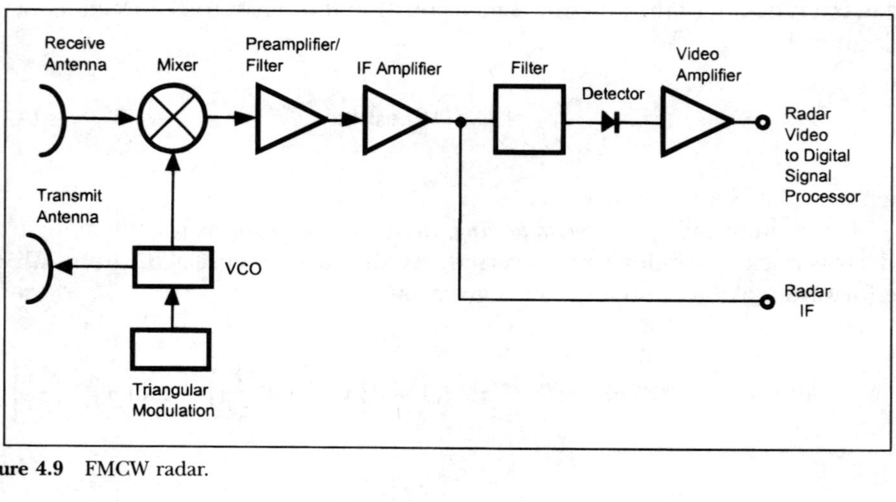

Figure 4.9 FMCW radar.

its output power to the mixer. Hence, this FMCW radar is an example of a homodyne radar, one that uses the same device as the transmit power source to the antenna and to the LO.

The portion of the transmitted energy that is backscattered from objects in the field of view of the radar enters the receive antenna and then the mixer, where it is nonlinearly combined with the VCO signal and translated in frequency into the IF signal. After amplification and filtering, the IF signal can be processed in its own right, as discussed in Chapter 5, or it can be processed noncoherently by passing it through an envelope detector. In the latter case, the signal is amplified, filtered, and digitized and then input to a digital signal processor where targets or objects are detected and classified and their range computed using algorithms developed for the particular application.

The IF FMCW radar signal, f_{IF}, contains the radar-to-target range information. The intermediate frequency at the output of the mixer is written as

$$f_{IF} = f\{t + \Delta t\} - f\{t\} \tag{4.22}$$

where $f\{t + \Delta t\}$ is the output frequency of the VCO at the time the backscattered signal reaches the receive antenna and $f\{t\}$ represents the frequency of the VCO at the time t when the FMCW signal is originally transmitted. By expressing the frequency of the FMCW signal as the sum of an instantaneous component f_0 and a component calculated from the rate of change of frequency with respect to time, (4.22) can be rewritten as

$$f_{IF} = [f_0 + \dot{f}(t + \Delta t)] - [f_0 + \dot{f}t] = \dot{f}\,\Delta t \tag{4.23}$$

$$= \dot{f}\,2R/c \tag{4.24}$$

where $\dot{f}$ = rate of change of frequency with respect to time and the relationship between R and Δt is established by (4.15).

Since $\dot{f}$ is equal to $2(\Delta F/T_m)$, as shown in Figure 4.10, and $T_m = 1/f_m$, (4.24) becomes

$$f_{IF} = 2\Delta F f_m(2R/c) = 4\Delta F f_m R/c \tag{4.25}$$

where

ΔF = RF modulation bandwidth of the transmitted waveform;
f_m = RF modulation rate corresponding to an up and down chirp cycle;
R = range between sensor and target or object of interest.

The factor of 2 in (4.25) that multiplies the $\Delta F f_m$ factor expresses the presence of both up and down frequency chirps in one modulation cycle T_m.

When the received signal backscattered from the target contains a Doppler-shifted component as shown in Figure 4.11, two difference frequency waveforms

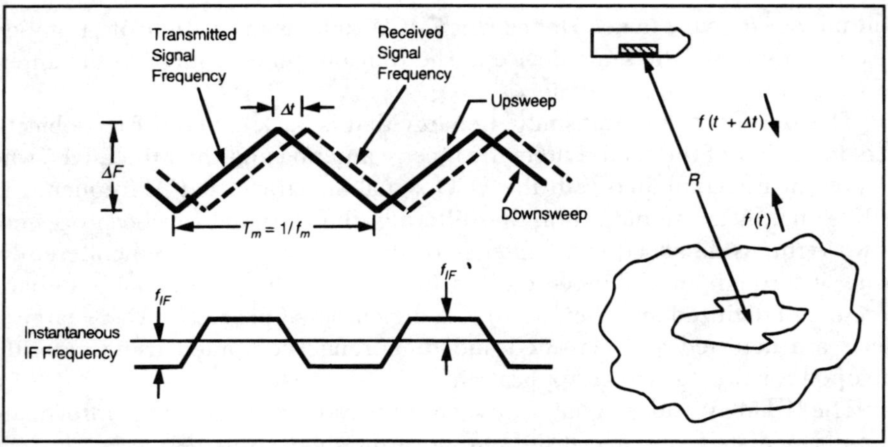

Figure 4.10 Generation of FMCW radar intermediate frequency.

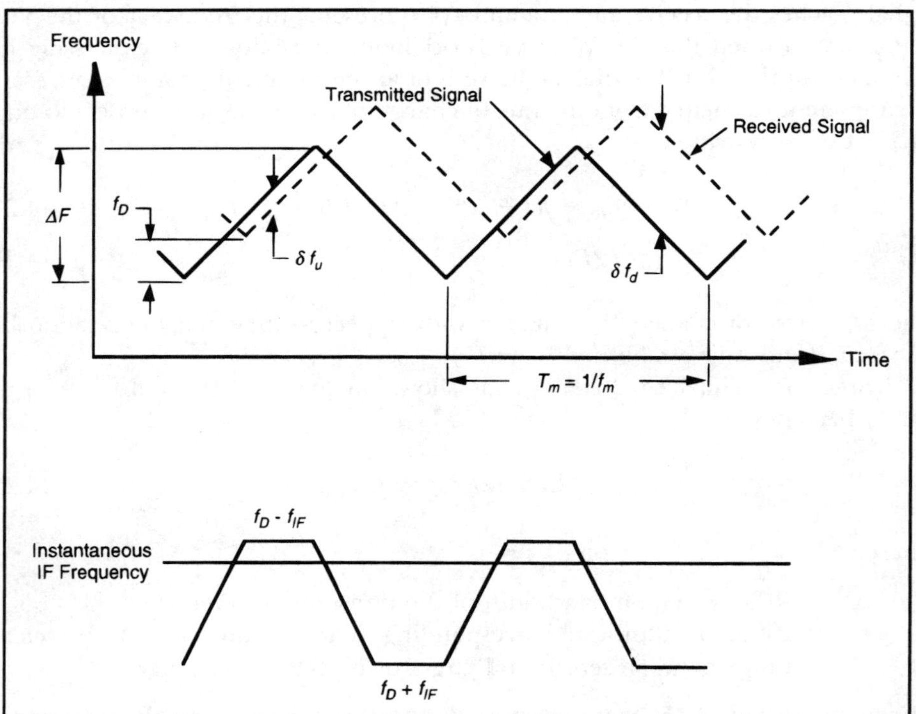

Figure 4.11 Doppler-shifted received signal from FMCW waveform.

are present: one for the illumination upsweep and one for the downsweep [7]. These waveforms, δf_u and δf_d, contain both the Doppler shift frequency f_D due to relative target motion and the frequency shift f_{IF} produced by the range to the target. When the Doppler shift is larger than the range-produced shift, δf_u represents the difference between the Doppler and range frequencies and δf_d represents the sum of the two frequencies. Assuming the slopes of the upsweep and downsweep are additive inverses (equal magnitudes, different signs) leads to the calculation of the Doppler shift frequency and range shift frequency as

$$f_D = (\delta f_d + \delta f_u)/2 \tag{4.26}$$

and

$$f_{\mathrm{IF}} = (\delta f_d - \delta f_u)/2 \tag{4.27}$$

When the Doppler shift is smaller than the range-induced frequency shift, the algebraic signs in the δf_d and δf_u terms enclosed by parentheses are reversed. The instantaneous intermediate frequency illustrated in Figure 4.11 is based on the convention used in (4.22).

FMCW radars may be designed to measure range in two ways: either (1) automatically decrease f_m with increasing range using a closed loop system that maintains f_{IF} constant, or (2) allow f_m to increase with range and, hence, also cause f_{IF} to increase. If the first option is employed, sufficient dynamic range must be designed into the system to accommodate the change in range over which the radar operates. For example, if the operating range varies between 50 and 1,000 feet (15 and 305m), then f_m must also change by a factor of 20 in order to maintain f_{IF} constant. A benefit of this design is the narrowband IF that imparts a measure of jamming immunity to the radar. If option two is used, then f_{IF} is 20 times as large as before and a large number of filters or the application of frequency-domain techniques must be used to find the range to the target. In this case, the use of small frequency intervals established by the filters helps to ameliorate the effects of jamming.

4.6 CW DOPPLER RADAR

Continuous wave (CW) Doppler radars transmit a constant frequency signal that is capable of detecting relative motion between targets, clutter, and the radar through the Doppler frequency shift of the signal backscattered from the target or clutter. CW radars are comparatively simple to build and require minimal transmitted signal bandwidth. They can measure the speed of targets over great distances and velocities. Figure 4.12 illustrates one implementation of a CW Doppler radar [8].

Here, a signal of frequency f_0 is transmitted from the radar. Some time later, a frequency-shifted signal is received at $f_0 + f_D$ where f_D, the Doppler shift caused by the motion of a target relative to the radar, is

$$f_D = (2v \sin \theta)/\lambda \tag{4.28}$$

where $v \sin \theta$ is the relative speed between target and radar, and θ is the incidence angle with respect to nadir defined in Figure 4.13. The corresponding depression angle (measured from the horizontal) is denoted by β. The value ϕ is the angle of the target track in the x, y plane.

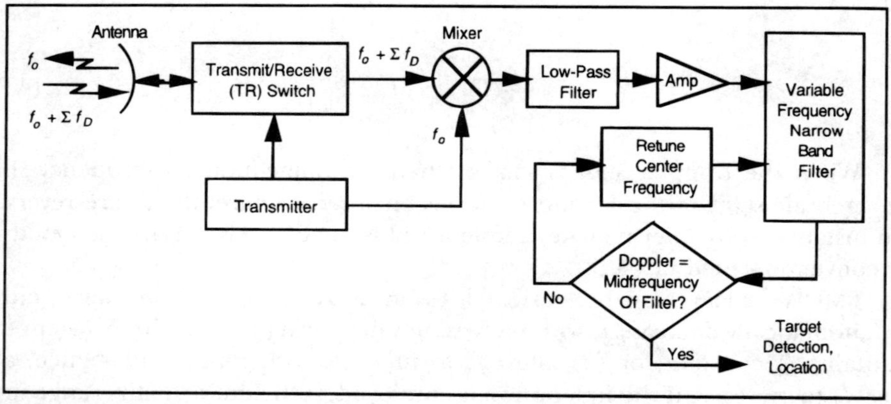

Figure 4.12 CW Doppler radar.

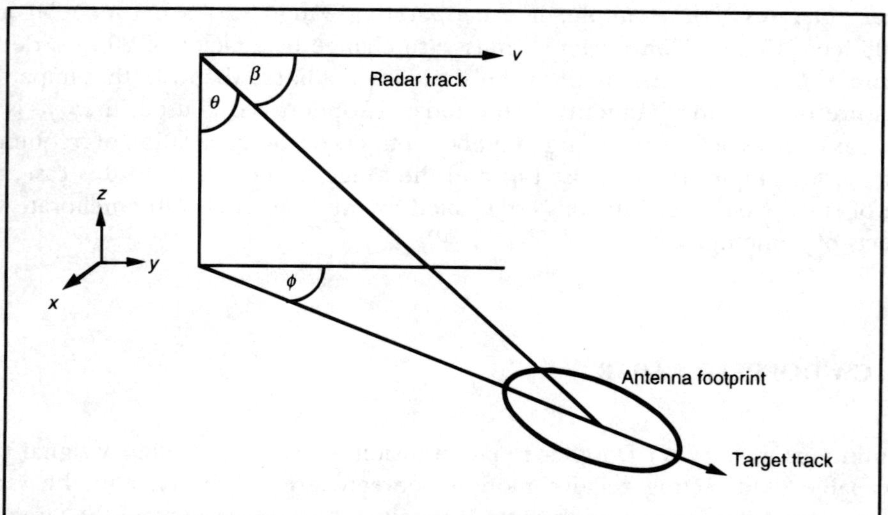

Figure 4.13 CW Doppler geometry.

If more than one moving object is present, a series of Doppler-shifted frequencies, represented as Σf_D in Figure 4.12, is returned. After passing through the TR switch, the backscattered signal enters the mixer where it is nonlinearly combined with the transmitted signal of frequency f_0, leaving only the Doppler in the IF. After filtering and amplification, the Doppler frequency is detected using a variable-frequency narrowband filter. By tuning the center frequency of the filter, a peak is detected at the Doppler frequency corresponding to the radial speed of the target.

4.7 PULSE DOPPLER RADAR

Pulse Doppler radars transmit one or more pulse waveforms that measure the relative speed and range of targets through coherent processing. Pulse Doppler waveform generation and processing is usually more expensive than that found in noncoherent radars and, hence, is applied to weapon applications against high-value targets or for Earth resource monitoring from space as with synthetic aperture radars.

As shown in Figure 4.14, the series of Doppler-shifted frequencies backscattered to the antenna passes through the TR switch and into a mixer. Here, they nonlinearly combine with a portion of the transmitted signal of frequency f_0, leaving only the Doppler frequencies in the IF. After filtering and amplification, the Doppler signals are processed by a range-gated bank of Doppler filters that outputs a target detection flag, speed, range, and cross-range information. Along-track and across-track resolutions for pulse Doppler radars are the same as those discussed for the noncoherent pulse radar.

4.7.1 Pulse Repetition Frequency

Low, medium, and high pulse repetition frequencies (PRFs) are used with pulse waveforms to determine the range to the target and its velocity. In a low-PRF waveform, the range to the target can be determined unambiguously by selecting the PRF as

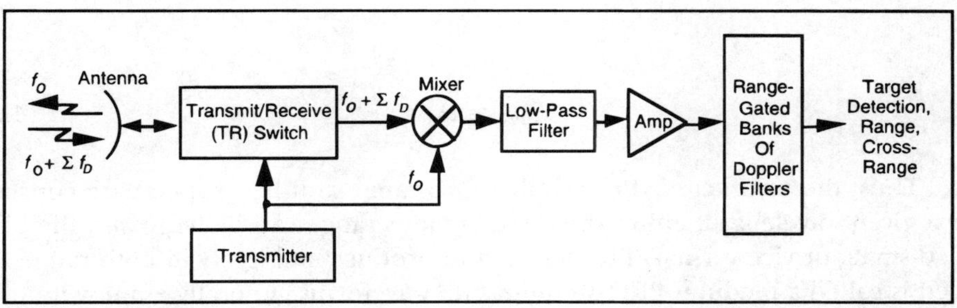

Figure 4.14 Pulse Doppler radar.

$$\text{PRF} = c/(2R_u) = 1/t_r \tag{4.29}$$

where

R_u = unambiguous range of the waveform;
t_r = pulse repetition interval.

When R_u is greater than the maximum expected range to the target, the waveform is called low PRF. Generally, low-PRF radars use time-multiplexed banks of Doppler filters to serve a large number of range gates [9,10].

A high-PRF waveform can be used to determine the unambiguous velocity of a target. The maximum speed that can be measured in an unambiguous manner is the blind speed v_b. The PRF that corresponds to v_b is

$$\text{PRF} = 2v_b/\lambda \tag{4.30}$$

where λ = wavelength of the transmitted energy.

When v_b is greater than the anticipated speed of the targets, the waveform is high-PRF. High-PRF waveforms generally employ one range gate with many Doppler filters to determine the range and speed of targets.

Medium-PRF waveforms are used when it is not possible to simultaneously satisfy the demands of unambiguous range and speed. In this case, the range and the speed are ambiguous and additional signal processing or transmission of different waveform patterns is required to resolve the ambiguity. The plot of the matched filter response to a particular waveform in the time-frequency plane, called the ambiguity function, is sometimes employed to select suitable waveforms for various applications [11].

Since by definition,

$$(\text{PRF})(t_r) = 1 \tag{4.31}$$

$$(2v_b/\lambda)(2R_u/c) = 1 \tag{4.32}$$

or

$$R_u v_b = c\lambda/4 \tag{4.33}$$

Thus, the product of the unambiguous range and blind speed are constant. For a given wavelength, either the unambiguous range can be large and the blind speed small, or vice versa. A PRF selected to produce ambiguity in both range and speed is called a medium PRF. Medium-PRF waveforms generally employ multiple range gates with multiple Doppler filters following each range gate.

As an example of how a medium-PRF waveform is designed, consider an X-band air-to-air radar with a requirement to detect Mach 2 targets at a range of 100 km. The design parameters for this example are given in Table 4.2.

The PRF consistent with an unambiguous range of 100 km is found from (4.29) as

$$\text{PRF} < c/2R = 1.5 \text{ kHz} \tag{4.34}$$

and thus specifies the maximum value of the PRF. The PRF, consistent with the blind speed of the Mach 2 target, is found from (4.30) as

$$\text{PRF} > 2v_t/\lambda = 80 \text{ kHz} \tag{4.35}$$

imposing a minimum value on the PRF. Since both of these constraints cannot be satisfied simultaneously, a medium-PRF waveform is selected for this application such that

$$1.5 \text{ kHz} \leq \text{PRF} \leq 80 \text{ kHz} \tag{4.36}$$

4.7.2 Resolving Ambiguities With a Medium PRF Waveform

Range ambiguities, produced by the uncertainty in associating a return with a particular transmitted pulse, may be resolved by transmitting multiple waveform patterns, each with a different PRF. The nonambiguous range is found by measuring (1) the number of elapsed pulses between the start of the transmission and the target return for each PRF and (2) the time T between the last transmitted pulse and the target return. Hovanessian [12] shows that these values allow the range to be calculated from the basic $R = cT/2$ relation using the known PRFs and the measured parameters described above. The transmission of multiple PRF waveforms is also used to reduce the blind zone caused by signal competition from sidelobe clutter.

Frequency ambiguities may also be resolved by transmitting multiple PRF waveforms. In addition to allowing the target speed to be found, they reduce the

Table 4.2
Design Parameters for Medium-PRF Radar Example

Parameter	*Value*
Target velocity v_t	1,200 m/sec
Wavelength λ	3 cm
Range R	100 km
Speed of light c	3×10^8 m/sec

blind zone caused by signal competition from mainlobe clutter. The technique is illustrated in the following example that employs three PRFs to resolve the frequency (speed) ambiguity in a medium PRF radar [12,13].

Assume that the basic PRF is 12 kHz and that Doppler filters are placed at 1-kHz intervals, as shown in Figure 4.15. The basic PRF is selected such that the Doppler shift from mainlobe clutter is located at the PRF line. Assume further that the clutter Doppler bandwidth is 3 kHz and thus occupies three filters. To prevent target returns from falling within the filters occupied by mainlobe clutter, select two additional PRFs separated from the basic PRF by the clutter Doppler bandwidth. Thus, the two added PRFs are at 9 kHz and 15 kHz (12 kHz ± 3 kHz).

If the Doppler shift due to the target (denoted as TGT in the figure) is at 24 kHz, it falls within the filters occupied by the clutter when the 12-kHz PRF is

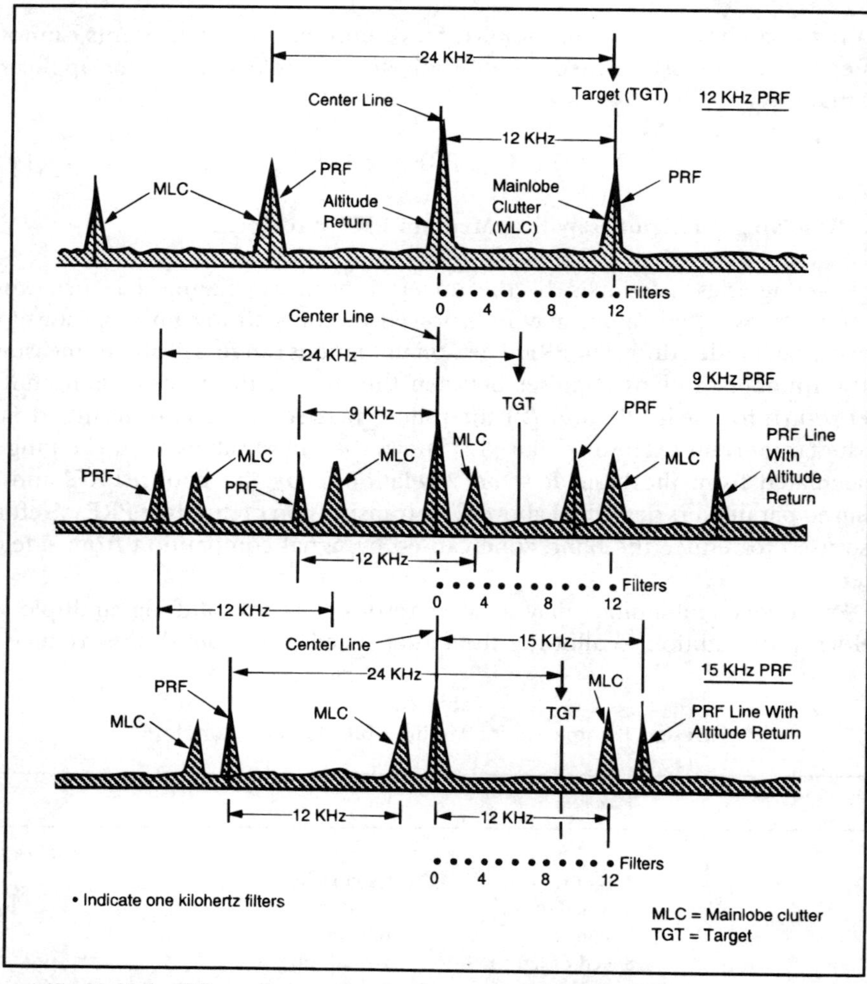

Figure 4.15 Multiple PRF waveforms to resolve frequency ambiguities [12].

transmitted as shown in the upper portion of Figure 4.15. However, the 9-kHz and 15-kHz PRFs remove the target Doppler from the clutter screen by placing the target in filter 6 and filter 9, respectively, as illustrated in the middle and lower parts of the figure. The true Doppler corresponding to the target is found by applying the Chinese remainder theorem [13]. PRF increments, in kilohertz, are added to the corresponding filter numbers until a match is obtained as described in Table 4.3. Thus, the correct Doppler shift is 24 kHz.

4.8 SYNTHETIC APERTURE RADAR

A synthetic aperture radar (SAR), operating in a side-looking geometry, increases the along-track resolution by keeping the target in the beam for a significant amount of time and observing the target from numerous locations along the radar flight path, as shown in Figure 4.16 [14–19]. Although early SAR applications were military in nature, SAR data are now processed to obtain images of geological features, ocean wave heights, sea ice, and crops [20–23].

Table 4.3
Resolving Frequency Ambiguity in a Medium-PRF Radar With Multiple Waveforms

PRF	*Doppler Shift*
9 kHz	6 + 9 + 9 = 24
15 kHz	9 + 15 = 24

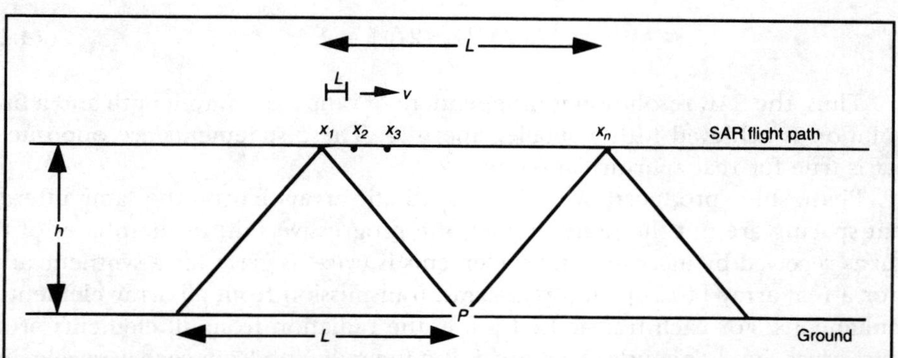

Figure 4.16 Observation points along flight path of a synthetic aperture radar.

In order to find the along-track resolution of a SAR, consider that it moves an antenna element of physical length L at a constant velocity v. The length of the synthetic antenna aperture L that is formed is given by

$$L = \theta_r h = k_a \lambda h / L \tag{4.37}$$

where

θ_r = 3-dB beamwidth of the antenna pattern generated by the real aperture of length L;
h = height or altitude at which the radar platform is located;
k_a = aperture illumination taper factor;
λ = wavelength of the transmitted energy.

As the radar moves along its flight path, successive echoes from an object at location P, shown in the middle of the ground reference line in Figure 4.16, are recorded at points $x_1, x_2, \ldots, x_n$ along the flight path. The amplitude and phase of the echoes are recorded coherently as a function of time. These represent signals that would have been received by an actual array of antenna elements if such existed. The echoes are combined in a signal processor to synthesize the synthetic antenna array whose corresponding 3-dB beamwidth θ_s is equal to

$$\theta_s = \lambda / (2L) \tag{4.38}$$

assuming the aperture illumination taper factor is unity.

The factor of 2 in the denominator of (4.38) arises because the phase difference between equally spaced elements of a synthetic array is twice that of a real array with the same spacing, as explained later in this section.

The resulting ground resolution r_a of the synthetic array is found by substituting h from (4.37) and θ_s from (4.38) into

$$r_a = h\theta_s \tag{4.39}$$

$$= [LL/\lambda][\lambda/(2L)] = L/2 \tag{4.40}$$

Thus, the SAR resolution is independent of range and wavelength and a finer resolution is achieved with a smaller aperture. These statements are opposite of what is true for real aperture antennas.

Phase shifts produced by real and synthetic arrays having the same interelement spacing are not the same. In fact, the progressive shift in the phases of the returns received by successive array elements is twice as great for a synthetic array as for a real array [10,24]. In a real array, transmission from all array elements is simultaneous. For each transmitted pulse, the radiation from all elements of the array arrives simultaneously at a point P. But the radiation from each array element is staggered in phase at point P as a result of the progressive differences in the distances from each successive element to *P*. For each pulse, the phase shift from the sum of the radiation from all the array elements is controlled by the distance from the central element to *P*. Therefore, if the position of the antenna is not changed, all the pulses reflected by *P* have the same phase. The energy in each reflected pulse received by the individual array elements similarly differs in phase as a result of the progressive difference in the distances from *P* to the elements. However, the phase differences are again due only to the differences in the one-way distances from *P* to the elements.

In a synthetic array, pulses are transmitted sequentially from the individual array elements (i.e., the first pulse is transmitted and received entirely by the first element, the second pulse is transmitted and received entirely by the second element, etc.). Consequently, the returns received by the successive elements differ in phase by amounts proportional to the differences in the round-trip distance from each element to *P* and back to the element again. Thus, for any angle off boresight, the progressive shift in the phases of the returns received by successive synthetic array elements is twice as great as for a real array.

Because of the doubling of the phase shift in a synthetic array, the null-to-null beamwidth of a synthetic array is one-half the null-to-null beamwidth of a real array of the same length. The two-way 3-dB beamwidth of a synthetic array is approximately 70% of the two-way 3-dB beamwidth of a real array. The doubling of the phase shift must be correctly accounted for when calculating such factors as phase corrections needed to focus a synthetic array and angles at which grating lobes occur.

4.8.1 Focused and Unfocused SAR Operation

In a focused SAR, each received echo is phase-shifted to account for the variable distance between SAR and target. During image processing, the phase shift ψ_i is added to the echo received from each position x_i. The phase shift value is

$$\psi_i = \frac{4\pi}{\lambda}(r_i - h) \tag{4.41}$$

where

h = height of SAR above the target;
r_i = distance between target and SAR position x_i.

Resolution of a focused SAR is independent of range because the length of the synthetic aperture is directly proportional to height as expressed by (4.37). Since a larger height gives a larger aperture, the corresponding 3-dB beamwidth is reduced, keeping the resolution, r_a, constant. Resolution is also wavelength-independent. This is shown by observing that longer wavelengths increase the potential synthetic aperture length given by (4.37). Since the synthetic beamwidth is proportional to wavelength divided by synthetic aperture length as in (4.38), beamwidth is also independent of wavelength. Therefore, resolution (equal to height multiplied by synthetic beamwidth as in (4.39)) is independent of wavelength.

In an unfocused SAR, the synthetic array length is shortened and signal processing is simplified as phase corrections less than $\pi/4$ radians are neglected. However, resolution now depends on height and wavelength. To find the resolution dependence on height and wavelength, we compare the maximum range at which unfocused SAR operation occurs with the maximum range at which focused SAR operation is possible.

Referring to Figure 4.17, the phase shift associated with the maximum operating range R_{mf} for a focused SAR is [19,24]

$$\psi_{mf} = 2k(R_{mf} - h) \tag{4.42}$$

and that associated with the maximum operating range R_{mu} for an unfocused SAR is

$$\psi_{mu} = 2k(R_{mu} - h) \tag{4.43}$$

By substituting $k = 2\pi/\lambda$,

$$\psi_{mu} = \frac{4\pi}{\lambda}(R_{mu} - h) \tag{4.44}$$

Since we want to find the conditions under which $\psi_{mu} \leq \pi/4$, we set

$$\frac{4\pi}{\lambda}(R_{mu} - h) \leq \frac{\pi}{4} \tag{4.45}$$

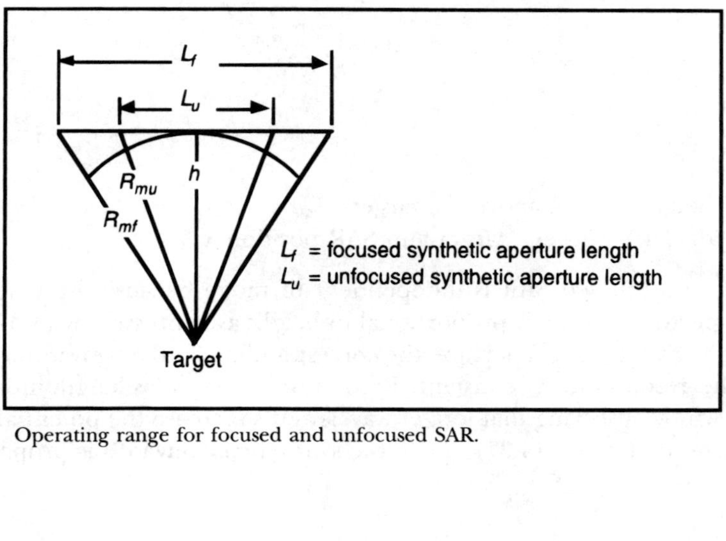

Figure 4.17 Operating range for focused and unfocused SAR.

or, rewriting,

$$R_{mu} - h \leq \lambda/16 \tag{4.46}$$

From Figure 4.17,

$$R_{mu}^2 = h^2 + (L_u/2)^2 \tag{4.47}$$

where L_u is the length of the unfocused synthetic aperture and L_f is the length of the focused synthetic aperture. Thus,

$$(R_{mu}/h)^2 = 1 + [L_u/(2h)]^2 \tag{4.48}$$

Since $h > L_u$,

$$R_{mu}/h = \sqrt{1 + \left(\frac{L_u}{2h}\right)^2} \approx 1 + \frac{1}{2}\left(\frac{L_u}{2h}\right)^2 \tag{4.49}$$

and

$$R_{mu} - h \approx L_u^2/(8h) \tag{4.50}$$

From (4.46), $R_{mu} - h \leq \lambda/16$ so that

$$L_u^2 \leq \frac{8\lambda h}{16} \leq \frac{\lambda h}{2} \tag{4.51}$$

and

$$L_u \leq \sqrt{\lambda h/2} \tag{4.52}$$

The resolution of the unfocused SAR, equal to the synthetic 3-dB beamwidth θ_s multiplied by height, becomes

$$r_a = \frac{\lambda h}{2L_u} \geq \sqrt{\frac{\lambda h}{2}} \geq L_u \tag{4.53}$$

Thus, the along-track resolution of an unfocused SAR is at best equal to its synthetic aperture length and, hence, is a function of height and wavelength.

However, the signal-processing requirements are relaxed with the unfocused SAR. The unfocused SAR views the resolution cell for all the time needed to generate the resolution cell image, but no longer. The data processor is turned over to the next resolution cell as soon as it has completed processing data from the previous cell. In focused SARs, the synthetic aperture length is much longer than a resolution cell. Hence data for many cells are processed simultaneously [24].

A comparison of the resolution of a real aperture radar with that of a focused and unfocused SAR is given in Figure 4.18 using a 10-GHz X-band radar having a real aperture length of 1 m as a model. As expected, the focused SAR has the best resolution, followed by the unfocused SAR and the radar with the real aperture.

4.8.2 Spotlight-Mode Operation

In spotlight-mode operation, shown in Figure 4.19, the SAR antenna is continuously aimed at the target located at point *P*. This allows the target to be viewed over a longer period from positions x_1 through x_n and improves along track resolution. If *P* is observed over a time *t* such that

$$t = L_s / v \tag{4.54}$$

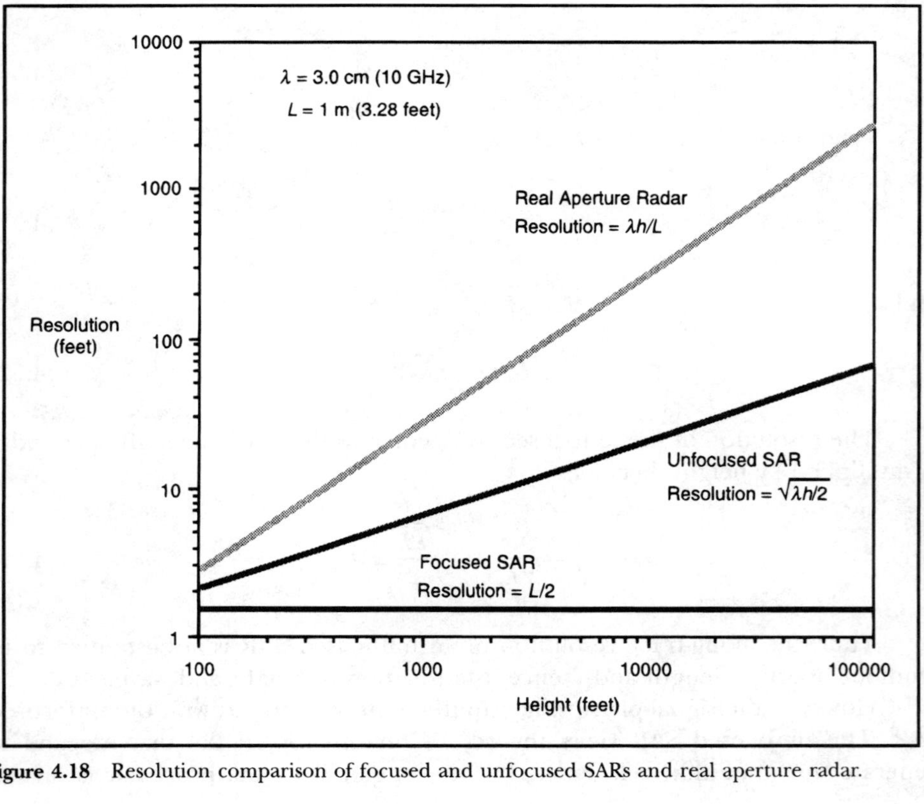

Figure 4.18 Resolution comparison of focused and unfocused SARs and real aperture radar.

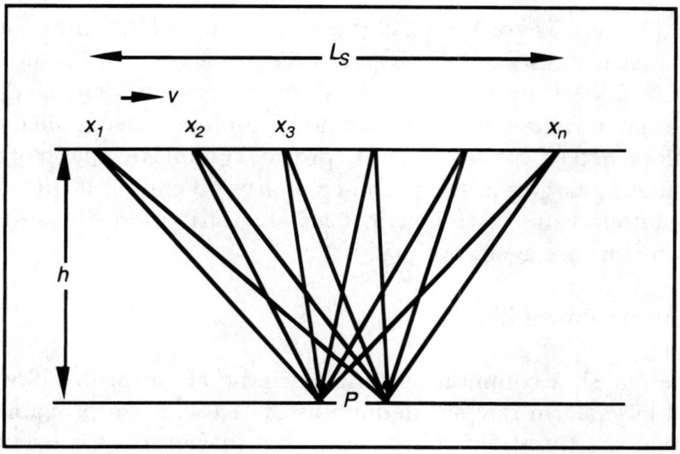

Figure 4.19 SAR spotlight mode.

then the resolution becomes

$$r_a = \lambda h/(2L_s) = \frac{\lambda h}{2vt} \tag{4.55}$$

where L_s = synthetic aperture length for the spotlight SAR.

Viewing time is limited by the curvature of the Earth and the requirement to minimize the lost opportunities to view other objects while staring at the target at point *P*.

4.9 OPTIMIZING RADAR TARGET DETECTION IN GROUND CLUTTER

Early radars were developed to detect targets in the presence of noise, as for example in the detection of air targets from ground-based radars that transmit a sequence of pulses. The optimal receivers for maximizing output signal-to-noise ratio (S/N) in the presence of white Gaussian noise for these noncoherent radars are envelope detectors and matched filters designed for the transmitted waveform [25]. For small-input signal-to-noise ratios, the envelope detector is operated in the square law region, whereas for large-input signal-to-noise ratios it is operated in the linear region. The performance of both detectors is approximately the same [25–27]. Models for nonfluctuating and fluctuating targets have been developed to describe their detection probability as a function of false alarm probability and S/N [27–30]. When radars are used to detect targets in ground clutter, however, the dominant interference is no longer Gaussian noise in most cases, but rather clutter whose amplitude statistics are more often described by Weibull or log normal distributions [31–34]. These imply a higher probability of large amplitude scatterers than with Gaussian statistics as illustrated in Figure 4.20.

Since the underlying statistic for the interference has changed, the square law and linear law envelope detectors are not optimum. Instead, binary integrators and log envelope detectors may be used to provide optimal detection of nonfluctuating and fluctuating targets in the presence of ground clutter. If the amplitudes of the ground-clutter scatterers follow a Rayleigh distribution, the square law and linear law detectors are again optimal.

4.9.1 False Alarm Probability

In order to establish a common terminology, false alarm probability, false alarm number, and false alarm rate are defined below. False alarm probability P_{fa} is the probability that the threshold voltage v_t, shown in Figure 4.21, is exceeded by a signal produced by noise or clutter when no target signal is present [35].

Fehlner's false alarm number [26] n_F is the number of independent samples after video integration that must be observed before one or more false alarms occur with probability of one-half. Fehlner's false alarm number and false alarm probability are related by

$$n_F = 0.693/P_{fa} \tag{4.56}$$

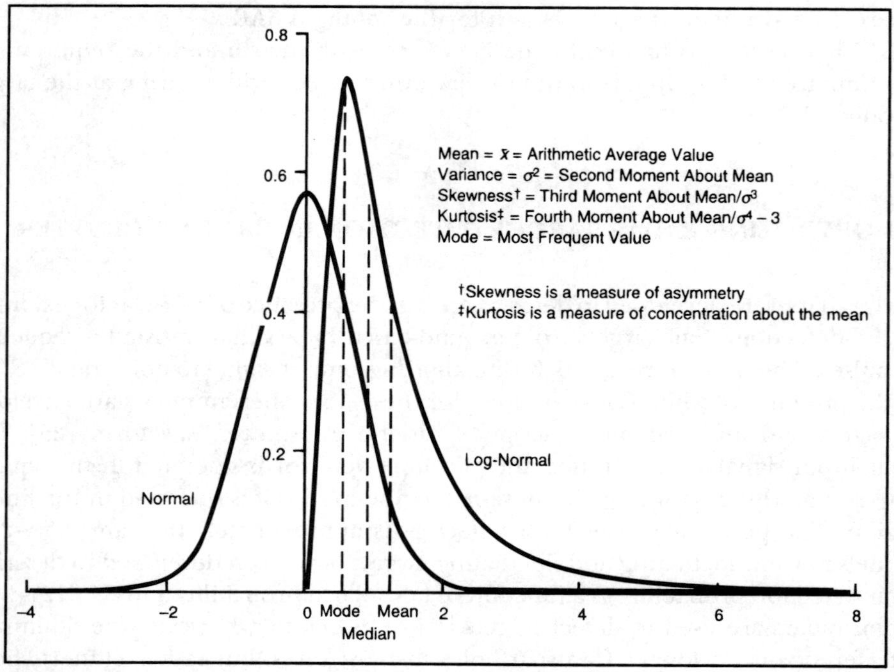

Figure 4.20 Normal and log-normal probability density functions [26].

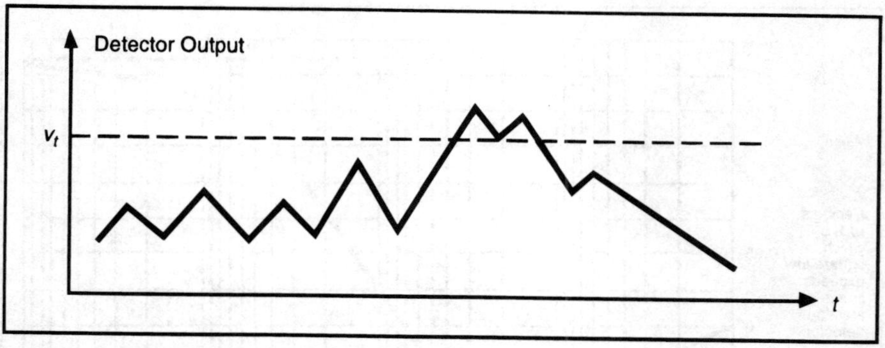

Figure 4.21 False alarm probability.

False alarm rate R_{fa}, equal to the reciprocal of the false alarm time, is defined as the average number of false alarms per unit time. False alarm rate is related to false alarm probability through the number of samples N noncoherently integrated and the integration time t_i by

$$R_{fa} = NP_{fa}/t_i \tag{4.57}$$

Marcum's false alarm number [26] n_M is related to the Fehlner false alarm number by

$$n_M = n_F N \tag{4.58}$$

In Figure 4.22, Marcum's false alarm number is used to illustrate the small differences in performance between linear law and square law detectors for nonfluctuating target detection in noise. Each value of N in the abscissa corresponds to a different false alarm probability, as shown in Table 4.4.

4.9.2 Target Fluctuation Models

In order to calculate the correct detection probability for a particular target, its fluctuation characteristics must be established. A summary of the features of popular nonfluctuating (Marcum) and fluctuating (Swerling) target models is given in Table 4.5. The nonfluctuating model was developed for spherical and nearly spherical objects many wavelengths in diameter. The Swerling 1 target was developed to model jet aircraft and Swerling 2 to model propeller aircraft and turbine blades. In general, small changes in the orientation of Swerling 2 targets produce significant changes in the echoing area.

Swerling 3 and 4 models are associated with stabilized missile bodies. They do not have as large an amount of inherent randomness as the group of random scatterers found in the Swerling 1 and 2 models. The chi-squared density function used in the Swerling 3 and 4 models represents a large reflector subject to small

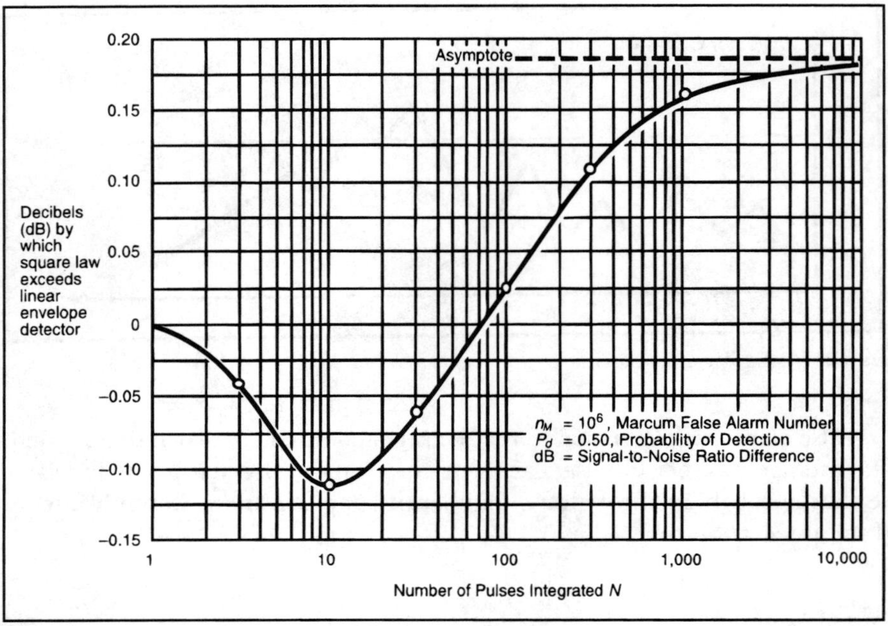

Figure 4.22 Linear law and square law envelope detector performance for nonfluctuating target detection in noise. [27].

Table 4.4
Comparison Among Fehlner and Marcum False Alarm Numbers and False Alarm Probability

N	n_M	n_F	P_{fa}
1	10^6	1.0×10^6	6.93×10^{-7}
3	10^6	6.3×10^5	2.10×10^{-6}
10	10^6	1.0×10^5	6.93×10^{-6}
30	10^6	6.3×10^4	2.10×10^{-5}
100	10^6	1.0×10^4	6.93×10^{-5}
300	10^6	6.3×10^3	2.10×10^{-4}
1,000	10^6	1.0×10^3	6.93×10^{-4}

changes in orientation. Swerling 1 and 3 models are also commonly applied to ground combat vehicles such as tanks. Detection probabilities for other target models, such as log normal often identified with ships and space vehicles, can be derived from the Swerling models as explained by Meyer and Mayer [26].

When a constant threshold voltage is selected to achieve the desired false alarm probability, the number of false alarms can still vary due to drifts in receiver gain, changes in receiver noise, or the introduction of external noise or clutter

Table 4.5
Nonfluctuating and Fluctuating Target Models

Target Model	*Probability Density Function for Target Scatterer Model*	*Physical Interpretation of Target Model*	*Fluctuation of Scatterer Amplitude*	*Application*
0 (Marcum)	Nonfluctuating	Steady reflector	No fluctuation	Spherical and nearly spherical objects such as balloons
1 (Swerling 1)	Rayleigh	Many independent random scatterers, where no single one is dominant	Slow (scan-to-scan fluctuation)	Jet aircraft and ground combat vehicles
2 (Swerling 2)			Rapid (pulse-to-pulse fluctuation)	Propeller aircraft, turbine blades, and long thin objects
3 (Swerling 3)	Chi-square with 4 degrees of freedom	One dominant plus many smaller independent scatterers	Slow (scan-to-scan fluctuation)	Ground combat vehicles and stabilized missile fuel tank or body
4 (Swerling 4)			Rapid (pulse-to-pulse fluctuation)	Stabilized missile fuel tank or body

signals. Excessive false alarms can cause signal-processing hardware to overload or algorithms to fail. Therefore, automatic approaches, referred to as constant false alarm rate (CFAR) techniques, are used to maintain the number of false alarms constant. Common approaches for CFAR observe the noise or clutter background in the vicinity of the target and adjust the threshold in response to the measured background backscatter statistics.

4.9.3 Target Detection in Non-Rayleigh Clutter

Consider the problem of maintaining a given detection probability at a constant false alarm rate without increasing the threshold voltage or requiring an increased signal-to-clutter ratio (S/C) (assuming clutter is the limiting interference) in the presence of non-Rayleigh clutter. It is undesirable to increase the threshold to maintain CFAR because valid target signals, in addition to clutter, will be suppressed.

The pervasiveness of non-Rayleigh clutter backgrounds is shown in Table 4.6 [31]. Terrain and sea state clutter are modeled with Weibull distributions that have probability density functions of the form

$$p_W(x; \alpha_W, \beta) = (x/\beta)^{\alpha_W - 1}(\alpha_W/\beta) \exp[-(x/\beta)^{\alpha_W}] \quad (4.59)$$

where

x = a random variable representing the clutter return;
α_W = skewness parameter;
β = scale parameter.

When $\alpha_W = 2$, the Weibull distribution reduces to Rayleigh, and when $\alpha_W = 1$, to an exponential distribution. As α_W decreases from 2, the probability of large amplitude scatterers increases.

4.9.4 Square Law Detector Performance

Suboptimal performance of a radar with a square law or linear law detector operating in Weibull clutter is illustrated in Figure 4.23(a) for a nonfluctuating target [31,32]. The signal-to-clutter ratios required to achieve relatively high detection probabilities are greater when $\alpha_W < 2$ (implying more skewing of the amplitude distribution from that of a Gaussian) than if the clutter were Rayleigh. Similarly, Figure 4.23(b) shows that larger signal-to-clutter ratios are needed to achieve high detection probabilities in log normal clutter than in Rayleigh clutter. In fact, Figure 4.23(c) shows that a radar designed to operate at a false alarm probability of 10^{-6} in Rayleigh clutter only achieves a false alarm probability

Table 4.6
Weibull Clutter Parameters for Distributed Clutter [31]

Terrain/Sea State	*Frequency*	*Beamwidth (degrees)*	*Depression Angle (degrees)*	*Pulse Width (μsec)*	α_W
Rocky mountains	S	1.5	—	2	0.512
Wooded hills	L	1.7	≈0.5	3	0.626
Forest	X	1.4	0.7	0.17	0.506–0.531
Cultivated land	X	1.4	0.7–5.0	0.17	0.606–2.0
Sea state 1	X	0.5	4.7	0.02	1.452
Sea state 3	Ku	5	1.0–30.0	0.1	1.160–1.783

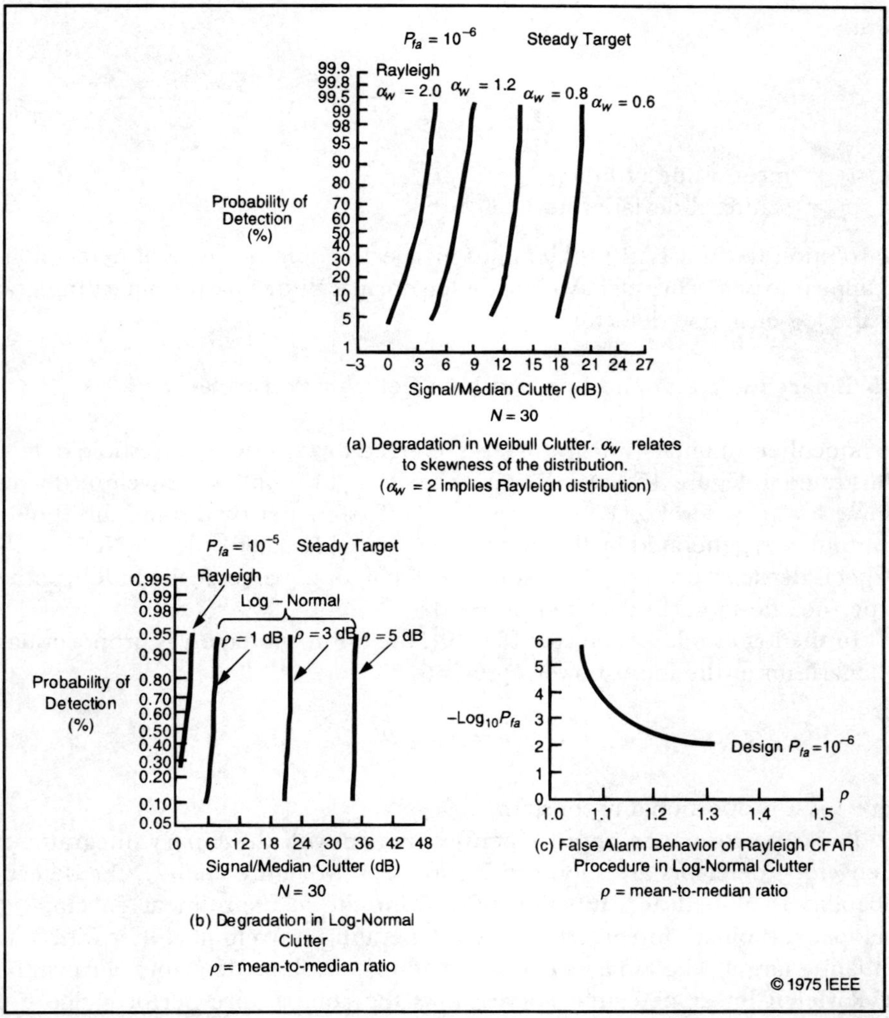

Figure 4.23 Radar performance in Weibull and log normal clutter using square and linear law detectors. (*Source:* [31,32]. Reprinted with permission.)

of 10^{-4} in log normal clutter characterized by a mean-to-median ratio ρ of approximately 1.1. Larger values of ρ correspond to a higher probability that large-amplitude scatterers will be encountered.

The probability density function for the log normal distribution is given by

$$p_L(x;\, m,\, \sigma) = \frac{1}{x\sigma\sqrt{2\pi}} \exp[-(\ln x - m)^2/2\sigma^2] \tag{4.60}$$

where

$$x > 0, \sigma > 0,$$

m = mean value of ln x;
σ = standard deviation of ln x.

The solution to the CFAR problem lies in selecting another type of detector. Two that appear to perform well in Weibull or log normal clutter are the binary integrator and the log envelope detector.

4.9.5 Binary Integrator and Log Envelope Detector Performance

The noncoherent binary integrator, also referred to as a double threshold detector, is illustrated in Figure 4.24. An IF signal consisting of N pulses is envelope detected yielding a voltage level v_i. When (sgn v_i + 1)/2 is greater than some threshold η_T, an output R_i is generated by the binary integrator [31,32,36].[2] If $\Sigma_{i=1}^{N} R_i > m$, then a target is declared present. If the sum is less than or equal to m, the null hypothesis is true (i.e., no target is declared present).

In the log envelope detector [37–39], the output voltage v is proportional to the logarithm of the input power P_i such that

$$v = b \log P_i \tag{4.61}$$

where b is a proportionality constant.

Radar performance improvements obtained with the binary integrator and log envelope detectors are shown in Figure 4.25. In Figure 4.25(a), the detection probability is plotted as a function of S/C for the median, linear envelope, log envelope, and binary integrator detectors operating in Weibull clutter with a nonfluctuating target. The binary integrator and log envelope detectors perform best. The Rayleigh linear detection curve shows the comparative performance if the radar were operating in Rayleigh clutter with a linear detector. The Chernoff bound represents a theoretical upper performance bound for the detection efficiency.

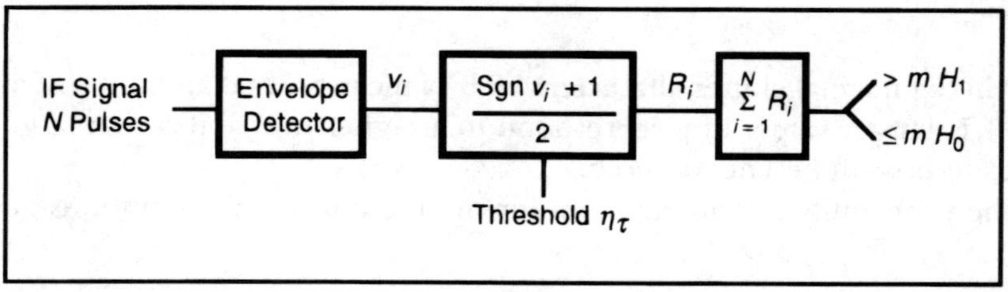

Figure 4.24 Binary integrator detector.

2. The signum of a real number a (abbreviated sgn a) is defined as sgn a = $a/|a|$.

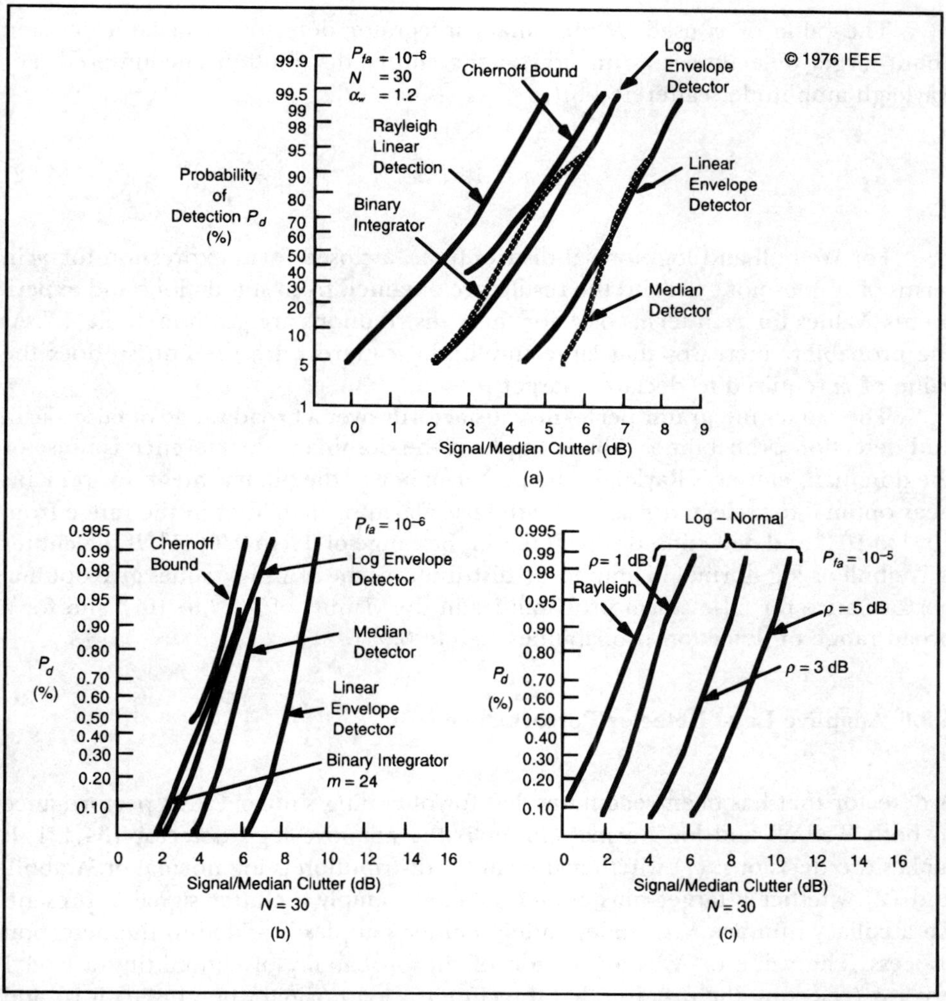

Figure 4.25 Radar performance in Weibull and log normal clutter using median, linear, binary integrator, and log envelope detectors: (a) probability of detection, Weibull clutter; (b) probability of detection, log normal clutter. (ρ = 1 dB); and (c) probability of detection, of log envelope detector, in Rayleigh and log normal clutter. (*Source:* [31,32]. Reprinted with permission.)

Figure 4.25(b) shows similar detector performance curves for operation in log normal clutter where the mean-to-median ratio ρ is 1 dB. Again the binary integrator and log envelope perform best. Figure 4.25(c) illustrates the performance of the log envelope detector as the mean-to-median ratio varies in magnitude. Larger values of the mean-to-median ratio require larger signal-to-clutter ratios to achieve the same detection probability. Although the binary integrator and log envelope detectors do not perform as well as a square law or linear law detector in Rayleigh clutter, they do outperform the square and linear envelope detectors in the more severe Weibull and log normal clutter.

The value of m used by the binary integrator detector to make a decision about target detection is a function of the clutter distribution encountered. For Rayleigh amplitude scatterers [40],

$$m = 1.5\sqrt{N} \tag{4.62}$$

For Weibull and log normal distributions, a closed form expression for m in terms of N does not exist and the results are obtained from simulations and experiments. Values for m in terms of N for these distributions are given in Table 4.7. As the probability increases that large-amplitude scatterers are present, so does the value of m required to declare a target present.

The binary integrator performs satisfactorily over a broad range of false alarm and detection probabilities. For example, if the dominant interference is noise or the dominant clutter is Rayleigh, the performance of the binary integrator remains near optimal for selected m and N with false alarm probabilities in the range from 10^{-4} to 10^{-8} and detection probabilities in the range of 0.5 to 0.9 [41]. If the clutter is Weibull or log normal in amplitude distribution, the m and N values give optimal performance for false alarm probabilities in the vicinity of 10^{-5} to 10^{-6} and for a broad range of detection probabilities (0.5 to 0.95).

4.9.6 Adaptive Log-*t* Detector Performance

A detector that has been recommended for obtaining similar CFAR performance in both Weibull and log normal clutter is the adaptive log-t detector [34,42]. It makes two decisions: (1) whether the clutter distribution is log normal or Weibull, and (2) whether a target plus clutter signal or simply a clutter signal is present. An auxiliary number N_c of independent clutter samples is added to the detection process. The value of N_c is a function of the probability of committing a type 1 error α (rejecting the premise that the clutter is log normal when in fact it is) and the probability P of rejecting the premise that the clutter is log normal distributed

Table 4.7
Binary Integrator Design for Rayleigh, Weibull, and Log Normal Clutter [31,32]

N	*m Rayleigh*	*m Weibull* ($\alpha_W = 1.2$)	*m Log Normal* ($\rho = 1$ dB)
3	2	2	2
10	4	7	7
30	8	23	24

when it is Weibull distributed. P is related to the probability of committing a type 2 error β by

$$P = 1 - \beta \tag{4.63}$$

where a type 2 error occurs when the clutter distribution is in fact Weibull and the decision is made to accept the premise that the clutter distribution is log normal. The probability of a type 1 error is usually set at a small value that determines the critical value of the test statistic used to make the first decision that differentiates log normal from Weibull clutter. The value N_c increases as the type 1 error decreases and as P increases.

The first decision is based on the value of the d-test statistic defined by Szajnowski as [42]

$$d(z) = -(1/N) \sum_{i=1}^{N} [(z_i - \hat{m})/\hat{\sigma}]^3 \tag{4.64}$$

where

$z = \ln x$;

$N =$ number of samples;

$\hat{m} =$ estimate of the mean value of the random variable given by

$$\hat{m} = (1/N) \sum_{i=1}^{N} z_i \tag{4.65}$$

and

$$\hat{\sigma}^2 = \text{estimate of the variance of the random variable}$$

$$= (1/N) \sum_{i=1}^{N} (z_i - \hat{m})^2 \tag{4.66}$$

The decision to reject or accept the log normal clutter distribution depends upon whether $d(z)$ is greater than or less than a predetermined critical value d_c. The transformation $z = \ln x$ converts the log normal probability density function associated with a clutter distribution into a normal density function, and the Weibull probability density function into a type 1 extreme value distribution, allowing better separation and identification of the underlying clutter distribution by the $d -$ test statistic.

The second decision, made using the log-t detector, is based on the value of the test statistic

$$t(z) = (z_0 - \hat{m})/\hat{\sigma} \tag{4.67}$$

where z_0 = a sample from the radar resolution cell being analyzed.

The log-t detector decides a target signal is present if t exceeds a detection threshold whose value controls the false alarm probability. The name of the statistic comes from (1) the log transformation of the maximum likelihood ratio formed by the probability density functions that describe the probability that a clutter sample belongs to a log normal distribution when the clutter is log normal distributed, and the probability that a clutter sample belongs to a Weibull distribution when the clutter is Weibull distributed; and (2) the Student-t distribution given by the statistic when the clutter is log normal.

In general, a larger threshold is required in log normal clutter to maintain a given false alarm probability than in Weibull clutter. Goldstein [34] points out, however, that for false alarm probabilities less than 0.22, a number of samples N_0 exists such that the threshold values for log normal and Weibull clutter are the same. The required value of N_0 increases with decreasing false alarm probability. For example, when $P_{fa} = 10^{-6}$, $N_0 = 30$ and for $P_{fa} = 10^{-4}$, $N_0 = 18$. To ensure an acceptable target signal detection probability, Szajnowski recommends that $P \geq 0.8$ to achieve satisfactory false alarm control with only moderate loss in target detectability.

4.10 INCREASING DETECTION PROBABILITY WITH MULTIPLE INDEPENDENT SAMPLES

In order of decreasing efficiency, the methods for increasing detection probability using multiple independent samples are (1) coherent integration, (2) video (noncoherent) integration, (3) binary integration, and (4) cumulative (multipulse) detection. Coherent integration has been briefly discussed in the sections on the CW Doppler, pulse Doppler, and synthetic aperture radars. Binary integration has been described above with respect to efficient target detection in clutter. Video integration and cumulative detection are discussed below.

4.10.1 Video Integration of Samples

In video or noncoherent integration of independent samples, the gain achieved is generally less than that obtained with coherent integration of the same number of samples. If N samples are noncoherently integrated after envelope detection, then the gain is approximated by [36]

- $\sqrt{N}$ for small signal-to-noise ratio;
- N for N small;
- $N^{0.8}$ over a wider range of conditions.

These regions of gain, alternatively described as the reciprocal of integration loss L_i, are illustrated in Figure 4.26.

Video integration works by summing radar returns and dividing by the number of returns summed. This process causes the variance of the noise and the variance of the signal plus noise to decrease by a factor proportional to N, but keeps their mean values the same as shown in Figure 4.27 [36,43]. Therefore, a given detection and false alarm probability are achieved with a smaller signal-to-noise ratio.

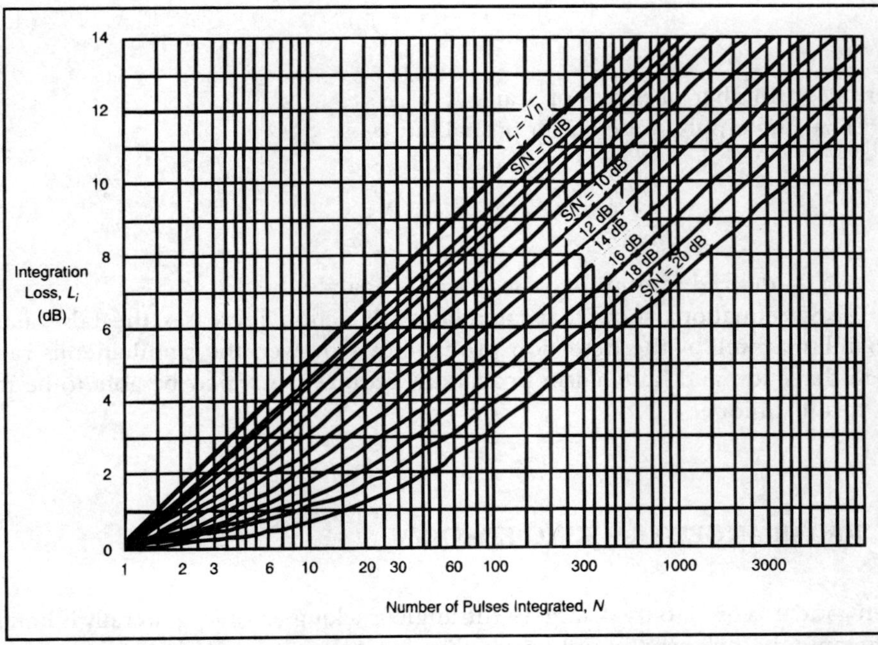

Figure 4.26 Integration loss with noncoherent integration of independent samples. (*Source:* [36]. Reprinted with permission.)

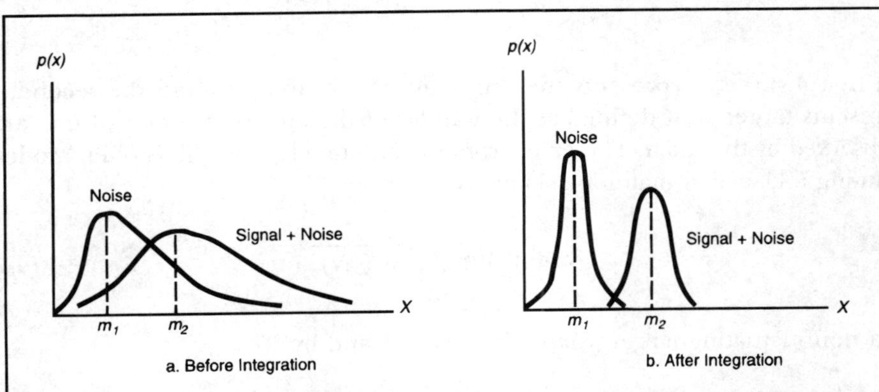

Figure 4.27 Noncoherent video integration of independent signal samples.

4.10.2 Cumulative Detection

Cumulative detection, sometimes referred to as multipulse and multiscan detection, is effective for increasing the detection probability of fluctuating targets. The multipulse detection probability P_d is defined in terms of the single pulse or single scan probability p_d as

$$P_d = 1 - (1 - p_d)^N \tag{4.68}$$

where N = number of pulses or scans.

The single pulse false alarm probability p_{fa} is

$$p_{fa} = P_{fa}/N \tag{4.69}$$

where P_{fa} = desired multipulse false alarm probability.

These equations show that cumulative detection increases the false alarm probability as well as the detection probability. However, the simultaneous radar system detection and false alarm probability requirements may be able to be met with this technique.

4.11 RADAR ANGLE-TRACKING ERROR

When a radar is used to track targets, the angle-tracking error σ_a generally is limited by the antenna beamwidth and target glint [11,44] to a value given by

$$\sigma_a = (\sigma_\theta^2 + \sigma_g^2)^{1/2} \tag{4.70}$$

The first term, σ_θ, represents the finite antenna beamwidth and the second, σ_g, represents target glint defined as the wander of the apparent center of the target as observed by the radar. The error from the antenna beamwidth is often modeled assuming a Gaussian mainlobe shape as

$$\sigma_\theta \approx 0.49\theta_a/\sqrt{N(S/N)_a} \tag{4.71}$$

for a nonfluctuating target when $(S/C)_a >> 1$ and by

$$\sigma_\theta \approx 0.58\theta_a/\sqrt{N(S/N)_a} \tag{4.72}$$

for a nonfluctuating target when $(S/C)_a << 1$, where

θ_a = two-way 3-dB antenna beamwidth in rad;
$(S/N)_a$ = signal-to-noise ratio as seen by the antenna;
N = number of pulses between the half-power points of the one-way antenna pattern, assuming coherent integration; with noncoherent integration, $N(S/N)_a$ is reduced by the integration loss L_i.

The error contribution from glint due to target scintillation can be modeled as

$$\sigma_g = 0.35\ L/R \tag{4.73}$$

where

L = target extent across the beam in m;
R = sensor-to-target range in m.

Although (4.71) and (4.72) were developed for noise-limited operation, they can be generalized to incorporate clutter-limited operation and signal-to-clutter ratios by modifying the value of the coefficient in front of θ_a. When the amplitude of the clutter signal is comparable with the return from the target, the angular resolution may be no better than the half-power beamwidth [44].

Target scatterer amplitude fluctuations are interpreted by the radar as deviations in target location from the on-axis track that would be provided by a nonfluctuating target. The glint error represented by (4.73) assumes the target scatterers are uniformly distributed and, as written, gives a worst case estimate of the tracking error induced by target fluctuations. In most situations, its magnitude is reduced by signal processing and the dynamic response of the tracker when the electronics cannot follow the sharp peaks of glint when the target fades. The magnitude of the glint error is also somewhat dependent on the antenna beamwidth relative to the subtended angle L/R of the target. As the target extent approaches the antenna beamwidth, the target edges are weighted less by the off boresight antenna pattern gain. Hence, contributions to glint from scatterers on target edges are reduced. When the target extent approaches the antenna beamwidth, the beamwidth is comparable to the width of the target radar cross-section lobe arriving at the radar. This provides smoothing of the phase front arriving from scatterers over the face of the antenna [44].

In sensors employing conical scan antennas, the magnitude of the angle-tracking error added by target fluctuations is inversely proportional to the product of conical scan frequency and error slope, and directly proportional to the square root of the scan servo bandwidth divided by the decorrelation time of the fluctuation. Barton [36] approximates this source of tracking error as $\theta_a/40$. When an antenna is used to scan a sector area, the contribution to the tracking error from target fluctuations is proportional to the time on target divided by the decorrelation time. Barton estimates this error source in the range $0.05\theta_a$ to $0.1\theta_a$. Monopulse radars are essentially immune from target fluctuation errors, especially if the desired signals are averaged over at least two fluctuation cycles [36]. The angle-tracking error due to target fluctuations, shown in Table 4.8, was modeled by Swerling for a rapidly fluctuating target [44]. The normalized standard deviation shown in columns 2 and 3 expresses the error as a function of the signal-to-noise ratio S/N for nonfluctuating and rapidly fluctuating targets, respectively. For a broad variation in S/N, the difference in error between nonfluctuating and rapidly fluctuating targets is confined to a small range. The modeled results predict that the normalized standard deviation for the rapidly fluctuating target approaches 0.152 for S/N greater than 25 dB, whereas the standard deviation for the nonfluctuating target continually decreases as the S/N increases, as shown by (4.71) and (4.72).

4.12 TARGET RESOLUTION

Proper specification of the antenna beamwidth helps optimize the performance of the radar for a particular application. If the beamwidth produces a footprint at the target that is smaller than the observed target dimension, then the target is termed *over-resolved*, as shown in Figure 4.28(a). The advantages of over-resolution are a smaller clutter return, because of the smaller footprint, and decreased angle-tracking error because of the smaller beamwidth. A major disadvantage is that the radar may not receive the energy backscattered by all of the target scatterers or from the largest scatterers because they may not be included within the footprint. Figure 4.28(b) shows an *under-resolved* target, where a large beamwidth produces a footprint that is equal to or larger than the target dimension of interest. Now, all the target scatterers are enclosed by the footprint to maximize the received

Table 4.8
Degradation in Angle-Tracking Error With Rapidly Fluctuating Targets [44]

	$\sigma\sqrt{N}/\theta$ *(Normalized Error)*		
S/N (dB)	*Nonfluctuating Target*	*Rapidly (Pulse-to-Pulse) Fluctuating Target*	*Difference in Tracking Error Between Nonfluctuating and Rapidly Fluctuating Target*
0	0.81	0.91	0.10
5	0.34	0.48	0.14
10	0.18	0.28	0.10
15	0.09	0.20	0.11
20	0.05	0.18	0.13

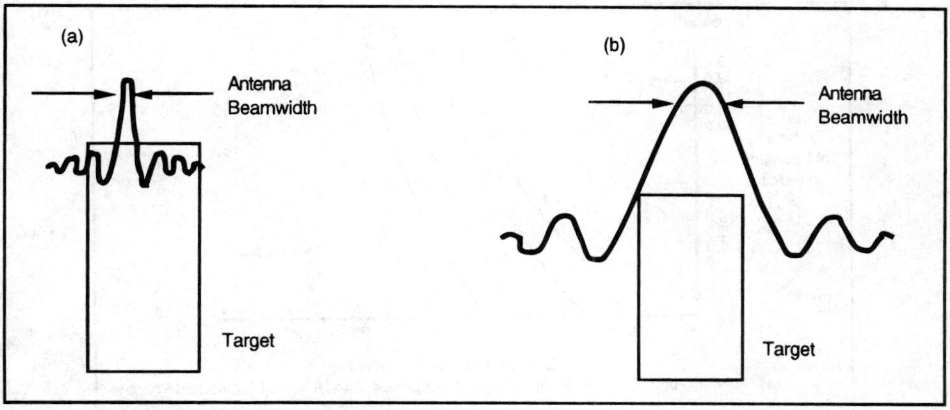

Figure 4.28 Resolution for (a) over- and (b) under-resolved targets.

backscattered signal. However, the larger beamwidth contributes to a larger angle-tracking error and a larger return from background clutter.

The effects of under- and over-resolved targets on detection probability are shown in Figure 4.29. These curves represent modeled returns from multiscatterer nonfluctuating and fluctuating targets [45]. Scatterers are arbitrarily distributed over M_S range cells. When the target occupies one range cell, $M_S = 1$ and the target is under or properly resolved. When the waveform over resolves the target, $M_S > 1$ and the target scatterer model is reconfigured into M_S subscatterers, each randomly located within one of the M_S range cells. The false alarm probability P_{fa}, total energy backscattered from the target, and the number of transmitted pulses N are held constant for each curve represented by the different value of M_S.

Each detection probability curve for the nonfluctuating amplitude target, shown in Figure 4.29(a), corresponds to a different target structure as represented by the number of subscatterers M_S. The under-resolved target ($M_S = 1$) produces the greatest detection probability with the smallest S/N. In the slow-fluctuation Swerling 1 target of Figure 4.29(b), the scatterer model for each M_S is represented by a complex Gaussian function that yields Rayleigh amplitude scatterers. Here, each curve represents the same scatterer since superposition of Gaussian random variables results in a Gaussian random variable. For large signal-to-noise ratios, the increase in detection probability with increasing M_S is due to the superposition of the returns from the scatterers. The particular way the returns from the M_S scatterers combine also gives rise to the loss that is exhibited at the lower signal-to-noise ratios. In the slow-fluctuating Swerling 3 target of Figure 4.29(c), the M_S dominant scatterers combine with Rayleigh subscatterers to produce a Rayleigh target. For large signal-to-noise ratios, the improvement in detection performance with increasing M_S is again due to the superposition of the returns from the scatterers. Detection probability deteriorates with increasing M_S for small values of S/N. Overall, the

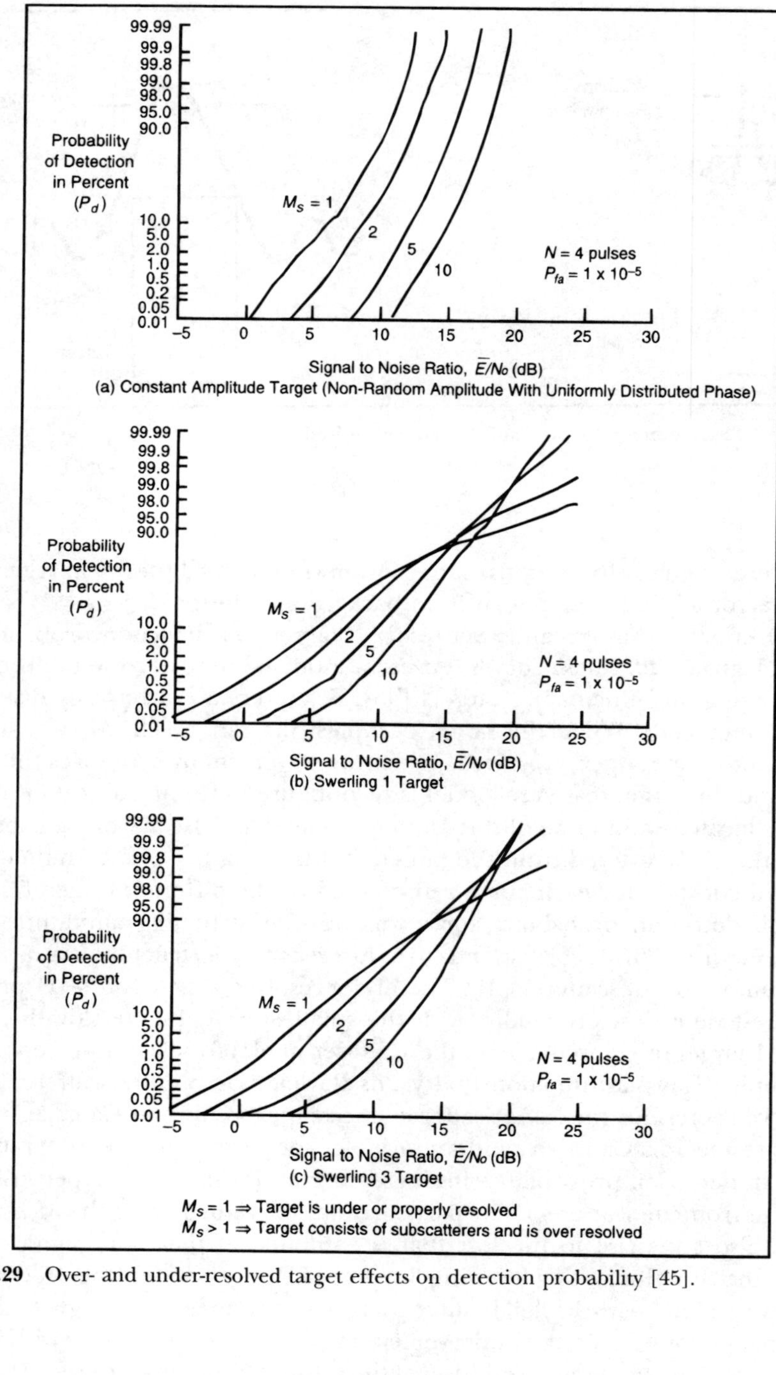

Figure 4.29 Over- and under-resolved target effects on detection probability [45].

detection performance of the Swerling 3 scatterer model falls somewhere between that of the nonfluctuating target and that of the Swerling 1 fluctuating target.

4.13 HIGH-CLUTTER BACKGROUNDS

Two high-clutter backgrounds that produce Weibull and log normal clutter distributions are forests and snow. The magnitude of the clutter is typically expressed in terms of a dimensionless backscatter coefficient σ^o that relates the radar cross-section of the clutter region to the antenna ground footprint such that

$$\sigma^o = \sigma_c / A_a \tag{4.74}$$

where

σ_c = radar cross-section of the clutter in m^2;
A_a = antenna ground footprint in square meters, as illustrated in Figure 4.30.

Another method to express backscatter coefficient is in terms of the projected area normal to the viewing direction. In this case, the coefficient is expressed as γ^o where

$$\gamma^o = \sigma_c / A_p \tag{4.75}$$

and A_p = antenna footprint projected normal to the viewing direction.

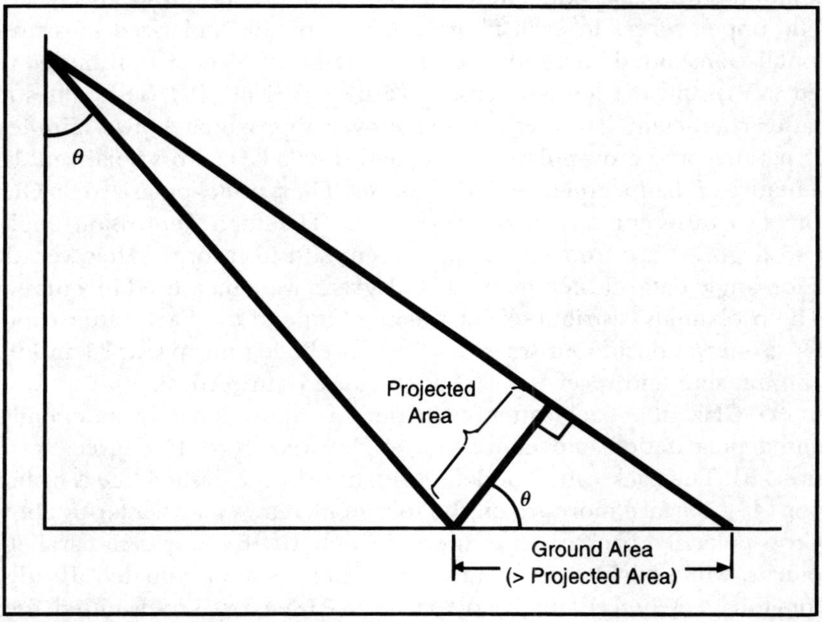

Figure 4.30 Backscatter coefficients.

The two backscatter coefficients are related by

$$\sigma^o = \gamma^o \cos\theta \tag{4.76}$$

where θ is the angle of incidence measured from nadir, as shown in Figure 4.30.

The actual radar cross-section of the clutter is the same no matter which average value of the scattering cross-section per unit area is used. The larger γ^o is multiplied by the smaller projected area, while the smaller σ^o is multiplied by the larger ground area.

4.13.1 Deciduous Trees

Deciduous tree clutter backscatter data measured at 35 and 95 GHz as a function of depression angle are given in Figures 4.31 and 4.32, respectively [46]. The depression angle is measured from the horizontal while the incidence angle used in (4.76) is measured with respect to nadir. Since the measured value of the clutter cross-section is dependent on the operating parameters of the measuring radar, it is important to note the pertinent characteristics of the radars. These data were measured with 35- and 95-GHz radars that transmitted 50-nsec duration pulses at a PRF of 2 kHz. The IF bandwidth was 20 MHz. The IF amplifiers used in both radars were logarithmic. The transmitted peak power of the 35-GHz radar was 40 kW and the antenna azimuth and elevation 3-dB beamwidths were each 1 degree. The 95-GHz radar transmitted a peak power of 6 kW into an antenna whose azimuth and elevation 3-dB beamwidths were 0.70 and 0.65 degrees, respectively.

The upper curves in each figure apply to parallel-polarized measurements (horizontal transmitted-horizontal received (HH) or vertical transmitted-vertical received (VV)) and the lower to cross-polarized (VH or HV). The values for the backscatter coefficients are averages made over approximately two minutes. The 35-GHz parallel- and cross-polarized data and the 95-GHz cross-polarized data are representative of damp, green, and lush forests. The parallel-polarized 95-GHz data are from two different sets of measurements. The high depression angle data (30 to 45 degrees) are from the damp, green, and lush forest. However, the low depression angle data, clustered about 10 degrees, were measured in a dryer forest [47]. The probability distribution for the amplitude of the backscatter coefficient obtained from wet deciduous trees was found to be log normal at 35 and 95 GHz when the transmit and receive polarizations were both vertical.

At 35 GHz, the backscatter coefficient is approximately independent of transmitted polarization and depression angle over 15 to 45 degrees, as shown in Figure 4.31. The backscatter model represented by the dashed line is from Hayes and Dyer [48]. Surface moisture on the foliage increases backscatter by about 3 to 5 dB. Cross-polarized backscatter is approximately 10 dB lower than parallel-polarized returns, while backscatter from a grassy field is approximately 10 dB lower than from foliage. When circularly polarized 35-GHz energy is transmitted, the cross-polarized backscatter components (RHC/LHC and LHC/RHC) are approximately

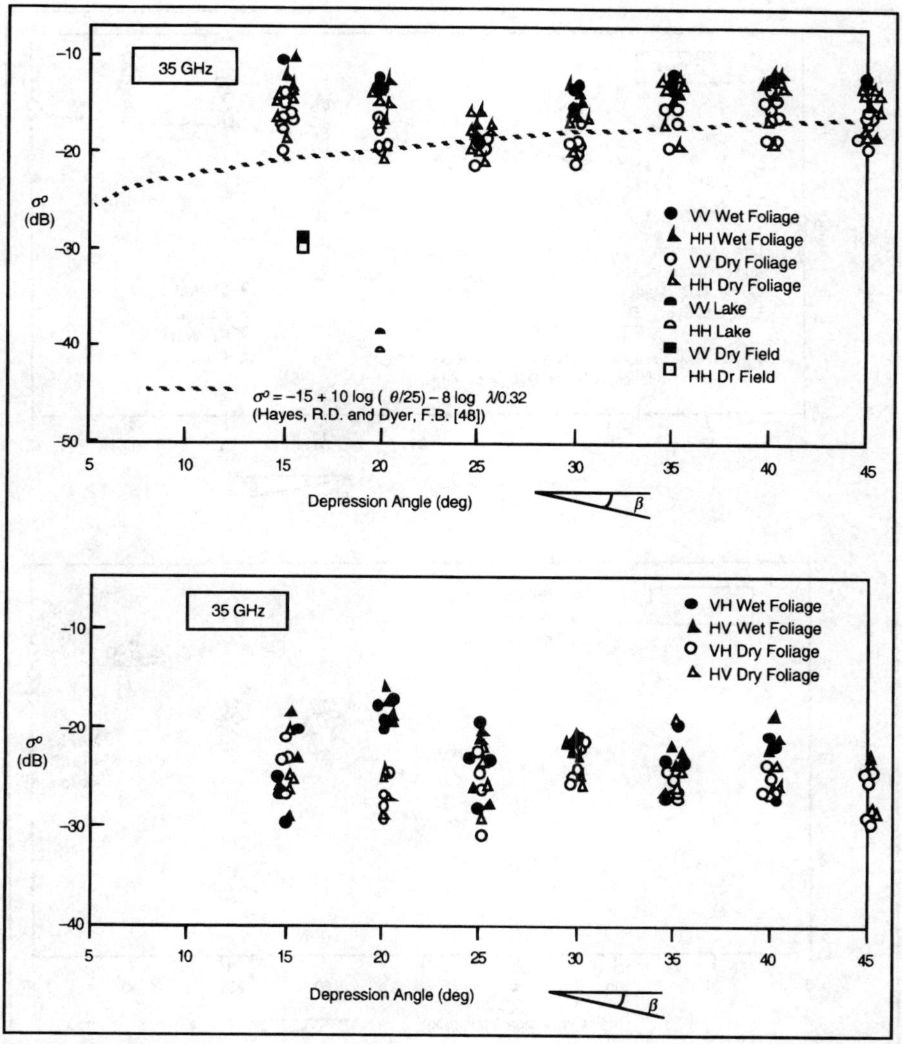

Figure 4.31 Radar cross-section measurements of deciduous trees at 35 GHz. Upper figure shows parallel polarized data and lower figure cross-polarized data [46].

equal to the parallel-polarized components (RHC/RHC and LHC/LHC), with each 3 to 5 dB less than those from linear polarizations.

Figure 4.32 shows that backscatter at 95 GHz is about 10 dB greater than at 35 GHz. The 95-GHz backscatter coefficient for wet foliage is no more than 2 to 3 dB greater than that of dry foliage. Cross-polarized backscatter is approximately 5 dB lower than parallel-polarized values. Higher depression angle parallel-polarized data appear 5 to 10 dB greater than the lower depression angle data clustered about 10 degrees. However, this effect may be due to the difference in the character of the foliage between the two test sites (i.e., wet versus dry), or indeed may represent

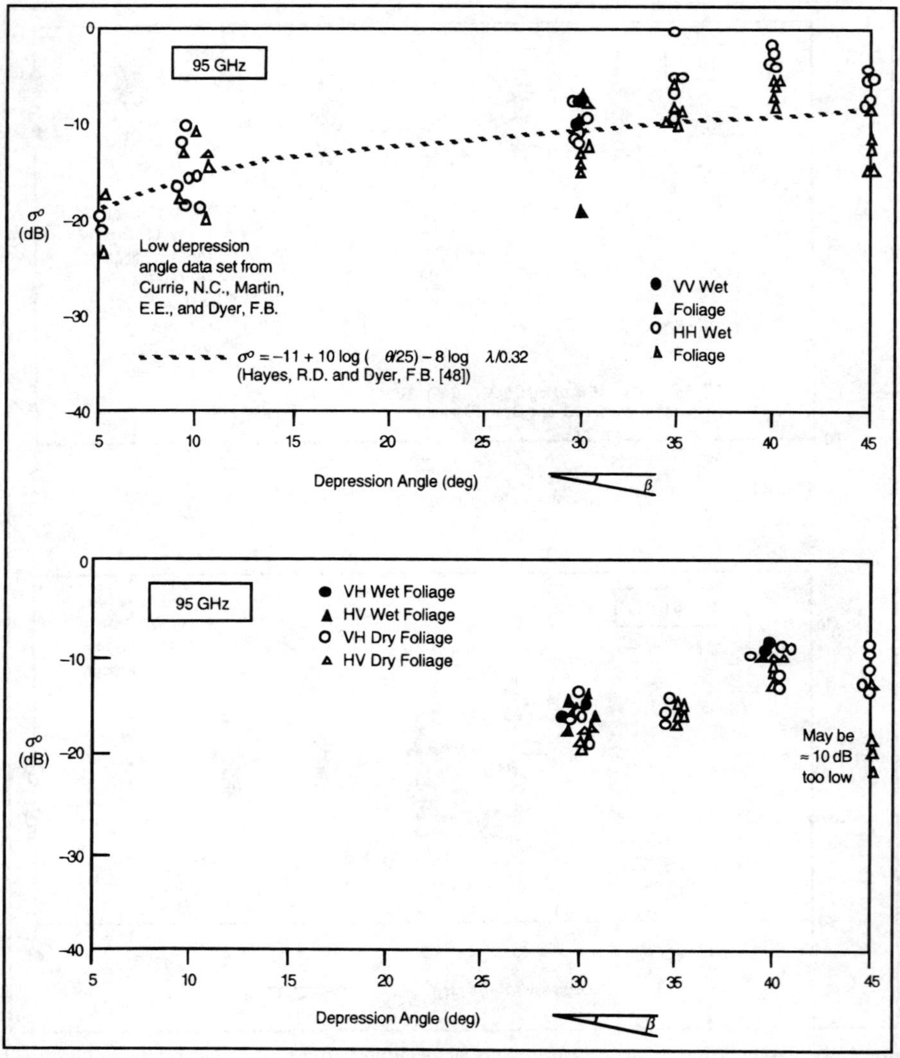

Figure 4.32 Radar cross-section measurements of deciduous trees at 95 GHz. Upper figure shows parallel polarized data and lower figure cross-polarized data [46].

a true depression angle dependency. The researchers conclude that more data are needed to resolve which of the two hypotheses is true.

Standard deviations of tree, grass, and rock clutter radar cross-sections, shown in Table 4.9, can be used to help characterize the clutter distributions at 9.5 to 95 GHz from narrowband radars (20-MHz IF bandwidth) [49,50]. The 9.5-GHz radar transmitted a peak power of 40 kW into an antenna having azimuth and elevation 3-dB beamwidths of 1.5 and 1.65 degrees, respectively. The 16.5-GHz radar transmitted a peak power of 50 kW into an antenna whose azimuth and

Table 4.9
Standard Deviation of Foliage and Rock Clutter From 9.5 to 95 GHz [49,50]

		Average Value of Standard Deviation (dB)			
Clutter Type	*Polarization*	*9.5 GHz*	*16.5 GHz*	*35 GHz*	*95 GHz*
Deciduous trees	Vertical	3.9	–	4.7	–
• Summer	Horizontal	4.0	–	4.0	5.4
	Average	4.0	–	4.3	5.4
Deciduous trees	Vertical	3.9	4.2	4.4	6.4
• Fall	Horizontal	3.9	4.3	4.3	5.3
	Average	3.9	4.2	4.3	5.0
Pine trees	Vertical	3.5	3.7	3.7	6.8
	Horizontal	3.3	3.8	4.2	6.3
	Average	3.4	3.7	3.9	6.5
Mixed trees	Vertical	4.3	–	4.0	–
• Summer	Horizontal	4.6	–	4.2	–
	Average	4.4	–	4.1	–
Mixed trees	Vertical	4.1	4.1	4.7	6.3
• Fall	Horizontal	4.5	4.3	4.6	5.0
	Average	4.4	4.2	4.6	5.4
Grassy field	Vertical	1.5	–	1.7	2.0
	Horizontal	1.0	1.2	1.3	–
	Average	1.3	1.2	1.4	2.0
Rocky area	Vertical	1.1	2.2	1.8	1.6
	Horizontal	1.2	1.7	1.7	1.7
	Average	1.1	1.9	1.8	1.7
10-inch corner reflector located in grassy field	–	1.0	1.0	1.2	1.2

Source: Currie, N. E., F. B. Dyer, and R. D. Hayes, "Radar Land Clutter Measurements at Frequencies of 9.5, 16, 35, and 95 GHz," Report A-1485-003, AD A012 709, Engineering Experiment Station, Georgia Institute of Technology, Apr. 2, 1975. Also appears in Currie, N.C., R. D. Hayes, and R. N. Trebits, *Millimeter-Wave Radar Clutter,* Artech House, Norwood, MA, 1992. Reprinted with permission.

elevation 3-dB beamwidths were each 1.5 degrees. The pulse width was 50 nsec for the 9.5-GHz and 16.5-GHz radars. The specifications of the 35- and 95-GHz radars were given above.

Standard deviations for trees increase slightly with frequency, although the standard deviations for pine trees appear significantly larger at 95 GHz than at other frequencies. This may be indicative of log normal amplitude fluctuations at the higher radar frequency. Currie and others [49] point out that wideband radars can be expected to reduce the amplitude variations through frequency decorrelation. This is discussed further in Chapter 5. The corner reflector is included to show the return from a large stationary target.

4.13.2 Radar Backscatter Coefficient Data Set Inconsistencies

Figure 4.33 summarizes measurements of backscatter coefficients made of crops and the Earth's surface at X-band frequencies at incidence angles between 30 and 50 degrees [51]. The early backscatter measurements, performed by researchers at Ohio State University, are lower than later data obtained by others. After Peake re-examined the Ohio State measurements [52], he acknowledged that systematic calibration errors or the variability and inability of the experimenter to adequately specify the surface itself caused the early results to be about 4 to 5 dB too low. However, Bush and Ulaby [51] conclude that the later measurements support approximately a 7-dB increase in the early values. Since the Ohio State results are often quoted and reproduced [53–55], it is important to be aware that they apparently understate the backscatter values.

4.13.3 Snow

Radar energy backscattered from snow is a strong function of the liquid water content of the snow. There are two quantities used to specify liquid water content, the volume percentage of liquid water contained in a unit volume of snow, denoted

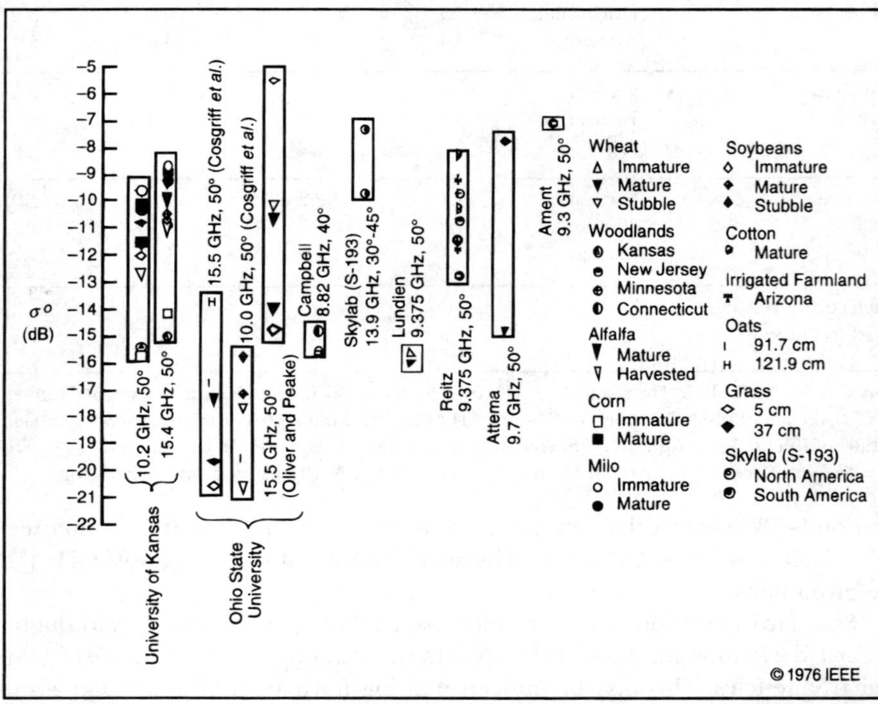

Figure 4.33 Comparison of vegetation radar backscatter data. (*Source:* [51]. Reprinted with permission.)

by m_v, and the snowpack water equivalent, denoted by W, equal to the height in centimeters of liquid water in a vertical column of snowpack having a horizontal cross-section of 1 cm^2. Mathematically, W is computed as

$$W = \rho_s d \tag{4.77}$$

where

ρ_s = snow density in g/cm^3;
d = snow depth in cm.

Figure 4.34 depicts the scattering coefficients of snow and several underlying surfaces at 35 GHz as a function of the incidence angle measured from nadir [56]. The data were acquired with the University of Kansas truck-mounted Microwave Active Spectrometer. The RF modulation bandwidth of this FMCW radar was

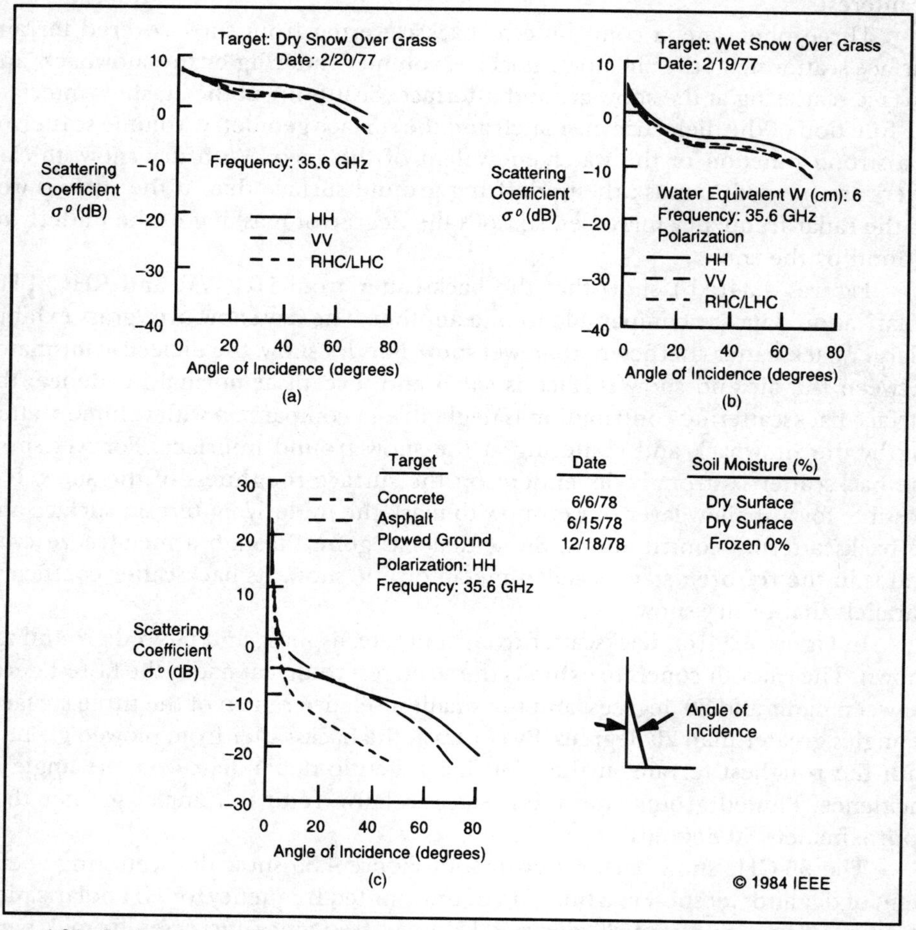

Figure 4.34 Backscatter coefficients of snow and other terrain at 35 GHz versus incidence angle and polarization: (a) dry snow over wet grass, (b) wet snow over grass, and (c) concrete, asphalt, and plowed ground. (*Source:* [56]. Reprinted with permission.)

800 MHz. It was operated at a height of 20m using an antenna with a 3-dB beamwidth of 3.05 degrees.

The backscatter data of Figure 4.34 represent *typical* angular responses between incidence angles of 0 (nadir looking) and 80 degrees. The backscatter measured from similar surfaces at other locations may differ from the values shown here. An asphalt surface found elsewhere may be rougher or smoother than the one used in these measurements and may thus affect the outcome of another set of measurements. Variations in soil moisture affect the return from plowed ground, as does water on concrete or asphalt. Finally, snow depth and water content affect the backscatter from snow at a given frequency, polarization, and incidence angle. Therefore, the backscatter data illustrated in Figure 4.34 express trends but not necessarily the precise values for snow on a specific surface. The latter can only be obtained through measurements made with the applicable sensor at the test site of interest.

Three phenomena contribute to backscattering from snow-covered terrain: surface scattering at the air-snow interface, volume scattering by the snowpack, and surface scattering at the snow-ground interface. Scattering at the air-snow interface is a function of the dielectric mismatch and the surface geometry. Volume scattering is a strong function of the water equivalent of the snow. With dry snow, thicker layers are needed to mask the underlying ground surface than if the snow is wet. As the radar frequency increases, so does the degree of masking of the underlying ground by the snow.

Figures 4.34(a,b) show that the backscatter from HH, VV, and RHC/LHC polarization data are comparable to one another. The dry snow over grass exhibits a larger backscatter coefficient than wet snow. For dry snow, the dielectric mismatch between the air and snow surface is small and, except at normal incidence, the surface backscattering contribution is negligible in comparison with volume scattering by the snowpack and scattering at the snow-ground interface. For wet snow, the backscatter is strongly dependent on the surface roughness of the snow. It is possible for a shallow layer of wet snow to mask the underlying terrain surface and its backscattering contributions. Snow that has gone through a melt-freeze cycle and is in the refrozen state is called metamorphic snow. Its backscatter coefficient parallels that of dry snow.

In Figure 4.34(c), backscatter from concrete, asphalt, and plowed ground are shown. The smooth concrete exhibits the strongest return at nadir, the fastest decay between nadir and 20 degrees, and the smallest relative return of the three surfaces at angles greater than 20 degrees. By contrast, the backscatter from plowed ground, with the roughest terrain in the plot, has a gentle dependence on the angle of incidence. Plowed ground has the greatest relative return at angles greater than approximately 30 degrees.

The 35-GHz snow backscatter data of Figure 4.35 show the scattering coefficient of dry and wet snow as a function of transmitted frequency for HH polarization at an incidence angle of 20 degrees [57]. As frequency increases from L-band through Ka-band, so does the backscatter, especially for dry snow. The larger back-

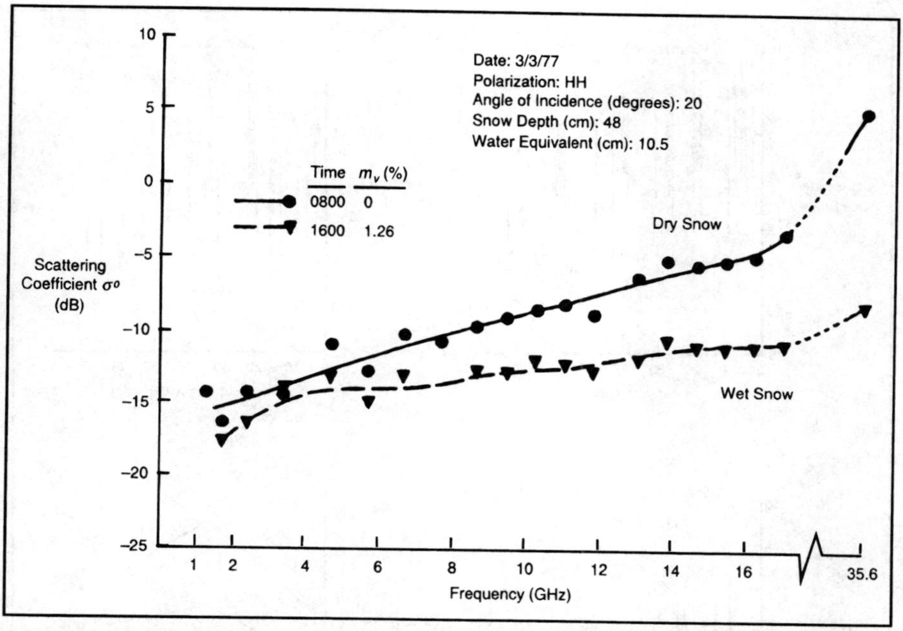

Figure 4.35 Spectral response of backscatter coefficient of wet and dry snow at 20-degree incidence angle [57].

scatter shown by dry snow at all frequencies is due to volume scattering by the snowpack.

At 94 GHz, snow backscatter exhibits similar behavior as demonstrated in Figure 4.36 [58]. These observations were made from a helicopter at a height of 60m using a Marconi 94-GHz monostatic radar containing a coherent pulsed transmitter and a dual-channel double-superheterodyne receiver. The beamwidth of the antenna was 1.5 degrees. The transmitted signal was selectable for any plane-polarized wave or RHC and LHC polarization. The two receiver channels recorded the amplitude and phase of the two orthogonally polarized received signal components.

The first three frames show data for VV polarization as a function of incidence angle. The length of the vertical bars corresponds to the range of measured values, while the symbols centered on the means are proportional to surface snow grain size and shape. The first frame shows that backscatter data for dry snow do not appear to exhibit a strong dependence on grain size and shape, but a larger dependence on incidence angle. For moist snow, shown in the second frame, the backscatter mean value is lower and there is a stronger dependence on grain size and shape. For wet snow, shown in the third frame, the mean backscatter values continue their downward trend and their dependence on grain size and shape.

The fourth frame demonstrates the diurnal behavior of the backscatter as the snow goes through a dry to wet cycle over 24 hours. As the snow progresses from a dry state, characterized by 0% water by volume to a moist state where the change

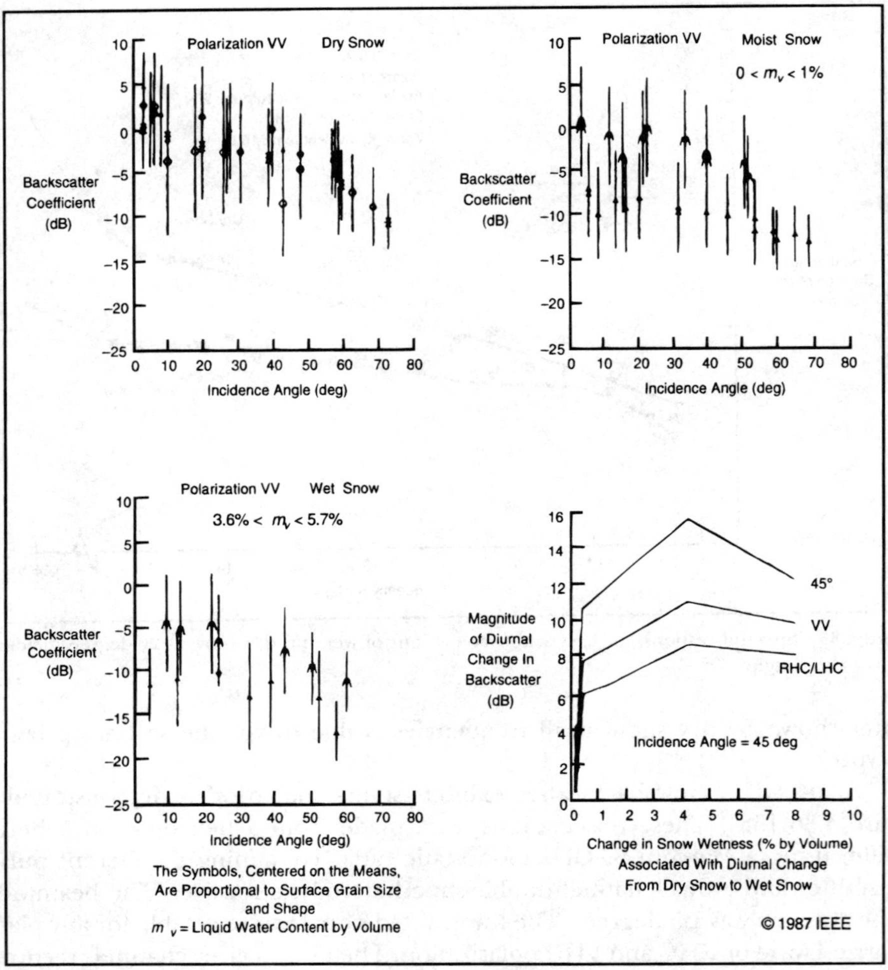

Figure 4.36 Backscatter coefficient of snow at 95 GHz versus incidence angle, wetness, and grain size (*Source:* [58]. Reprinted with permission.)

in snow wetness is less than 1%, there is a dramatic change in the backscatter, approaching 11 dB for 45-degree parallel polarization. It is apparent that just a small amount of liquid water can change the snow backscatter by a large amount. Data such as these help explain early observations made by radar developers about the detectability of targets in snow during sunny afternoons, when the liquid water content was relatively high, and the lack of detection during early morning hours when the snow was still dry and had a large backscatter coefficient that masked the target. More recent measurements of snow backscatter at 95 GHz have been used to derive a relatively simple model that predicts the polarimetric scattering behavior for dry metamorphic snow cover [59].

In summary, the backscatter coefficient from dry and metamorphic snow is greater than that from wet snow. A relatively small percentage of liquid water is sufficient to convert snow from its high backscatter state to its lower backscatter state. Backscatter increases as frequency increases and decreases as incidence angle increases.

4.14 BACKSCATTER FROM TARGETS

Radar cross-section measurements of cargo trucks and tanks are shown in Figure 4.37. These were obtained using a 94-GHz pulsed IMPATT at a 30-degree incidence angle and VV polarization [60]. The radar was scanned over the vehicles in the directions indicated by the arrows at the bottom of the figure. Most manmade targets exhibit large specular returns at the cardinal angles (front, back, and orthogonal side views), as shown by the large spikes. In fact, the specular values of the radar cross-sections of many targets often exceed even those shown and have magnitudes of several hundred square meters. The large cross-section is due to retroreflection from angles and surfaces that are shaped as dihedral and trihedral reflectors. At other viewing angles, the radar cross-section is much lower. By comparison,

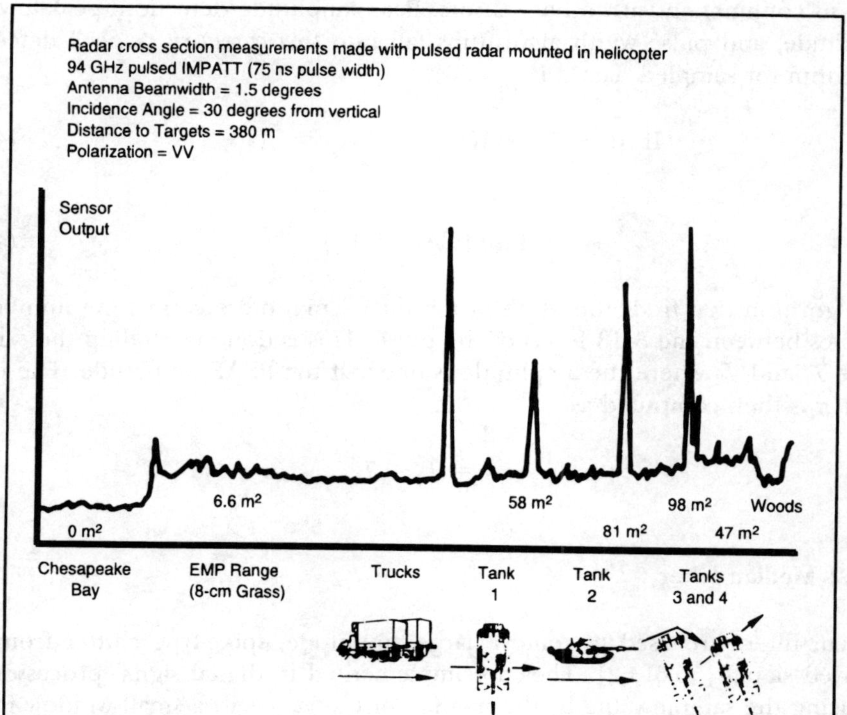

Figure 4.37 Typical target cross-sections at 94 GHz obtained from a pulsed radar [60].

the radar cross-section of the background grass is 6.6 m^2. Typical radar design specifications include the radar cross-sections of targets, averaged over 360 degrees of viewing, that it must detect. These average cross-sections are typically between 20 and 30 m^2 for uncamouflaged ground-vehicle targets.

4.15 SIGNAL-PROCESSING TECHNIQUES FOR TARGET DETECTION AND CLASSIFICATION

As was shown in Table 4.1, there are a variety of signal-processing algorithms and processing domains that can be applied to MMW radar signals. The selection depends on the sensor resolution, field of view, relative motion of the radar and target, processing time available, cost goals, degree of discrimination desired, and external factors such as clutter type and density, artifacts such as glint, atmospheric and weather effects, and countermeasures.

4.15.1 Peak and Pulse Width Time-Domain Algorithms

Relatively simple types of time-domain processing can be effective in low-cost radars used in conjunction with other sensors. Peak amplitude detection, peak-to-valley amplitude, and pulse width algorithms fall into this category. A peak detection algorithm for sampled data A is

$$\text{If } A(i+1) > A(i), \qquad \text{PEAK} = A(i+1) \tag{4.78}$$

$$\text{Else PEAK} = A(i) \tag{4.79}$$

An algorithm that finds the width of a return signal pulse counts the number of samples between the 3-dB levels of the pulse. This is done by finding the sample times T_1 and T_2 where the amplitude is one-half the PEAK amplitude. The pulse width τ is then computed as

$$\tau = T_2 - T_1 \tag{4.80}$$

4.15.2 Median Filter

Median filters are used to remove large amplitude, spike-type clutter from the received signal [56,61,62]. They are implemented in digital signal processors by replacing the sample value by the median of the values in a small window in the neighborhood of the sample value. The window size is selected based on the expected target size. If the window is too large, target features will be eliminated as well as the clutter. Conversely, if the window is too small, the clutter will not be removed. Consequently, some experimentation with the window size selection is necessary. The window size is always odd to make calculation of the median value easy. For example, if the window size is 3, median filtering replaces the value at each sample by the median value of that sample and its two neighbors. Thus, if the 3 samples are a, b, c, where $a \leq b \leq c$, then the median is b.

Before the signal is passed through the median filter, a number of zero-valued samples are appended to the beginning and end of the sample string to minimize edge effects. The number of appended samples s_0 is related to the window size W_M by

$$s_0 = (W_M - 1)/2 \tag{4.81}$$

Figure 4.38 shows the application of a median filter to a digitized signal. The input to the filter is represented by the waveform in the upper left corner of the figure. Suppose a window of size 3 is chosen. The first step in applying the filter is to select the first three samples of the waveform (0, 7, 0) and order them (0, 0, 7). Since the median value is 0, the output of the filter is 0 and replaces the original sample number 2 value of 7 in the filter output waveform. By sliding the 3-window filter to the right so that it acts on samples 2, 3, and 4, the waveform samples are ordered as 0, 1, 7. Hence the median value of 1 replaces the sample number 3 value of 0 in the filter output waveform. As the window continues to be shifted to the right, the input sample values at the center window of the filter are replaced by the median value output of the filter. Applying the filter to all the samples of the input waveform results in the output waveform shown in the upper right corner of the figure. The median filter has removed the spike-type clutter and retained much of the shape from the rest of the original waveform.

4.15.3 Feature-Based Inference for Target Classification

Feature-based inference techniques perform classification or identification by mapping data, such as statistical knowledge about an object or recognition of object features, into a declaration of identity. Feature-based inference can be divided into parametric and nonparametric techniques. Parametric classification is used when the probability density function that describes the random variables is known. The parametric approaches include maximum likelihood, Bayesian, Dempster-Shafer, and generalized evidence processing. When the density function is unknown, nonparametric classification techniques are applied. Examples of these are artificial neural networks, probability density function estimators, and Kth nearest neighbor classifiers [63].

Parametric Techniques

In *maximum-likelihood estimation*, features have fixed but unknown values. The best estimate of the feature value is defined as the one that maximizes the probability of obtaining the samples actually observed.

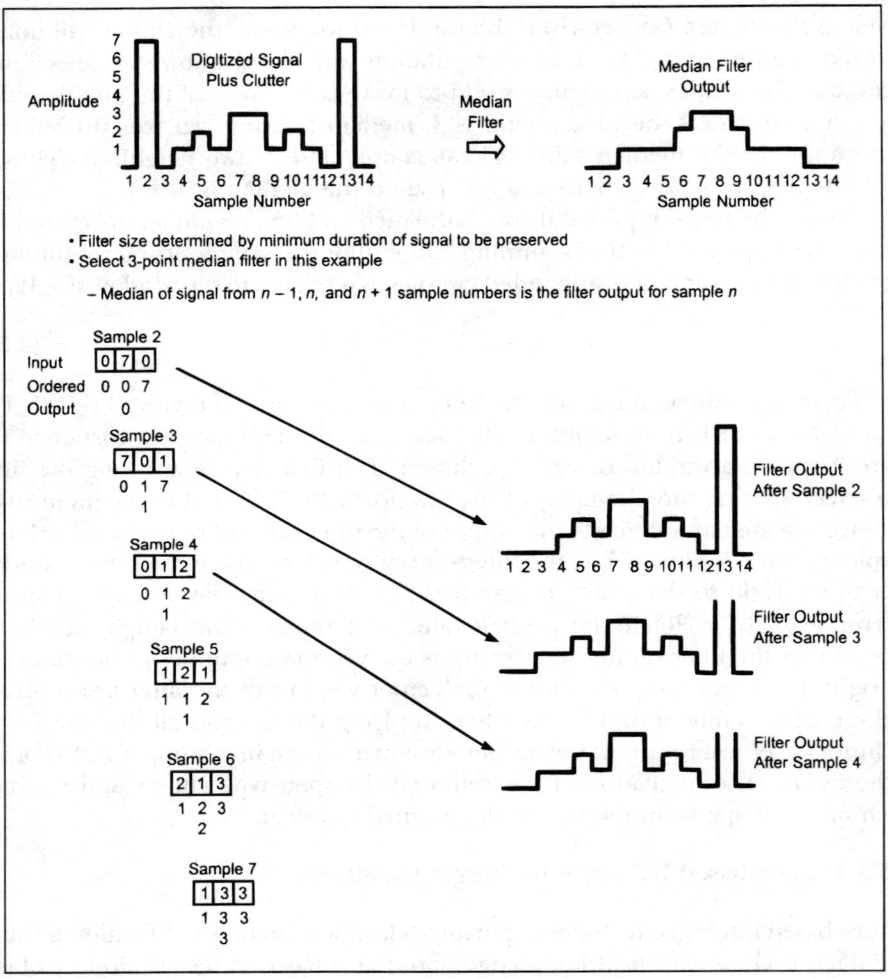

Figure 4.38 Median filter application to digitized signal plus clutter.

In *Bayesian estimation,* the feature values are random variables having an unknown *a priori* distribution. The sample observation converts the *a priori* distribution into an *a posteriori* distribution that revises our opinion about the true values of the features. Observing additional samples sharpens the *a posteriori* density function, causing it to peak near the true value of the feature. The distance measures, shown in Table 4.10, are used in Bayesian target classification when the feature vector consists of normally distributed random variables. The feature characteristics that control the choice of the metric are listed.

The *Dempster-Shafer* method generalizes Bayesian theory to allow for uncertainty by distributing support for a proposition (e.g., that an object is of a particular type) not only to the proposition itself, but also to the union of propositions (disjunctions) that include it. Any support that cannot be directly assigned to a

Table 4.10
Distance Measures for Feature Vectors Whose Values Are Normally Distributed

Characteristics	*Distance Measure*	*Classifier*
• Features are statistically independent • Each feature has the same variance such that $[\mathbf{C}]_i = \sigma^2\mathbf{I}$	Euclidean $\|\mathbf{x} - \boldsymbol{\mu}_i\|$	Minimum distance or template matching
• Covariance matrices for all classes are identical such that $[\mathbf{C}]_i = [\mathbf{C}]$	Squared Mahalanobis distance $(\mathbf{x} - \boldsymbol{\mu}_i)^t\,\mathbf{C}^{-1}(\mathbf{x} - \boldsymbol{\mu}_i)$	Decision boundary is a hyperbola

Note: $\mathbf{x}$ = feature vector for unknown target class; $\boldsymbol{\mu}_i$ = feature vector containing mean values for class i; $\mathbf{C}$ = covariance matrix = $E[(\mathbf{x} - \boldsymbol{\mu}_i)(\mathbf{x} - \boldsymbol{\mu}_i)^t]$; $\mathbf{I}$ = identity matrix; t = transpose operation; -1 = inverse operation.

proposition, or its negation, is assigned to the set of all propositions making up the hypothesis space (i.e., uncertainty). Bayesian and Dempster-Shafer produce identical results when all the singleton propositions are mutually exclusive and there is no support assigned to uncertainty [64–66].

Generalized evidence processing (GEP), which separates the hypotheses (propositions) from the decisions, allows Bayesian decisions to be extended into a hypothesis space (called the frame of discernment in Dempster-Shafer theory) that incorporates multiple hypotheses (as with Dempster-Shafer). With GEP, evidence from nonmutually exclusive propositions can be combined in a Bayesian formulation to reach a decision [67].

Nonparametric Techniques

Artificial neural networks are hardware or software systems that give desired responses to specified inputs. Different types of neural networks exist, each with its own capabilities as listed in Table 4.11. They include single perceptrons, multilayer perceptron, Kohonen networks, Grossberg adaptive resonance networks, and Hopfield networks [68–70].

Nonparametric techniques that estimate the probability density function $p_N(x)$ require the estimate to converge to the true density function $p(x)$ as the number of samples N grows large. One such approach that meets this requirement estimates the density function with *Parzen windows.* Parzen windows design an optimal classifier using estimates of the density function. The region over which the random samples x are selected is reduced until the estimate of the density function converges to the true value. As the region size decreases, the number of samples increases to ensure that enough samples fall within the region. Examples of Parzen windows

Table 4.11
Properties of Artificial Neural Networks

Type	*Key Operating Principles*	*Applications*
• Multilayer perception	• A back-propagation network that uses an error-correcting learning process, in the form of adjustable weights, to map inputs into outputs • Errors are minimized with backward error propagation that updates the weights applied to the input data by the hidden layer neurons • Number of weights is equal to the number of hidden layer neurons	• Accommodates complex decision regions in the feature space • Target classification • Speech synthesis
• Kohonen network	• Feed-forward network that works with an unsupervised learning paradigm (processes unlabeled data, i.e., data where the desired classification is unknown) • Uses a mathematical transformation to convert input data vectors into output vectors	• Speech recognition
• Hybrid network	• Uses supervised and unsupervised paradigms • The lower unsupervised layer organizes data making for easier target classification • Example is a counterpropagation network that allows propagation from the input to a classified output, as well as propagation in the reverse direction	• Target recognition
• Grossberg adaptive resonance network	• Unsupervised learning paradigm that employs feed-forward and feedback computations • Teaches itself new categories and continues storing information without rejecting pieces of information that are temporarily useless as they may be needed later • Uses two layers—an input layer and an output layer. The output layer itself has two sublayers: a comparison layer for short-term memory and a recognition layer for long-term memory	• Pattern recognition • Target classification
• Hopfield network (also called a relaxation network)	• Learning does not occur as fixed interconnection weights are used, making the network easy to implement in hardware • Each node receives inputs that are processed through a hard limiter. The outputs of the nodes (±1) are multiplied by the weight assigned between the nodes connected by the weight • The minimum number of nodes is seven times the number of memories to be stored	• Data association • Optimization problems

from Duda and Hart [71] are shown in Figure 4.39 for $p(x)$ equal to a zero mean, unit variance, univariate normal density and a window function of the same Gaussian form.

When the number of samples $N = 1$, $p_N(x)$ is a single Gaussian hill centered on the first sample. For $N = 16$ and $h_1 = 0.25$ (h_1 is a parameter that specifies the window width h_N as a function of the number of samples from the relation $h_N = h_1/\sqrt{N}$), the contributions from individual samples are discernible; however, for $h_1 = 1$ and $h_1 = 4$, the individual sample contributions are smoothed out. As N increases, $p_N(x)$ is better able to follow variations in $p(x)$. As N approaches infinity, $p_N(x)$ converges to the smooth normal distribution curve.

The *Kth nearest neighbor* nonparametric classifier bypasses the estimation process that generates the probability density function and goes directly to the decision function that classifies the data. The data assignment rule assigns the unknown object to class C if a majority of the Kth nearest neighbors are labeled class C. To proceed, the stored prototype signatures of target classes are sorted until K are found whose distance measure from the unknown are smallest, as illustrated in Figure 4.40. Here, the unknown is assigned to class 1 since two out of three nearest neighbors belong to class 1. The sort time is a function of the nearest neighbor density near the unknown samples. As the number K approaches infinity, the large sample, two class error rate for the Kth nearest neighbor classifier approaches that of the optimal Bayesian classifier, as shown in Figure 4.41 with K odd [71].

4.15.4 Spectral Feature Extraction

In spectral feature extraction, the frequency of the transmitted waveform changes with time and thereby excites resonance modes in the target that produce return signals that are unique to that target. Features identified in advance are extracted from the signals and are input to an algorithm that detects and classifies the object. Practical waveform sources include chirp pulses produced by a hyperabrupt IMPATT diode, a train of pulses each of a different frequency, and highly linear FMCW transmitters. The return signal can be processed either coherently using the IF signal or noncoherently using the video signal produced after detection by a diode.

Figure 4.42 depicts the radar system and illustrative return signals as might be processed coherently and noncoherently. A tank target is shown in the upper left of the figure, followed by the transmitted chirp pulse waveform. The pulses emit a time varying frequency. The return signal is received by an antenna and converted to IF in the mixer. The high resolution range profile, shown at the mixer output, represents the positions of individual scatterers on the target that are resolved through coherent signal processing. The effect of the processing is to increase the resolution ΔR normally obtained from the transmitted pulse of width τ by an improvement factor I as shown in the figure. Disadvantages of the coherent processing technique are the requirements for a phase stable signal source and a coherent receiver [72].

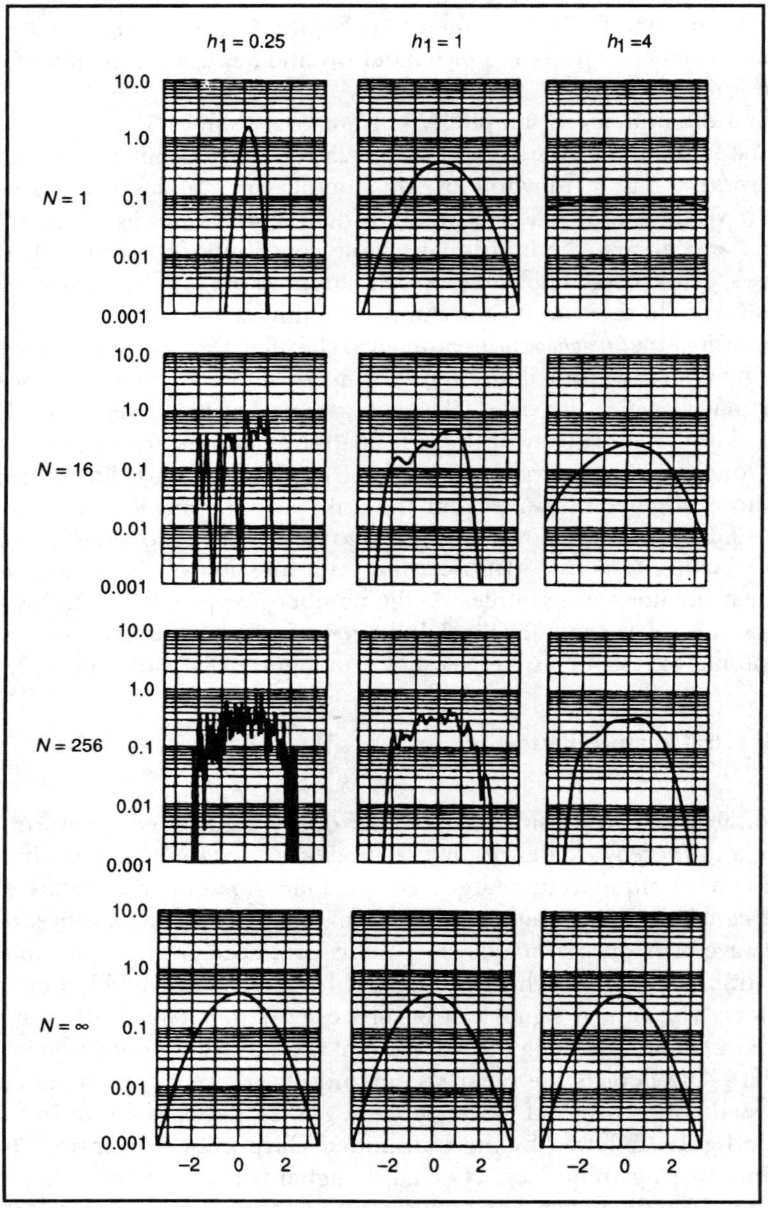

Figure 4.39 Parzen windows estimate of probability density function (*Source:* [71]. Reprinted with permission of John Wiley & Sons, Inc.)

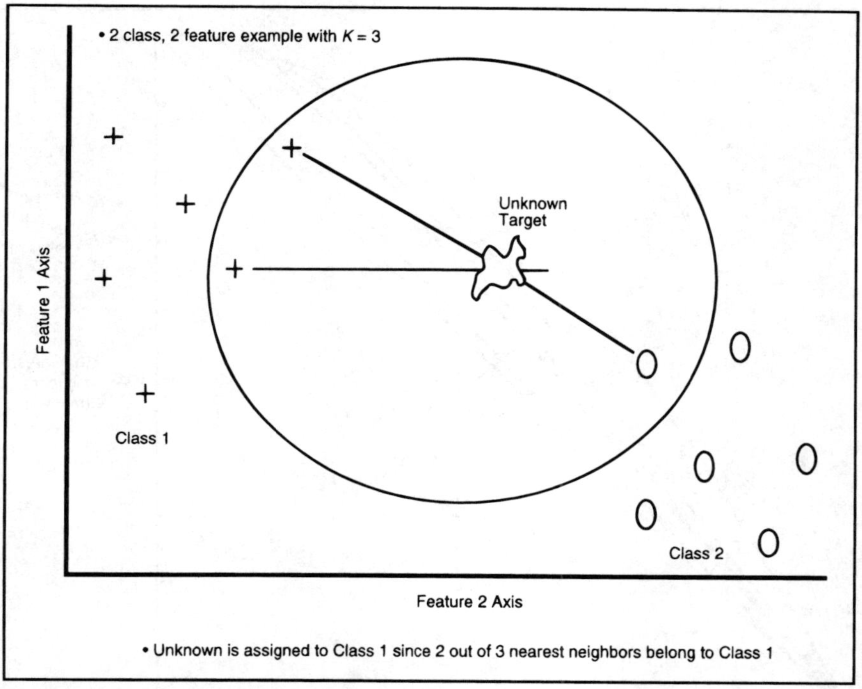

Figure 4.40 *K*th nearest neighbor classification.

If the IF signal is passed through an envelope detector, the resulting signal can be noncoherently processed to obtain the autocorrelation function of the high-resolution range profile as shown in the lower portion of Figure 4.42. Noncoherent processing does not require as stable a signal source nor motion compensation. However, the specification imposed on the amount of nonlinearity that can be tolerated in the change in frequency with respect to time in the transmitted waveform is still small when high resolution is required. The relation between the allowable waveform nonlinearity and the desired resolution r is

$$\text{Nonlinearity} = r/R \tag{4.82}$$

where R is the range to the target.

The autocorrelation waveform is proportional to the responses from pairs of scatterers. If the resolution after processing is 1 ft, as shown by the tick marks on the range delay separation axis in the lower part of Figure 4.42, then peak AB represents the return from scatterers 3 ft apart, peak BC represents the return from scatterers 5 ft apart, and peak AC represents the return from scatterers 8 ft apart. The magnitude of the peaks is proportional to scatterer size. Thus, scatterer C is larger than scatterer A since peak BC is greater than peak AB. Scatterer B is larger than scatterer A since peak BC is larger than peak AC. The magnitude of the main peak is proportional to the total energy in all the range bins.

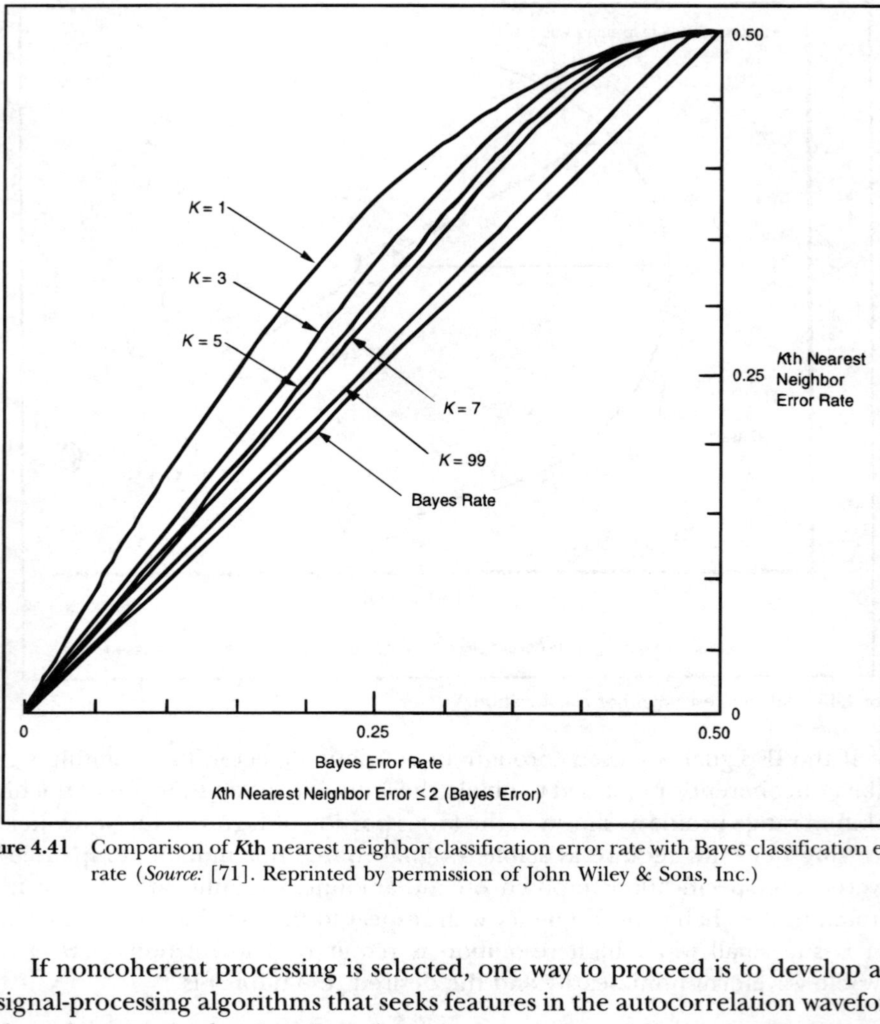

Figure 4.41 Comparison of *K*th nearest neighbor classification error rate with Bayes classification error rate (*Source:* [71]. Reprinted by permission of John Wiley & Sons, Inc.)

If noncoherent processing is selected, one way to proceed is to develop a set of signal-processing algorithms that seeks features in the autocorrelation waveform, such as the magnitude of each peak or the separation of the peaks. The values of the features are then used to define orthogonal surfaces or distance measures that separate the targets of interest from background clutter and nontarget objects and, perhaps, even one type of target from another (e.g., tank versus truck).

A technique that adds to the target discrimination information, and ultimately increases the number of correlation waveforms available for noncoherent signal processing, is the use of a polarimetric radar that receives both parallel-polarized and cross-polarized signals. In this manner, up to three correlation waveforms can be generated and used to extract feature information. These are the two autocorrelation waveforms, one corresponding to the parallel-polarized signal and the other corresponding to the cross-polarized signal, and one cross-correlation waveform corresponding to the parallel- and cross-polarized signals.

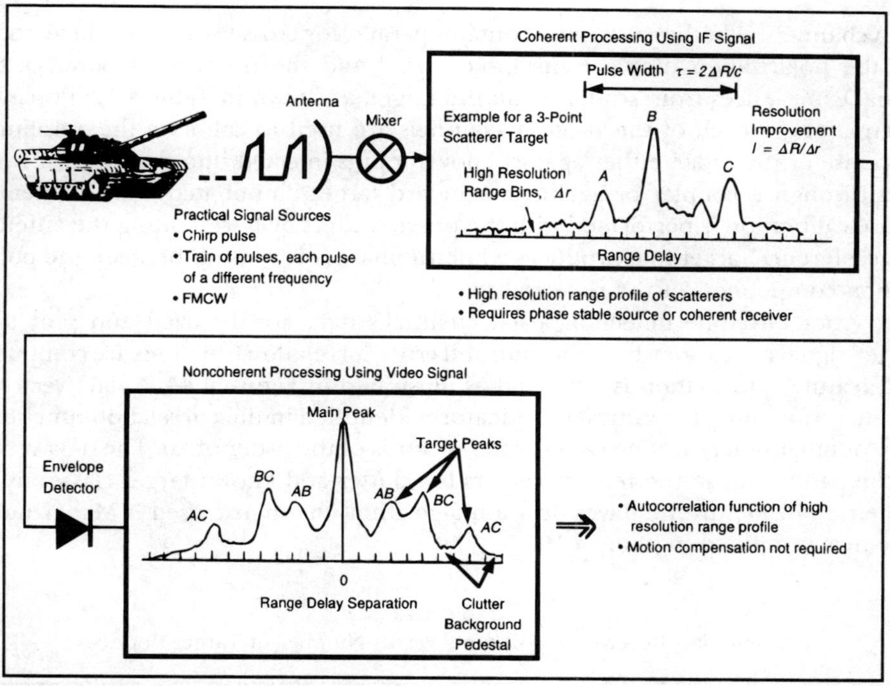

Figure 4.42 Spectral feature extraction [72].

A polarimetric instrumentation radar that transmits and receives multiple polarization signals is shown in Figure 4.43. The waveform generator has the capability to change the transmitted polarization from one burst of pulses to the next [73–75]. It also supplies the local oscillator signal. The dual-mode coupler and circulators separate the receive and transmit signals into parallel- and cross-polarized

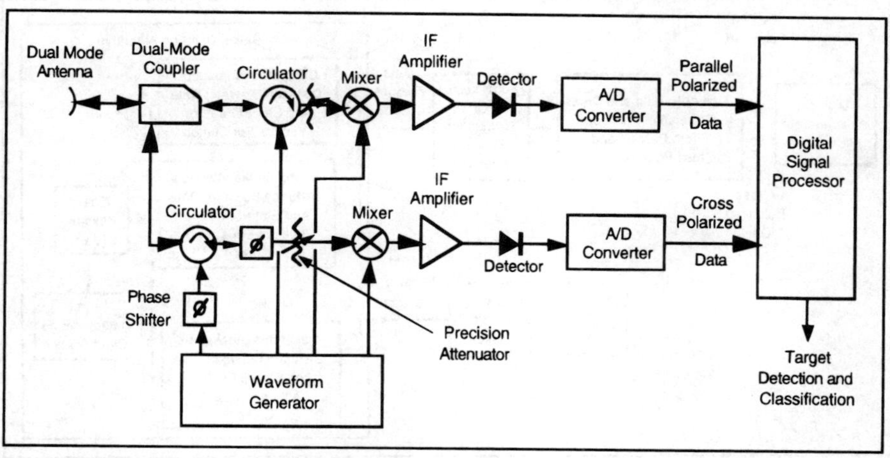

Figure 4.43 Polarimetric instrumentation radar.

data channels. Whether a channel contains parallel- or cross-polarized data depends on the polarization of the transmitted signal and the number of bounces the signal experiences from scatterers on the target, as shown in Table 4.12. Precision attenuators in each of the receiver channels are used to calibrate the amplitude response of the radar, either against known inputs injected into the receiver front end through a coupler or against a standard target illuminated by the antenna. Phase calibration is performed with the phase shifters by boresighting the antenna on a reference target that produces a known phase relationship between the polarization components of the reflected signal.

After envelope detection, each channel's data are digitized and sent to a digital signal processor where the auto and cross-correlation functions are computed and feature information is extracted as illustrated in Figure 4.44. A data vector is formed containing the values of the features identified in the correlation functions. The number of features needed for the vector is of the order of 20. The data vector is compared with stored feature vectors for clutter and known target classes using a metric [76,77]. In the Bayes decision algorithm, the metric used is Mahalanobis distance as defined in Table 4.10.

Table 4.12
Transmit and Receive Polarizations Versus Number of Target Bounces

Reflector	*Bounce*	*Transmit/Receive Polarization*
Trihedral	Odd	V/V, H/H, RHC/LHC, LHC/RHC
Dihedral	Even	RHC/RHC, LHC/LHC, V/H, H/V

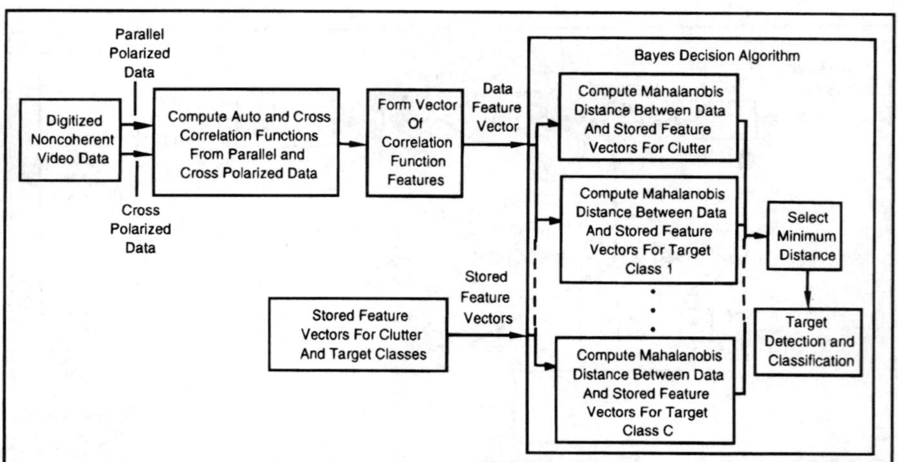

Figure 4.44 Polarimetric processing.

An underlying premise of polarimetric signal processing is that targets have about equal amounts of scattered energy from odd and even bounce scatterers, whereas clutter has more scattered energy from odd bounce than from even bounce scatterers [54]. The features that are selected attempt to measure the relative amounts of odd to even bounce scattered energy. A "target space" is defined when the outputs from a number of these features match those expected from a target. Otherwise, no target or a specific nontarget is declared. Other signal-processing approaches that use optimal polarimetric detectors to detect and classify objects in ground clutter have also been proposed [78–80]. A type of polarimetric detector that utilizes polarization ratios that relate the magnitudes of the returns from one polarization to those from another has been applied to multipolarization SAR data from forests. It has been able to classify stands of hardwoods and pine trees, wet and dry clearcut areas, and swamps [81].

The Bayes algorithm provides an optimal detection and classification solution if the values of the features are drawn from normally distributed random variables. If the values are not normally distributed, processes that normalize the data (such as using the logarithms of the feature values) can be applied to improve the performance of the Bayes decision algorithm. The data are declared to belong to the class represented by the stored feature vector that produces the minimum distance with the data vector. Thus, if the minimum Mahalanobis distance is between the data vector and a stored feature vector for a tank, the data are declared as belonging to a tank. Bhattacharyya distance has been used to estimate the expected accuracy of trial classifier designs [76,77].

Values of two features that help differentiate targets from clutter and that can be used as part of a feature vector in noncoherent processing are shown in Table 4.13. The features are defined as

$$\text{Feature 1} = \left[\sum \text{Energy (odd bounce)}\right] \Big/ \left[\sum \text{Energy(even bounce)}\right] \tag{4.83}$$

Table 4.13

Autocorrelation Function Feature Values for Target and Clutter Objects

Feature	*Target*	*Clutter*	*Explanation*
1	≈1	>1	Scattered target energy is about evenly distributed between odd and even bounce scatterers, whereas clutter return contains more odd bounce scattered energy
2	≈1	<1	Scattered energy from target of finite size will be contained in a number of range bins that represent the maximum target scatterer separation, whereas clutter return will not be constrained in range

and

$$\text{Feature 2} = \left[\sum \text{Energy } (n \text{ range bins})\right] \Big/ \left[\text{Total energy}\right] \qquad (4.84)$$

where

Σ Energy (odd bounce)	=	total energy in range bin from features obtained from the odd bounce channel;
Σ Energy (even bounce)	=	total energy in range bin from features obtained from the even bounce channel;
Σ Energy (n range bins)	=	sum of the energy in the first n fine resolution range bins;

n is proportional to the target size.

The following example conveys the performance capability of the autocorrelation feature vector and the Bayes decision algorithm. The scenario and radar characteristics are defined in Table 4.14.

The confusion matrix obtained after processing the data is shown in Table 4.15. Approximately 10 features were used in each feature vector. Each of the three targets was correctly classified 80 to 85% of the time. The false classifications were about equally distributed between the two remaining targets. After processing, approximately one false alarm per square kilometer remained [82].

If coherent processing is employed, data feature vectors and stored feature vectors representing known target and clutter classes can also be constructed from the high-resolution range profiles and be input to a Bayes decision algorithm. The types of features used with the range profiles differ from those found to be optimal

Table 4.14
Scenario and Radar Characteristics Used to Evaluate Autocorrelation Feature Vector and Bayes Decision Algorithm

Scenario	*Radar*
• ERIM tower test data	• Frequency = 35 GHz ± 1 GHz
• Targets on turntable	• Transmit polarization = RHC
—Target A	• Receiver polarization = RHC, LHC
—Target B	• Transmit waveform = FMCW
—Target C	• Resolution = 5 cm
• Depression angles = 23, 33, 62 degrees	• Waveform nonlinearity = 0.03% when range is 500 ft (152m)
• Slant range ≤ 500 ft (data corrected for non-far field effects when processed)	

Table 4.15
Confusion Matrix Showing Target Classification Results Using the Bayes Decision Algorithm

	Target A	*Target B*	*Target C*
Target A	80–85%	7–10%	7–10%
Target B	7–10%	80–85%	7–10%
Target C	7–10%	7–10%	80–85%

for the correlation functions. Examples of features used to form vectors from range profile data are given in Table 4.16 [83].

4.16 SUMMARY

In radar design, parameters such as the transmitted waveform, receive and transmit signal polarizations, and signal-processing technique are selected in accordance with the application and operating environment of the radar (e.g., long- versus short-range and high-value versus low-value target detection or classification; noise-

Table 4.16
High-Resolution Range Profile Features Used to Form Feature Vectors [83]

Range Profile Feature	*Transmit/Receive Polarization*
• Distance from central to largest peak	RHC/V + RHC/H
• Distance from central to largest peak	V/V + V/H
• Distance from central to largest peak	V/V
• Distance from first to largest peak	RHC/V + RHC/H
• Distance from largest to last peak	RHC/H
• Number of peaks greater than X m^2	RHC/H
• Number of peaks greater than X m^2	RHC/V
• Number of peaks greater than X m^2	V/H
• Maximum distance between three largest peaks	V/H
• Average distance to two largest peaks	V/H
• Maximum adjacent peak distance	V/H
• Average peak-to-peak distance	V/H
• Ratios of selective features, such as first to largest peak distance divided by maximum adjacent peak distance	–

Source: Olson, R. R., D. L. Halliday, J. R. Gannaway, and B. L. Masteller, "An Armored Target Recognition System Based on Polarimetric Signatures," Bendix Guidance Systems Division, Mishawaka, IN, 1983.

versus clutter-limited operation; target attributes desired, such as range and cross-range position, speed, size, and shape; weather and countermeasure effects; and packaging and power supply constraints). While the relatively simple pulse and FMCW radars are less expensive to build, their pulse Doppler and synthetic aperture counterparts provide target location, speed, clutter rejection, and imagery not available in the other designs.

When clutter-limited operation is of concern, the square law detector (optimal for noise-limited operation) must be replaced by another type of detector, such as a binary or log envelope detector, to maintain relatively high detection probability with a given signal-to-clutter ratio and false alarm probability. With each of these detectors, CFAR performance is achieved by adjusting the detector design parameters to the clutter distribution. Other types of detectors, such as the adaptive log-*t* detector, allow optimal performance in both Weibull and log normal clutter without the need for detector parameter adjustment. Other methods for increasing detection probability are to integrate multiple independent signal samples through coherent, video, binary, and cumulative techniques. Detection probability is also affected by the antenna beamwidth through the over and under resolution of targets. Over-resolved targets generally require increased signal-to-interference ratios to achieve the same detection probabilities as under-resolved targets. Radar target tracking accuracy is determined by the antenna beamwidth, the target scatterer amplitude fluctuations that cause the apparent center of the target to wander, and the type of antenna search or scan that is employed. When the antenna beamwidth approaches the target extent, the effects of glint or target scintillation are reduced.

Measured tree and snow radar backscatter data show high clutter characteristics at 35 and 95 GHz. Surface moisture, leaf structure, and polarization affect the backscatter coefficients of trees. Snow backscatter is primarily affected by water content, depth, underlying ground composition, and radar transmitted frequency. Care should be exercised when searching the literature for vegetation backscatter data as some of the early measurements may be 4 to 7 dB too low in value.

Several signal-processing algorithms were introduced. These include time-domain amplitude and pulse width measurements, and frequency-domain approaches such as spectral feature extraction. Target features can be enhanced through median filtering that removes spike-type clutter. Additional features obtained by processing multiple polarization data may potentially increase the detection and classification probabilities of targets. Several feature-based inference algorithms that perform target classification or identification were described. Bayesian decision algorithms, useful when the target features are represented by Gaussian-distributed random variables, are an example of a parametric classifier that compares values of features that are calculated in real time with those belonging to known objects that are stored in a microprocessor. Nonparametric techniques such as artificial neural networks and *K*th nearest neighbor classification can be used to classify objects when the probability density function is unknown.

References

[1] Willis, N. J., *Bistatic Radar,* Norwood, MA: Artech House, 1991.

[2] Corson, D. R., and P. Lorrain, *Introduction to Electromagnetic Fields and Waves*, San Francisco, CA: W. H. Freeman, 1962, p. 320.

[3] Taylor, Jr., J. W., and J. Mattern, "Receivers," Chap. 5 in *Radar Handbook*, M. I. Skolnik, (Ed.), New York, NY: McGraw-Hill, 1970.

[4] Tsui, J. B-Y, *Microwave Receivers and Related Components*, Ph.D. Thesis, Avionics Laboratory, Air Force Wright Aeronautical Laboratories, Report PB84-108711, 1983.

[5] Ulaby, F. T., R. K. Moore, and A. K. Fung, *Microwave Remote Sensing Active and Passive: Vol. I Microwave Remote Sensing Fundamentals and Radiometry*, Norwood, MA: Artech House, 1981.

[6] Mumford, W. W., and E. H. Scheibe, *Noise Performance Factors in Communication Systems*, Norwood, MA: Artech House, 1968.

[7] Edde, B., *Radar: Principles, Technology, Applications*, Englewood Cliffs, NJ: PTR Prentice Hall, 1993.

[8] James, D. A., *Radar Homing Guidance for Tactical Missiles*, London, U.K.: Macmillan Education Ltd., 1986.

[9] Williams, F. C., and M. E. Radant, "Airborne Radar and the Three PRFs," *Microwave Journal*, July 1983.

[10] Stimson, G. W., *Introduction to Airborne Radar,* El Segundo, CA: Hughes Aircraft Company, 1983.

[11] Skolnik, M. I., *Introduction to Radar Systems*, New York, NY: McGraw-Hill, 1980.

[12] Hovanessian, S. A., *Introduction to Sensor Systems*, Norwood, MA: Artech House, 1988.

[13] Chang, C. Y., and J. C. Curlander, "Application of the Multiple PRF Technique to Resolve Doppler Centroid Estimation Ambiguity for Spaceborne SAR," *IEEE Trans. on Geosci. and Rem. Sens.*, Vol. 30, No. 5, Sept. 1992, pp. 941–949.

[14] Cutrona, L. J., et al., "A High-Resolution Radar Combat-Surveillance System," *IRE Trans. on Military Electronics*, Vol. MIL-5, Apr. 1961, pp. 127–131.

[15] Develet, Jr., J. A., "Performance of a Synthetic-Aperture Mapping Radar System," *IEEE Trans. on Aerospace and Navigational Electronics*, Vol. ANE-11, No. 3, Sept. 1964, pp. 173–179.

[16] Harger, R. O., "An Optimum Design of Ambiguity Function, Antenna Pattern, and Signal for Side-Looking Radars," *IEEE Trans. on Military Electronics*, Vol. MIL-9, Jul.-Oct. 1965, pp. 264–271.

[17] Brown, W. M., "Synthetic Aperture Radar," *IEEE Trans. on Aerospace and Electronic Systems*, Vol. AES-3, No. 2, Mar. 1967, pp. 217–229.

[18] Harger, R. O. , "The Side-Looking Radar Image of Time-Variant Scenes," *Radio Science*, Vol. 15, No. 4, Jul.-Aug. 1980, pp. 749–756.

[19] Elachi, C., *Spaceborne Radar Remote Sensing: Applications and Techniques*, New York, NY: IEEE Press, 1988.

[20] Vesecky, J. F. "The Observation of Ocean Surface Phenomena Using Imagery From the SEASAT Synthetic Aperture Radar," *J. Geophysical Research*, Vol. 87, No. C5, Apr. 30, 1982, pp. 3397–3430.

[21] Curlander, J. C., "Utilization of Spaceborne SAR Data for Mapping," *IEEE Trans. on Geosci. and Rem. Sens.*, Vol. GE-22, No. 2, Mar. 1984, pp. 106–112.

[22] Holmes, Q. A., D. R. Nuesch, and R. A. Shuchman, "Textural Analysis and Real-Time Classification of Sea-Ice Types Using Digital SAR Data," *IEEE Trans. on Geosci. and Rem. Sens.*, Vol. GE-22, No. 2, Mar. 1984, pp. 103–120.

[23] Durand, J. M., B. J. Gimonet, and J. R. Perbos, "SAR Filtering for Classification," *IEEE Trans. on Geosci. and Rem. Sens.*, Vol. GE-25, No. 5, Sept. 1987, pp. 629–637.

[24] Ulaby, F. T., R. K. Moore, and A. K. Fung, *Microwave Remote Sensing Active and Passive: Vol. II Radar Remote Sensing and Surface Scattering and Emission Theory*, Norwood, MA: Artech House, 1982.

[25] DiFranco, J. V. and W. L. Rubin, *Radar Detection*, Englewood Cliffs, NJ: Prentice-Hall, 1968.

[26] Meyer, D. P., and H. A. Mayer, *Radar Target Detection: Handbook of Theory and Practice*, San Diego, CA: Academic Press, 1973.

[27] Marcum, J. I., "Studies of Target Detection by a Pulsed Radar: Mathematical Appendix," *IRE Trans. on Information Theory*, Vol. IT-6, Apr. 1960, pp.145–268.

[28] Fehlner, L. F., "Marcum's and Swerling's Data on Target Detection by a Pulsed Radar," AD 602 121, Silver Springs, MD: The Johns Hopkins University, July 1962.

[29] Trunk, G., and S. George, "Detection of Targets in Non-Gaussian Sea Clutter," *IEEE Trans. on Aerospace and Electronics Systems*, Vol. AES-6, Sept. 1970.

[30] Mitchell, R. L., and J. F. Walker, "Recursive Methods for Computing Detection Probabilities," *IEEE Trans. on Aerospace and Electronic Systems*, Vol. AES-7, No. 4, Jul. 1971, pp. 671–676.

[31] Scheler, D. C., "Radar Detection in Weibull Clutter," *IEEE Trans. on Aerospace and Elect. Systems*, Vol. AES-12, No. 6, Nov. 1976, pp. 736–743.

[32] Scheler, D. C., "Radar Detection in Log-Normal Clutter," *Proc. IEEE International Radar Conference*, Washington, D.C., Apr. 1975, pp. 262–267.

[33] Scheler, D. C., "Detection Performance of Logarithmic Receivers Employing Video Integrators," *IEEE Trans. on Aerospace and Elect. Systems*, Vol. AES-15, No. 6, Nov. 1979, pp. 831–839.

[34] Goldstein, G. B., "False-Alarm Regulation in Log-Normal and Weibull Clutter," *IEEE Trans. on Aerospace and Elect. Systems*, Vol. AES-9, No. 1, Jan. 1973, pp. 84–92.

[35] Brookner, E., *Radar Technology*, Norwood, MA: Artech House, 1985.

[36] Barton, D. K., *Modern Radar System Analysis*, Norwood, MA: Artech House, 1986.

[37] Green, Jr., B. A., "Radar Detection Probability with Logarithmic Detectors," *IRE Trans. on Information Theory*, Mar. 1958, pp. 50–52.

[38] Hansen, V. G., "Studies of Logarithmic Radar Receiver Using Pulse-Length Discrimination," *IEEE Trans. on Aerospace and Electronic Systems*, Vol. AES-1, No. 3, Dec. 1965, 246–253.

[39] Hansen, V. G., and H. R. Ward, "Detection Performance of the Cell Averaging Log/CFAR Receiver," *IEEE Trans. on Aerospace and Electronic Systems*, Vol. AES-8, No. 5, Sept. 1972, 648–652.

[40] Schwartz, M., "A Coincidence Procedure for Signal Detection," *IRE Trans. on Info. Theory*, Dec. 1956, pp. 135–139.

[41] Caspers, J. W., "Automatic-detection Theory," Chap. 15 in *Radar Handbook*, M. I. Skolnik, (Ed.), New York, NY: McGraw-Hill, 1970.

[42] Szajnowski, W. J., "Discrimination Between Log-Normal and Weibull Clutter," *IEEE Trans. on Aerospace and Elect. Systems*, Vol. AES-13, No. 5, Sept. 1977, pp. 480–485.

[43] Echard, J. D., "Detection in Noise," Chap. 9 in *Principles of Modern Radar*, J. L. Eaves and E. K. Reedy, (Eds.), New York, NY: Van Nostrand, 1987.

[44] Barton, D. K., *Radar System Analysis*, Norwood, MA: Artech House, 1979.

[45] Vannicola, V. C., and K. G. Hillman, "Detection Performance for Over Resolved Targets," RADC-TR-84-61, AD-B089 870, Rome Air Development Center, Griffiss Air Force Base, NY, Sept. 1984.

[46] Currie, N. C., et al., "Radar Millimeter Backscatter Measurements, Vol. I, Snow and Vegetation," Report AFATL-TR-77-92, AD BO 30190L, Eglin Air Force Base, FL, July 1977.

[47] Currie, N. C., E. E. Martin, and F. B. Dyer, "Radar Foliage Penetration Measurements at Millimeter Wavelengths," Tech. Report No. 4 on Contract DAAA25-73-C-0256, Engineering Experiment Station, Georgia Institute of Technology, Dec. 31, 1975.

[48] Hayes, R. D., and F. B. Dyer, "Land Clutter Characteristics for Computer Modeling of Fire Control Radar Systems," Tech. Report No. 1 on Contract DAAA25-73-C-0256, Engineering Experiment Station, Georgia Institute of Technology, May 15, 1973.

[49] Currie, N. C., F. B. Dyer, and R. D. Hayes, "Radar Land Clutter Measurements at Frequencies of 9.5, 16, 35, and 95 GHz," Report A-1485-003, AD A012 709, Engineering Experiment Station, Georgia Institute of Technology, April 2, 1975.

[50] Currie, N. C., R. D. Hayes, and R. N. Trebits, *Millimeter-Wave Radar Clutter*, Norwood, MA: Artech House, 1992.

[51] Bush, T. F., and Ulaby, F. T., "Variability in the Measurement of Radar Backscatter," *IEEE Trans. Ant. and Prop.*, Nov. 1976, pp. 896–899.

[52] Cosgriff, R. L., W. H. Peake, and R. C. Taylor, "Terrain Scattering Properties for Sensor System Design," Engineering Experiment Station Bulletin 181, Ohio State University, Columbus, OH, 1960.

[53] Barton, D. K., (Ed.), *Radars, Volume 5: Radar Clutter*, Norwood, MA: Artech House, 1975.

[54] Long, M. W., *Radar Reflectivity of Land and Sea*, Norwood, MA: Artech House, 1983.

[55] Peake, W. H., and T. L. Oliver, "The response of terrestrial surfaces at microwave frequencies," Report 2440-7, Ohio State University Electroscience Laboratory, May 1971.
[56] Ulaby, F. T., W. H. Stiles, and M. Abdelrazik, "Snowcover Influence on Backscattering from Terrain," *IEEE Trans. on Geosci. and Rem. Sens.*, Vol. GE-22, No. 2, March 1984, pp. 126–133.
[57] Ulaby, F. T., and W. H. Stiles, "The Active and Passive Microwave Response to Snow Parameters," Report RSL 340-2, Remote Sensing Laboratory, University of Kansas Center for Research, Dec. 1978.
[58] Williams, L. D., and J. G. Gallagher, "The Relation of Millimeter-Wavelength Backscatter to Surface Snow Properties," *IEEE Trans. on Geosci. and Rem. Sens.*, Vol. GE-25, No. 2, March 1987, pp. 188–194.
[59] Chang, P. S., et al., "Polarimetric Backscatter from Fresh and Metamorphic Snowcover at Millimeter Wavelengths," *IEEE Trans. on Ant. and Prop.*, Vol. 44, No. 1, Jan. 1996, pp. 58–73.
[60] Knox, J. E., "Terrain and Target Signature Measurements at 94 GHz," BRL Memorandum Report 2708, Dec. 1976.
[61] Rabiner, L. R., M. R. Sembur, and C. E. Schmidt, "Applications of a Nonlinear Smoothing Algorithm to Speech Processing," *IEEE Trans. on Acoustics, Speech, and Sig. Proc., ASSP-23*, No. 6, Dec. 1975, pp. 552–557.
[62] Sonderegger, J. F., Ph.D. Thesis, Dep't. of Electrical Engineering, University of Minnesota, 1985.
[63] Fukunaga, K., *Introduction to Statistical Pattern Recognition*, New York, NY: Academic Press, 1972.
[64] Shafer, G, *A Mathematical Theory of Evidence*, Princeton, NJ: Princeton University Press, 1976.
[65] Barnett, J. A., "Computational Methods for a Mathematical Theory of Evidence," *Proc. Seventh International Joint Conf. on Artificial Intelligence*, Vol. II, IJCAI-81, 1981, pp. 868–875.
[66] Buede, D. M., "Shafer-Dempster and Bayesian Reasoning: A Response to 'Shafer-Dempster Reasoning with Applications to Multisensor Target Identification Systems'," *IEEE Trans. on Systems, Man, and Cybernetics*, Vol. 18, No. 6, Nov./Dec. 1988, pp. 1009–1011.
[67] Thomopoulos, S.C.A., "Theories in distributed decision fusion: comparison and generalization," *Sensor Fusion III: 3-D Perception and Recognition*, SPIE Vol. 1383, 1990, pp. 623–634.
[68] Wasserman, P. D., *Neural Computing: Theory and Practice*, New York, NY: Van Nostrand Reinhold, 1989.
[69] Widrow, B., and M. A. Lehr, "30 Years of Adaptive Neural Networks: Perceptron, Madaline, and Backpropagation," *Proc. of the IEEE*, Vol. 78, No. 9, Sept. 1990, pp. 1415–1442.
[70] Rogers, S. K., and M. Kabrisky, *An Introduction to Biological and Artificial Neural Networks for Pattern Recognition*, SPIE Optical Engineering Press, Bellingham, WA, 1991.
[71] Duda, R. O. and P. E. Hart, *Pattern Classification and Scene Analysis*, New York, NY: John Wiley & Sons, 1973.
[72] Einstein, T. H., "Generation of High Resolution Radar Range Profiles and Range Profile Auto-Correlation Functions Using Stepped-Frequency Pulse Trains," Project Report TT-54, AD A149 242, MIT Lincoln Laboratory, Oct. 18, 1984.
[73] Currie, N. C., J. A. Scheer, and W. A. Holm, "mm-Wave Instrumentation Radar Systems," *Microwave Journal*, Aug. 1978, pp. 35–44.
[74] Gent, M., I. M Hunter, and N. P. Robinson, "Polarization of Radar Echoes, Including Aircraft, Precipitation, and Terrain," *Proc. IEEE*, Vol. 110, No. 12, Dec. 1963, pp. 2139–2148.
[75] W. A. Holm, "Polarimetric Fundamentals and Techniques," Chapter 20 in *Principles of Modern Radar*, J. L. Eaves and E. K. Reedy, (Eds.), New York, NY: Van Nostrand, 1987.
[76] Lee, C., and D. A. Landgrebe, "Fast Likelihood Classification," *IEEE Trans. on Geosci. and Rem. Sens.*, Vol. 29, No. 4, Jul. 1991, 509–517.
[77] Kim, B., and D. A. Landgrebe, "Hierarchical Classifier Design in High Dimensional, Numerous Class Cases," *IEEE Trans. on Geosci. and Rem. Sens.* Vol. 29, No. 4, Jul. 1991, 518–528.
[78] Klein, L. A., and E. C. Thiede, "Polarimetric Radar Signal Processing Comparison," *Polarimetric Radar Workshop*, U.S. Army Missile Command, Redstone Arsenal, AL, 1980.
[79] Huddleston, G. K., "Simplified Scattering Models for Polarimetric Processing Radar," *Second Workshop on Polarimetric Technology*, Redstone Arsenal, AL, May 3–5, 1983, GACIAC PR-83-01, pp. 125–151.

[80] Novak, L. M., M .B. Sechtin, and M. J. Cardullo, "Studies of Target Detection Algorithms That Use Polarimetric Radar Data," Project Report TT-71, Lincoln Laboratory, Massachusetts Institute of Technology, Oct. 12, 1988.

[81] Wu, S-T., and S. A. Sader, "Multipolarization SAR Data for Surface Feature Delineation and Forest Vegetation Characterization," *IEEE Trans. on Geosci. and Electronics*, Vol. GE-25, No. 1, Jan. 1987, pp. 67–76.

[82] Glynn, T., Glynn Scientific Incorporated, Annapolis, MD.

[83] Olson, R. R., et al., "An Armored Target Recognition System Based on Polarimetric Signatures," Bendix Guidance Systems Division, Mishawaka, IN, 1983.

CHAPTER 5

Processing of Intermediate Frequency Data in an FMCW Radar

A frequency-modulated continuous wave (FMCW) radar not only provides a detected video signal (one without phase information), but also an intermediate frequency (IF) signal at the mixer output that contains relative target phase information and higher resolution data corresponding to objects in the radar's field of view. Since the IF is proportional to the range from sensor to target, the IF can be used to calculate the range. Furthermore, the measurement of the IF is limited by noise, potentially providing less interference than clutter. One application of the range data is to construct range profiles of objects in the scene and thereby create images from which the desired targets or objects can be identified. These data allow targets to be discriminated from clutter and nontargets through the use of algorithms that utilize features such as a height variation greater than some threshold, a backscattered pulse width characteristic of a target, and a rate of change of height with respect to time that matches a known target shape, size, and speed. This chapter develops the theoretical foundation for the design and operation of an FMCW radar whose intermediate frequency signal can be used to generate range imagery. Methods of increasing the signal-to-noise ratio, and hence detection probability, by obtaining additional independent samples through temporal and frequency decorrelation techniques are described. Fractal models are shown to generate range profile imagery suitable for development of signal-processing algorithms that can later be applied to actual range profile data.

5.1 RANGE RESOLUTION

In Chapter 4, the intermediate frequency f_{IF} in an FMCW radar was shown to be related to range R by

$$f_{IF} = 4\Delta F f_m R/c \tag{5.1}$$

where

ΔF = radio frequency (RF) modulation bandwidth of the transmitted waveform in Hz;
f_m = RF modulation frequency corresponding to an up and downsweep cycle in Hz;
R = range between sensor and target or object of interest in m;
c = speed of light = 3×10^8 m/sec.

Referring to Figure 5.1, the intermediate frequency can be measured over an interval T by counting the number of cycles N as

$$f_{IF} = N/T = 4\Delta F f_m R/c \tag{5.2}$$

allowing the range R to be expressed as

$$R = \frac{c}{4\Delta F f_m}\frac{N}{T} \tag{5.3}$$

If T is limited to $t_m/2$ equal to $1/(2f_m)$ to avoid having to count cycles in regions of distortion that occur near the turnaround time of upsweep to downsweep, (5.3) becomes

$$R = \frac{c}{4\Delta F f_m} 2f_m N = \frac{cN}{2\Delta F} \tag{5.4}$$

Rewriting (5.4), the range resolution per cycle δR is

$$\delta R = \frac{R}{N} = \frac{c}{2\Delta F} \tag{5.5}$$

Either a full-wave or a half-wave counter can be configured to count the IF [1].

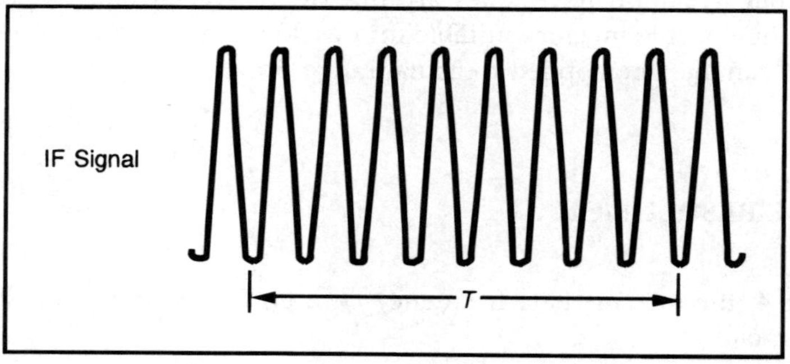

Figure 5.1 IF signal.

The full-wave counter produces two counts per cycle, whereas the half wave counter has one count per cycle. With a full-wave counter, the range resolution per count ΔR is

$$\Delta R = \delta R/2 = \frac{c}{4\Delta F} \tag{5.6}$$

If a half-wave counter is used, the range resolution per count is

$$\Delta R = \delta R = \frac{c}{2\Delta F} \tag{5.7}$$

If the desired range resolution per count is 1 ft (0.3 m) and a full-wave counter is selected, the required RF modulation bandwidth is

$$\Delta F = 250 \text{ MHz} \tag{5.8}$$

This bandwidth is not difficult to achieve with modern monolithic microwave integrated circuits (MMICs) at 35 GHz [2,3].

There are several sources of error that distort the measurement of the intermediate frequency. These are due either to effects manifested in the return signal or effects generated in the receiver [4,5]. They are summarized in Table 5.1.

Table 5.1
Error Sources in Intermediate Frequency Measurement

Return Signal Error Sources	*Receiver Error Sources*
• Doppler spread across antenna beam, caused by returns from scatterers at different azimuth locations, introduces several intermediate frequencies. • Range depth of antenna beam allows returns from scatterers at different elevation angles (ranges) to produce more than one intermediate frequency. • Sidelobes allow returns from scatterers not in the antenna's mainlobe to introduce extraneous frequency information and increase signal distortion.	• Thermal noise competes with returns from weak targets or clutter. • Voltage standing wave ratio (VSWR) permits reflected signals between the mixer and antenna to introduce extraneous intermediate frequencies. • Spurious frequencies caused by nonlinear processes in the mixer produce high frequency harmonics of the RF and LO sum and difference frequencies. These intermodulation products distort the IF signal and may mask the signal features of interest. • Low values of LO-to-RF isolation allow the LO signal to leak into the antenna and be reflected back into the mixer, producing unwanted IF signals. • Nonlinearities shift returns from scatterers to an IF region corresponding to an incorrect range.

Insufficient local oscillator (LO)-to-RF isolation distorts the IF waveform as shown in Figure 5.2. The waveform on the left is representative of an intermediate frequency signal produced by a mixer with high LO-to-RF isolation of the order of 20 dB, such as an MMIC mixer. The cycles are relatively easy to identify and count, although some distortion of the IF is evident in the vicinity of the turnaround time of the up and downsweeps. The waveform on the right is representative of intermediate frequencies produced by a mixer with lower LO-to-RF isolation of approximately 14 dB. Here, the cycles are not distinct and frequency counting is subject to larger errors. The relatively low LO-to-RF isolation found in the particular MIC (microwave integrated circuit) mixer that produced this waveform is not typical of all mixers with this construction, but rather is due to the poor matching of design parameters in the diodes selected for the mixer.

A range profiler can provide several types of information including the range to the background, range to target, and range or height profiles of the target that are useful in object classification. The implications these measurements have on the design of the FMCW radar are discussed below.

5.2 RADAR RANGE EQUATION

The requirements and performance of a radar can be understood by analyzing the radar range equation. The power P_R received by a monostatic radar is given by the range equation as

$$P_R = \frac{P_T G_T A_R \sigma_T}{(4\pi)^2 R^4} \tag{5.9}$$

where

P_T = transmitted power in W;
G_T = gain of transmitting antenna;
A_R = effective area of receiving antenna aperture in m^2;

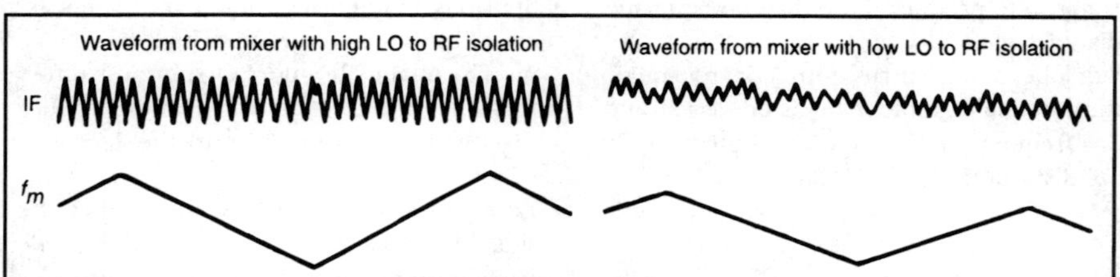

Figure 5.2 IF waveforms produced by mixers with high and low LO-to-RF isolation and corresponding RF modulation frequency. The waveform on the left was produced by an MMIC mixer, while the one on the right was produced by an MIC mixer constructed with diodes whose properties were not well-matched.

σ_T = radar cross section of the target in m^2;
R = sensor-to-target or sensor-to-ground range in m.

The antenna gain G can be written in terms of the effective area of the antenna aperture A_e as

$$G = \frac{4\pi A_e}{\lambda^2} \tag{5.10}$$

where λ = wavelength of transmitted energy in m. When the aperture is uniformly illuminated, the gain can be expressed in terms of the solid angle captured by the beam as

$$G = \frac{9.84}{\theta_a \theta_e} \tag{5.11}$$

where θ_a and θ_e are the azimuth and elevation two-way 3-dB antenna beamwidths in radians, respectively. When the aperture illumination is tapered as a function of the distance from the center of the aperture to reduce sidelobes, the gain is nearly constant at

$$G = \frac{11.3}{\theta_a \theta_e} \tag{5.12}$$

for all practical illuminations [6].

By substituting the receive antenna gain from (5.10) for the effective aperture of the receive antenna, the received power given by (5.9) becomes

$$P_R = \frac{P_T G_T G_R \lambda^2 \sigma_T}{(4\pi)^3 R^4} \tag{5.13}$$

Losses in the atmosphere or receiver reduce P_R to a value given by

$$P_R = \frac{P_T G_T G_R \lambda^2 \sigma_T}{(4\pi)^3 R^4 L} \tag{5.14}$$

where L = sum of atmospheric and receiver losses. The value $L = 1$ represents lossless transmission and reception and $L > 1$ represents a loss.

The signal-to-noise ratio S/N of the IF signal is written as

$$S/N = \frac{P_R}{k_B TBF_{SSB}} \quad (5.15)$$

$$= \frac{P_T G_T G_R \lambda^2 \sigma_T}{(4\pi)^3 R^4 L k_B TBF_{SSB}} \quad (5.16)$$

where

$k_B TBF_{SSB}$ = receiver noise power in W, (5.17)

k_B = Boltzmann's constant = 1.380662×10^{-23} J/K;

T = absolute temperature of receiver noise emitting components = 290K;

B = equivalent noise bandwidth of the radar in Hz;

F_{SSB} = single sideband noise figure of the radar.

5.3 RANGE TO BACKGROUND

The ability to range to the background is governed by the backscatter coefficient of the clutter, assuming the design of the radar, including transmitted power, antenna gains, wavelength, and noise figure is fixed. To examine the impact of the clutter on the signal received by the radar, we rewrite the S/N as a clutter-to-noise ratio C/N where the clutter return is now the backscattered signal of interest. Thus,

$$C/N = \frac{P_T G_T G_R \lambda^2 \sigma_c}{(4\pi)^3 R^4 L k_B TBF_{SSB}} \quad (5.18)$$

where σ_c = radar cross section of the clutter in m^2.

Values for the parameters in (5.18) are a function of the specific radar application. A set of values is given in Table 5.2 for a 35-GHz radar that might be used to detect targets from a low-flying missile.

With these values, the C/N becomes

$$C/N = \frac{(0.02)(10^{2.5})(10^{2.5})(8.6 \times 10^{-3})^2 \sigma_c}{(4\pi)^3 (50)^4 (1)(1.38 \times 10^{-23})(290)(5 \times 10^6)(10^{0.9})} \quad (5.19)$$

$$= 18.7 \text{ dB} + 10 \log \sigma_c \quad (5.20)$$

Figure 5.3 shows that to achieve a detection probability of 0.90 at a false alarm

Table 5.2
Radar Parameters Used in Radar Range Equation for Clutter-to-Noise Ratio Calculation

Parameter	*Value*
P_T	0.02 W
G_T	25 dB = $10^{2.5}$
G_R	25 dB = $10^{2.5}$
λ	8.6×10^{-3} m
R	50 m
k_B	1.38×10^{-23} J/K
T	290 K
B	5×10^6 Hz
F_{SSB}	9 dB = $10^{0.9}$
L	0 dB = 1

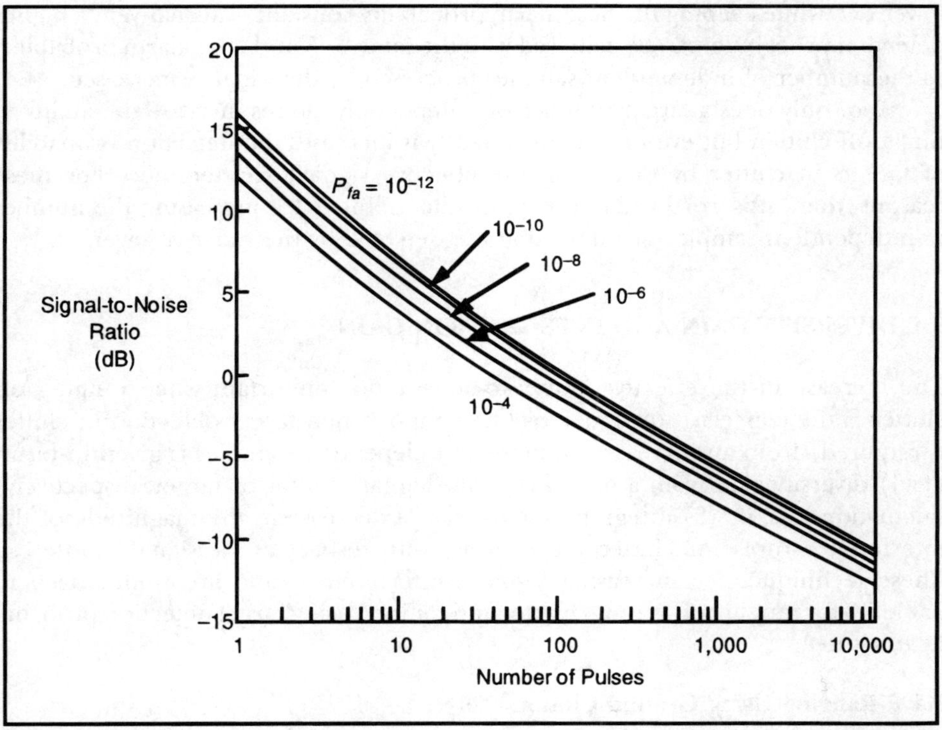

Figure 5.3 Signal-to-noise ratio as a function of number of pulses noncoherently integrated for a detection probability of 0.90 using a linear detector with a nonfluctuating target. (*Source:* [7]. Reprinted with permission.)

probability of 10^{-6} requires a signal-to-noise ratio of 13.2 dB when a single return pulse is detected [7]. Setting (5.20) equal to 13.2 dB and solving for the required value of σ_c yields

$$\sigma_c = 10^{-5.5/10} = 0.28 \text{ m}^2 \tag{5.21}$$

If the area of the ground footprint A_a formed by the antenna is 10 m^2, then the clutter backscatter coefficient σ^o is

$$\sigma^o \geq 0.28/10 = -15.5 \text{ dB m}^2/\text{m}^2 \tag{5.22}$$

where σ^o is equal to σ_c/A_a as defined in Section 4.13.

Not all background clutter regions produce this large a value of σ^o, especially those corresponding to smooth surfaces such as asphalt or concrete and cross-polarized returns from clutter [8]. How, then, can the probability of ranging off a background be increased when the backscatter coefficient is low and the transmitted power of the radar is limited? The answer is found in Figure 5.3 by observing the abscissa. When the number of independent backscattered pulses that are noncoherently integrated is increased, a given detection probability can be achieved with a lower S/N while keeping the false alarm probability constant. Alternatively, a higher detection probability can be achieved with the same S/N and false alarm probability as the number of independent samples processed by the radar is increased.

Not only does a larger number of independent pulses increase the ability to range off clutter, but even more importantly it increases the detection probability of targets in clutter by increasing the effective signal-to-clutter ratio. For these reasons, then, it is worthwhile to explore the methods for increasing the number of independent samples or pulses that are detected by the radar receiver.

5.4 DIVERSITY GAIN AND INTEGRATION GAIN

The increase in the effective clutter-to-noise ratio (important when ranging off clutter is the concern), or signal-to-clutter ratio (when target detection in clutter is required) from an increased number of independent samples is given in terms of (1) diversity gain from a decrease in the signal (clutter or target, respectively) fluctuation loss, or (2) integration gain from a decrease in the magnitude of the interference (noise or clutter, respectively) with respect to the signal of interest. These techniques for increasing signal-to-interference ratio are summarized in Table 5.3 for ranging off-ground clutter and for increasing target detection probability in clutter.

5.4.1 Ranging Using Ground Clutter

Ranging from ground clutter offers the potential for *diversity gain* when clutter behaves as a fluctuating target and the radar system suffers an associated fluctuation

Table 5.3
Increasing Effective Signal-to-Interference Ratio Through Diversity and Integration Gain

Ranging Off-Ground Clutter	*Target Detection in Clutter*
• Noise-limited operation • Technique 1: Diversity gain —Applies to clutter that fluctuates —Wide bandwidth of transmitted pulse reduces fluctuation loss by providing independent clutter samples from frequency decorrelation —Large integration time reduces fluctuation loss by providing independent samples at intervals equal to decorrelation time of the clutter • Technique 2: Integration gain —Applies if fluctuation rate of clutter is less than the frequencies of the dominant noise sources —Integrate clutter signal over many independent noise samples	• Clutter-limited operation • Technique 1: Diversity gain —Applies to targets that fluctuate —Wide bandwidth of transmitted pulse reduces fluctuation loss by providing independent target samples from frequency decorrelation —Large integration time reduces fluctuation loss by providing independent samples at intervals equal to decorrelation time of the target • Technique 2: Integration gain —Applies if fluctuation rate of the target is less than the fluctuation frequency of the clutter —Integrate target signal over many independent clutter samples that are produced by temporal or frequency decorrelation

loss [6]. In this case, the number of independent clutter samples is increased through temporal and frequency decorrelation. This increases the effective clutter-to-noise ratio and, hence, detection probability.

The diversity gain $G_d(N_d)$ is given by

$$G_d(N_d) = [L_f(1)]^{1-1/N_d} \tag{5.23}$$

$$= [1 - 1/N_d][L_f(1)] \text{ dB} \tag{5.24}$$

where N_d = number of independent clutter samples from temporal and frequency decorrelation that are integrated during n pulses (N_d can be less than n) and $L_f(1)$ = fluctuation loss for Swerling 1 target given by Figure 5.4.

Diversity gain is effective only when the nondiverse system has a fluctuation loss and a detection probability greater than 0.33. The latter constraint is imposed by the nature of the $L_d(1)$ curves, which are negative for detection probabilities less than 0.33.

Diversity gain can be accomplished in four ways: (1) time diversity, in which independent samples are received at intervals equal to or greater than the decorrelation time of the target; (2) frequency agility, in which independent samples are obtained by changing transmitter frequency; (3) frequency diversity, in which parallel channels produce independent samples; and (4) combined diversity, in which both time and frequency effects are incorporated to increase the number of samples. Samples from time diversity are limited when the decorrelation time of the target

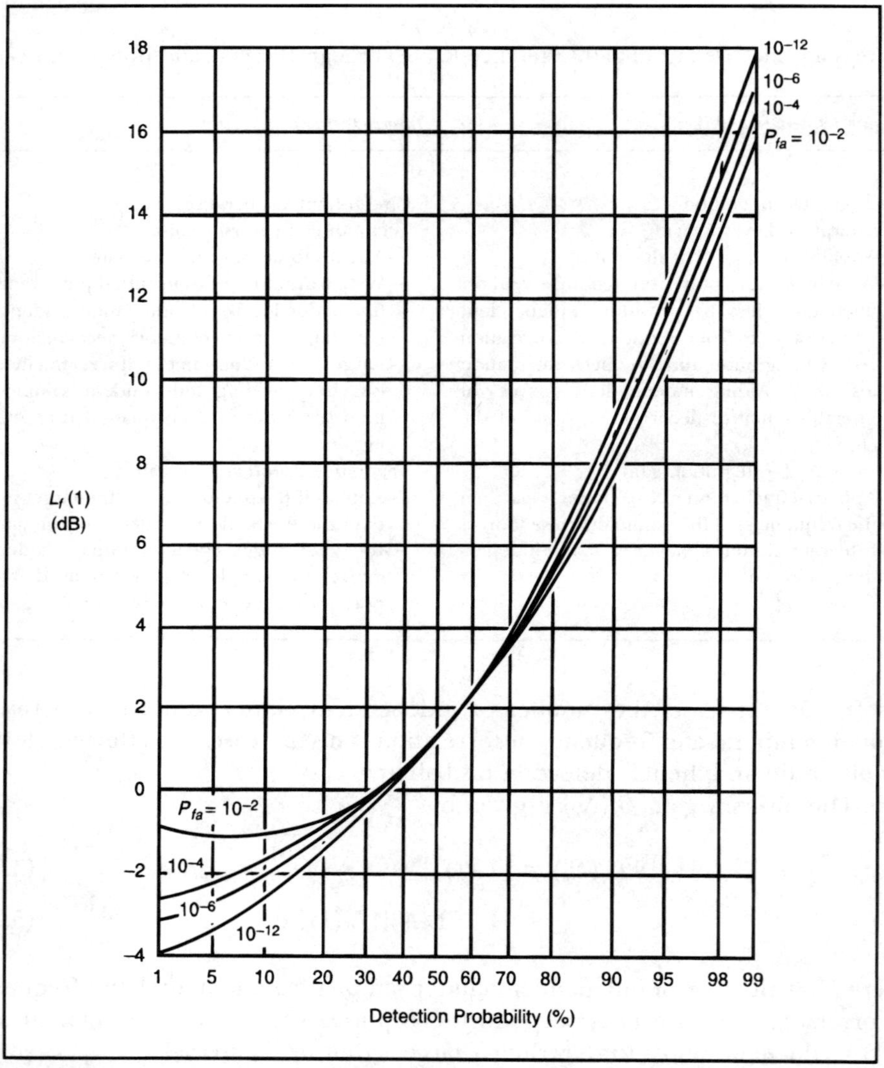

Figure 5.4 Fluctuation loss for a Swerling 1 target. (*Source:* [6]. Reprinted with permission.)

is large compared with the allowable integration interval as constrained by the mission scenario. Samples from frequency agility are limited by the transmission bandwidth that must be large compared with the decorrelation frequency of the target. This constraint may be more severe if requirements such as extreme waveform linearity are imposed as in FMCW radar applications to target identification. The transmitter also becomes more complex if circuits are built to vary the transmitted frequency from pulse to pulse. The frequency diversity radar is effective when there is a large fluctuation loss. Here, two or more channels are operated in parallel, with their center frequencies spaced by at least the decorrelation frequency of the target. Pulses are received in each of the N_d channels. There is an obvious cost and packaging volume impact as the number of samples (equal to the number of channels) increases. Time diversity, frequency agility, and combined diversity are discussed further in later sections of this chapter.

The effective clutter-to-noise ratio $(C/N)_e$ due to diversity gain becomes

$$(C/N)_e = G_d(N_d)\,(C/N)_{\text{single sample}} \tag{5.25}$$

and increases the detection probability accordingly.

If decorrelation of clutter samples does not occur, the clutter-to-noise ratio may still be increased through an *integration gain* $G_i(N_n)$ when the clutter return and internal receiver noise are integrated following detection. If the fluctuation rate of the clutter is less than the frequencies of the dominant noise sources, the effects of the noise are integrated or smoothed out and the effective clutter-to-noise ratio is increased. The gain is equal to

$$G_i(N_n) = \frac{N_n}{L_i(N_n)} \tag{5.26}$$

where

$$\begin{aligned} N_n &= \text{number of independent noise samples} \\ &= (\text{IF predetection bandwidth}) \cdot (\text{integration time}) \end{aligned} \tag{5.27}$$

The predetection bandwidth is discussed in Section 6.6.3 in conjunction with MMW radiometer operation. Integration loss $L_i(N_n)$ is obtained from the curves of Figure 4.26 as discussed in Section 4.10.1.

The effective clutter-to-noise ratio due to integration gain is

$$(C/N)_e = G_i(N_n)\,(C/N)_{\text{single sample}} \tag{5.28}$$

leading to an increase in detection probability.

When both diversity and integration gain are present, the net clutter-to-noise ratio is given by the product of the two gains and the single sample clutter-to-noise ratio as

$$(C/N)_e = G_d(N_d)\,G_i(N_n)\,(C/N)_{\text{single sample}} \tag{5.29}$$

5.4.2 Target Detection in Clutter

For target detection in clutter, diversity and integration gain are used to increase the effective target signal-to-clutter ratio and, hence, the detection probability of the target. The potential for diversity gain exists when the target signal fluctuates and the radar system suffers a fluctuation loss. The number of independent target samples is increased through the effects of temporal and frequency decorrelation, producing a diversity gain when the nondiverse system has a detection probability greater than 0.33. The diversity gain is given by (5.23) and (5.24) where N_d is now the number of independent target signal samples from temporal and frequency decorrelation that are integrated during n pulses. The effective signal-to-clutter ratio, given by

$$(S/C)_e = G_d(N_d)\,(S/C)_{\text{single sample}} \tag{5.30}$$

increases the detection probability of the target in clutter.

If time and frequency decorrelation of the target signal do not occur, the signal-to-clutter ratio may still be increased through integration gain when the returns from the target and clutter are integrated following detection. If the fluctuation rate of the target is less than the fluctuation rate of the clutter, the target-masking effects of the clutter are integrated or smoothed out and the effective target-to-clutter ratio is increased. The gain is equal to

$$G_i(N_c) = \frac{N_c}{L_i(N_c)} \tag{5.31}$$

where N_c = number of independent clutter samples that are integrated.

Integration gain increases the effective signal-to-clutter ratio to

$$(S/C)_e = G_i(N_c)\,(S/C)_{\text{single sample}} \tag{5.32}$$

When both diversity and integration gain are present, the net signal-to-clutter ratio is given by the product of the two gains and the single sample signal-to-clutter ratio as

$$(S/C)_e = G_d(N_d)\,G_i(N_c)\,(S/C)_{\text{single sample}} \tag{5.33}$$

5.5 TEMPORAL AND FREQUENCY DECORRELATION

Temporal decorrelation can increase the number of independent samples through three phenomena, as shown in Figure 5.5: (1) Doppler-induced signal fading, (2) scanning, and (3) movement of the clutter or target. More than one phenomena may be active at a given time. Frequency decorrelation produces independent samples when the RF modulation bandwidth of the transmitted signal is greater than the decorrelation bandwidth of the clutter or target. Signal transmission should be uniformly distributed over time-frequency space. This prevents correlation among samples obtained from temporal and frequency decorrelation that can

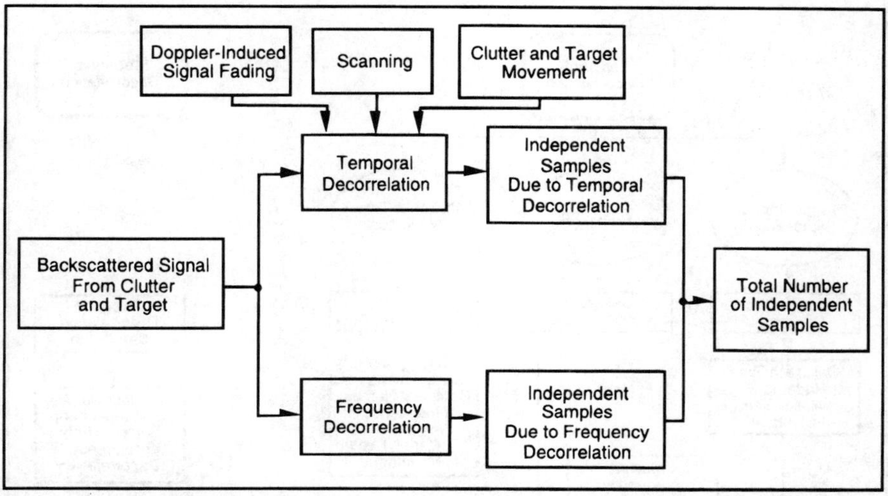

Figure 5.5 Increasing independent samples through temporal and frequency decorrelation.

otherwise reduce the effective number of independent samples that contribute to diversity and integration gain.

Figure 5.6 shows the details of the temporal and frequency decorrelation processes for increasing independent samples. The number of independent samples from Doppler-induced signal fading is equal to the product of the Doppler bandwidth, produced by scatterers in an antenna resolution element, multiplied by the time it takes the radar to move a distance equal to the resolution element in the along-track dimension.

Independent samples from scanning arise from the frequency spread produced by the mechanical motion or electrical scanning of the antenna. When both Doppler-induced signal fading and scanning effects are present, the two are combined to form a net frequency spread that is multiplied by the time it takes the radar to move a distance equal to the resolution element. The number of independent samples produced by clutter and target movement is a function of (1) the decorrelation time of the backscattered signal produced by the movement and (2) the integration time used in the signal processing.

5.5.1 Doppler-Induced Signal Fading

Doppler-induced signal fading is a phenomenon due to the spread in frequency of the returns backscattered from different parts of the antenna footprint [9,10]. The difference in range from each of the scatterers to the radar produces the spread in frequency.

Forward-Looking Pulse Doppler Radar

In a forward-looking pulse Doppler radar, the Doppler bandwidth Δf_D arises from the spread in Doppler frequency from leading edge and trailing edge scatterers in

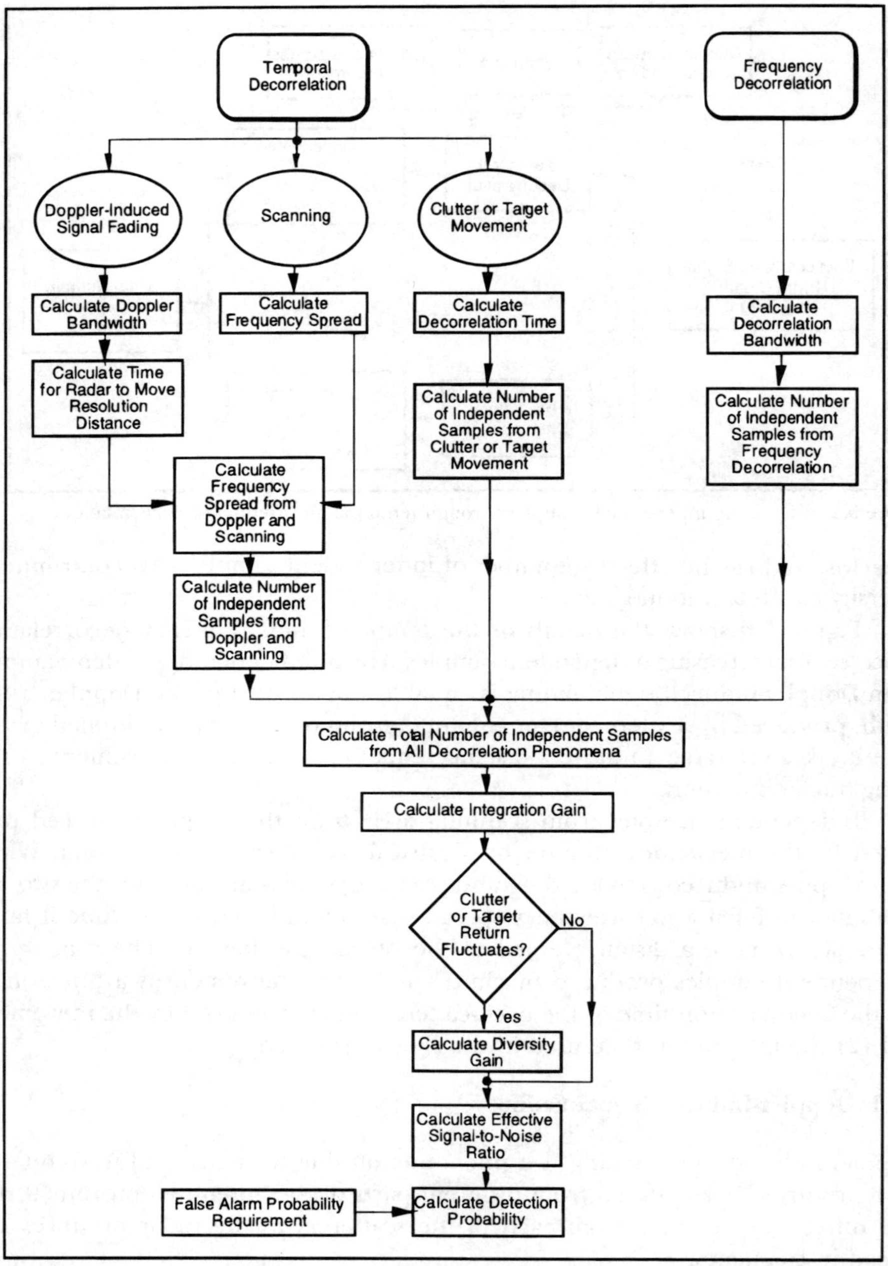

Figure 5.6 Calculating independent samples and effective signal-to-noise ratio from temporal and frequency decorrelation.

the along-track resolution distance x_a corresponding to the pulse width τ, where

$$x_a = \frac{c\tau}{2 \sin \theta} \tag{5.34}$$

as illustrated in Figure 5.7.

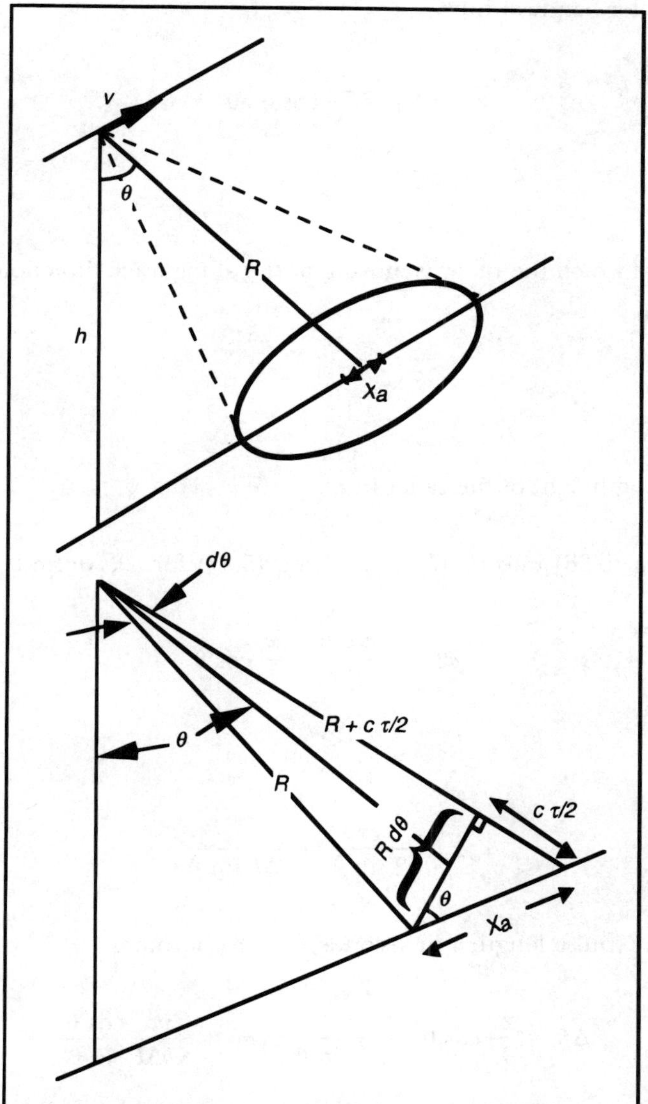

Figure 5.7 Doppler-induced signal fading geometry for a forward-looking pulse Doppler radar.

Trailing edge scatterers are located at an incidence angle of $\theta - d\theta/2$, while the leading edge scatterers are located at an incidence angle of $\theta + d\theta/2$. The Doppler frequency f_D is given by

$$f_D = \frac{2v}{\lambda} \sin \theta \tag{5.35}$$

and the Doppler bandwidth by

$$\Delta f_D = \frac{2v}{\lambda} \cos \theta \; d\theta \tag{5.36}$$

where

v = speed at which the radar is moving in the along-track direction in m/sec;

$$\cos \theta = \frac{R d\theta}{x_a}; \tag{5.37}$$

$$\mathrm{R} = \frac{\mathrm{h}}{\cos \theta}; \tag{5.38}$$

h = operating height of the radar in m.

By substituting (5.38) into (5.37) and solving (5.37) for $d\theta$, we find

$$d\theta = \frac{x_a \cos \theta}{R} = \frac{x_a}{h} \cos^2\theta \tag{5.39}$$

Since

$$x_a = \frac{c\tau}{2 \sin \theta} = \frac{c}{2\Delta F \sin \theta} \tag{5.40}$$

where $\Delta F = 1/(\text{pulse length } \tau \text{ in seconds})$, (5.36) becomes

$$\Delta f_D = \frac{2v}{\lambda} \cos \theta \frac{c}{2h\Delta F \sin \theta} \cos^2\theta = \frac{vc}{\lambda h \Delta F} \frac{\cos^3\theta}{\sin \theta} \tag{5.41}$$

Moore [9] provides another method for calculating the Doppler bandwidth for a forward-looking radar in terms of the incidence angles to the leading edge and trailing edge scatterers in the resolution element x_a. As shown in Figure 5.8,

$$\Delta f_D = \frac{2v}{\lambda}(\sin\theta_2 - \sin\theta_1) \tag{5.42}$$

where

$$\theta_1 = \theta - d\theta/2 \quad \text{and} \quad \theta_2 = \theta + d\theta/2$$

The number of independent samples N_T is equal to

$$N_T = \Delta f_D T \tag{5.43}$$

where

T = time for radar to move the distance x_a equal to $c\tau/(2\sin\theta)$

$$= \frac{x_a}{v} = \frac{c}{2v\Delta F\sin\theta} \tag{5.44}$$

Thus,

$$N_T = \frac{vc}{\lambda h\Delta F}\frac{\cos^3\theta}{\sin\theta}\frac{c}{2v\Delta F\sin\theta} = \frac{c^2}{2\lambda h(\Delta F)^2}\frac{\cos^3\theta}{\sin^2\theta} \tag{5.45}$$

Side-Looking Pulse Doppler Radar

In a side-looking pulse Doppler radar, the Doppler bandwidth arises from the spread in Doppler frequency from leading edge and trailing edge scatterers in

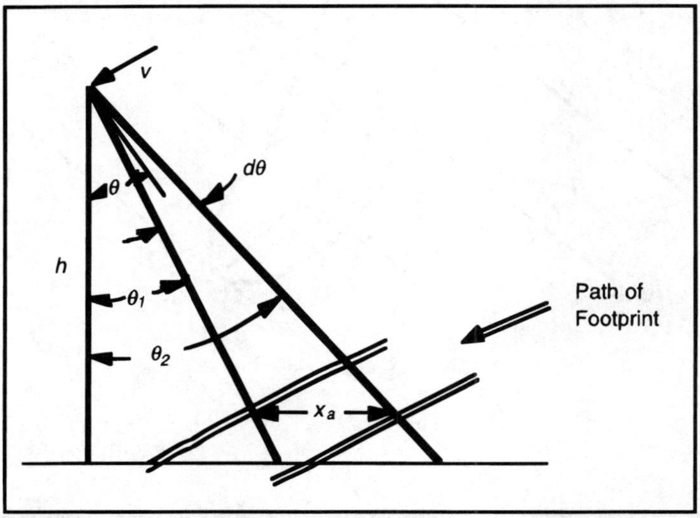

Figure 5.8 Geometry for calculating Doppler bandwidth for a forward-looking radar in terms of incidence angles to the leading edge and trailing scatterers in the resolution element.

the along-track footprint x_a corresponding to the antenna 3-dB beamwidth θ_a as illustrated in Figure 5.9. Doppler spread due to the antenna beamwidth is [10]

$$\Delta f_D = \frac{2v}{\lambda} \cos \theta_a \, d\theta_a \tag{5.46}$$

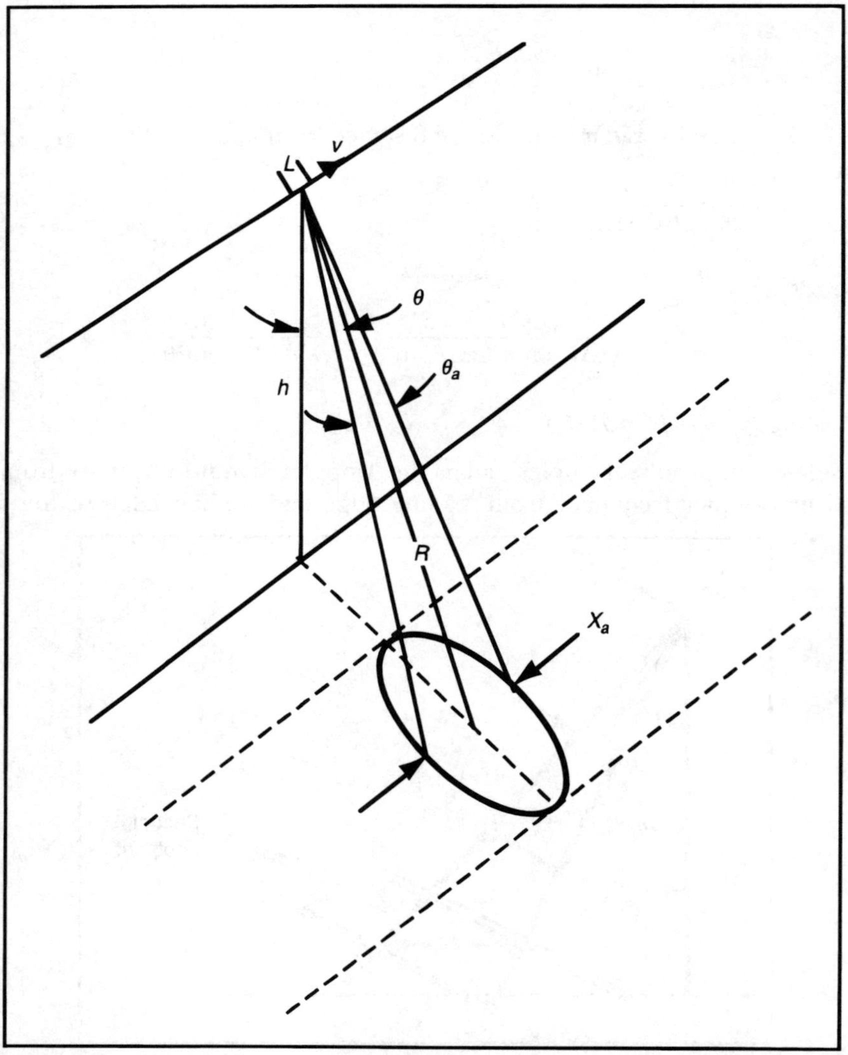

Figure 5.9 Doppler-induced signal fading geometry for a side-looking pulse Doppler radar.

For θ_a small,

$$\Delta f_D \approx \frac{2v}{\lambda} \Delta\theta_a \tag{5.47}$$

where $\cos\theta_a \approx 1$.

From the geometry shown in Figure 5.9,

$$\tan\frac{\theta_a}{2} = \frac{x_a/2}{R} = \frac{x_a/2}{h/\cos\theta} \tag{5.48}$$

For θ_a small,

$$\theta_a \approx \frac{x_a}{h}\cos\theta \tag{5.49}$$

and

$$\Delta\theta_a = \frac{\Delta x_a}{h}\cos\theta \tag{5.50}$$

Thus

$$\Delta f_D = \frac{2v}{\lambda}\frac{\Delta x_a}{h}\cos\theta = \frac{2v}{\lambda h} x_a \cos\theta \tag{5.51}$$

where the along-track antenna resolution is given by

$$x_a = \Delta x_a \tag{5.52}$$

Since

$$x_a = \frac{\lambda h}{L\cos\theta} \tag{5.53}$$

where L = effective antenna aperture length in the along-track direction,

$$\Delta f_D = \frac{2v}{\lambda h}\frac{\lambda h}{L\cos\theta}\cos\theta = \frac{2v}{L} \tag{5.54}$$

The number of independent samples N_T is again given by

$$N_T = \Delta f_D T$$

where T = time for radar to move the distance x_a as it was in (5.44). However, now x_a is equal to $\lambda h/(L \cos \theta)$ and

$$N_T = \frac{2v}{L} \frac{\lambda h}{vL \cos \theta} = \frac{2\lambda h}{L^2 \cos \theta} \tag{5.55}$$

The influence of wavelength, height, and RF bandwidth on the number of independent samples produced by Doppler-induced signal fading in forward-looking and side-looking radars is shown by comparing (5.45) and (5.55), respectively. In the forward-looking case, the number of samples is inversely proportional to wavelength, height, and the square of bandwidth. Thus a larger bandwidth, corresponding to a shorter pulse length, or longer wavelength radar produce fewer independent samples from Doppler-induced signal fading. By contrast, in the side-looking configuration, the number of samples increases directly as wavelength and height, and inversely as the square of the aperture. However, to maintain along-track resolution, the size of the aperture has to increase as wavelength increases. Here, the benefits of an increased number of samples from increased wavelength are more than offset by the decrease caused by the inverse relation to the square of aperture size.

5.5.2 Doppler Beam Sharpening

Doppler beam sharpening is a phenomenon related to signal fading. It diminishes the effects of surface clutter through the application of narrow-bandwidth Doppler filters that reduce the effective antenna beamwidth. As with Doppler-induced signal fading, the radar motion creates Doppler frequencies that are associated with the clutter return from each angular direction within the main beam of the antenna. This causes spectral spreading of the Doppler clutter return that is a function of the radar velocity, antenna beamwidth, and antenna boresight relative to the radar velocity vector.

Conversely, a target return has a unique Doppler frequency associated with its angular direction if the target occupies only a small portion of the footprint corresponding to the mainlobe of the antenna. If a narrowband Doppler filter whose bandwidth is inversely proportional to dwell time (time required to scan the portion of the footprint occupied by the target) is used to process the expected target return, only the clutter backscattered from angular directions associated with the filter's bandwidth competes with the target. Under favorable conditions, a dramatic increase in signal-to-clutter ratio can be achieved for either stationary or slowly moving targets. The processing is equivalent to reducing the antenna's beamwidth and thus gives rise to the term Doppler beam sharpening.

5.5.3 Temporal Decorrelation via Scanning

Scanning induces a frequency spread Δf_S in the backscattered signal through a velocity component produced by antenna motion [11] such that

$$\Delta f_S = \frac{1}{1.15TOT} \tag{5.56}$$

where TOT = time-on-target over which backscatter signals are collected during the scanning motion. A similar frequency spread phenomenon applies to electronically scanned antenna arrays, since the change in phase shape across the array is equivalent to that produced by mechanical scanning motion.

If both Doppler-induced signal fading and scanning contribute to the frequency spread, the total frequency spread Δf_T is given by

$$\Delta f_T = \sqrt{(\Delta f_D)^2 + (\Delta f_S)^2} \tag{5.57}$$

5.5.4 Temporal Decorrelation via Clutter and Target Movement

Clutter and target surfaces that are in motion also produce signal fading when illuminated by stationary radars due to movement of individual scatterers. Objects that may potentially induce signal fading from this phenomenon are oceans, vegetation, and moving targets and machines [11,12].

The number N_T of independent samples produced is

$$N_T = 1 + \frac{t_i}{t_d} \tag{5.58}$$

where

t_i = signal integration time;
t_d = signal decorrelation time.

The unity term indicates that at least one independent sample is always obtained.

Several definitions of decorrelation time are used in the literature. These are as follows:

1. Reciprocal of the noise bandwidth of the clutter frequency spectrum [6];
2. Time required for the normalized correlation coefficient to reach a value of 0.25 [13];
3. Time at which the modulus of the autocorrelation function of the backscattered signal decreases to 0.2 of its maximum value as given by

$$t_d = \frac{0.65}{\Delta f_T} \tag{5.59}$$

where Δf_T is defined in (5.57) [9,14];

4. Time at which the correlation function reaches e^{-1} or 37% of its peak value [14,15].

Decorrelation by Clutter Movement

Barton [11] expresses the clutter signal decorrelation time in terms of the standard deviation σ_v of the clutter velocity spectrum as

$$t_d = \frac{\lambda}{2\sqrt{2\pi}\sigma_v} = \frac{0.2\lambda}{\sigma_v} \tag{5.60}$$

Clutter decorrelation time may also be obtained directly from experimental measurements. Figure 5.10 shows the results of computing the autocovariance functions from windblown tree clutter backscatter data collected by noncoherent radars having pulse lengths of 50 nsec, narrow antenna beams of between 0.7 and 1.7

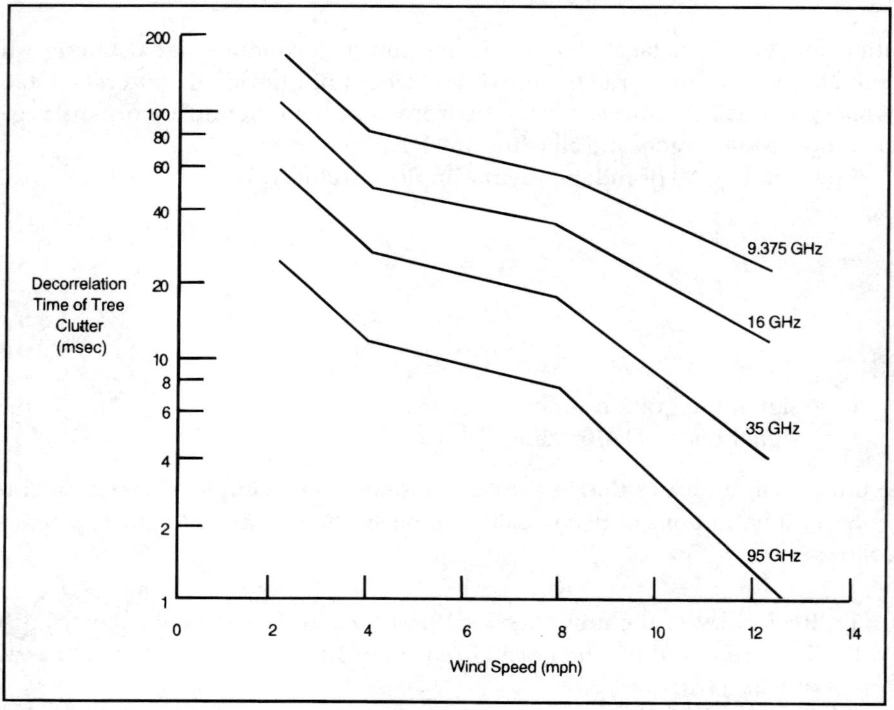

Figure 5.10 Decorrelation time of tree clutter as a function of wind speed. (*Source:* [13]. Reprinted with permission.)

degrees, and radar-to-target range of about 1 km [13–15]. Data points from which the curves were plotted represented the average of eight samples. The decorrelation time was set equal to the time required for the normalized correlation coefficient to decay to 0.25.

Decorrelation time decreases as the wind speed increases and as the radar transmission frequency increases. The decorrelation time curves appear to separate into several regions at the frequencies tested. The first, at wind speeds up to approximately 4 mi/h, is due to leaf rustling; the second, at speeds up to 8 mi/h, to movement of small twigs; and the final, up to speeds of about 12.5 mi/h, to movement of branches.

As an example of how to estimate the number of independent samples from temporal decorrelation using these data, consider a radar operating at 35 GHz in tree clutter with wind blowing at 6 mi/h. Figure 5.10 gives a decorrelation time of approximately 23 msec under these conditions. Therefore, to obtain an appreciable number of temporal diversity samples, the integration time needs to be much greater than 23 msec. The constraints of a particular mission do not always permit large integration times to be achieved and thus limit the number of independent samples available from temporal diversity.

Decorrelation by Rigid Target Movement

The decorrelation time for rigid targets produced by virtue of their movement is given by Barton [6] as

$$t_d = \frac{\lambda}{2\omega L} \tag{5.61}$$

where

- λ = wavelength of the transmitted energy in m;
- ω = rate of rotation of the target about the line of sight in rad/sec;
- L = target dimension normal to the line of sight in m.

Tables 5.4 and 5.5 show the impact of radar operating frequency and target rotation on the decorrelation time of the backscattered signal from tank and aircraft targets for radars operating at 35 and 95 GHz. Table 5.4 gives the target dimensions and rotation rates, while Table 5.5 shows the decorrelation times.

Under the conditions stated in Table 5.4, the tank target does not offer much possibility for temporal diversity from target movement because the integration times must be large (i.e., of the order of seconds) to obtain many independent samples and the restrictions imposed by target engagement times may not be compatible with the large integration time requirement. However, the decorrelation time of ground targets may be reduced substantially when the movement of the sensor platform, such as a missile, is added to the model. Such an analysis is beyond

Table 5.4
Rigid Target Dimensions and Rotation Rates for Temporal Diversity Example for Target Movement

Tank Target	Aircraft Target
L = 3.3m $\omega = 1.5 \times 10^{-4}$ rad/sec (based on a velocity of 6 inches per sec and a range of 1,000m)	L = 30m ω = 0.03 rad/sec

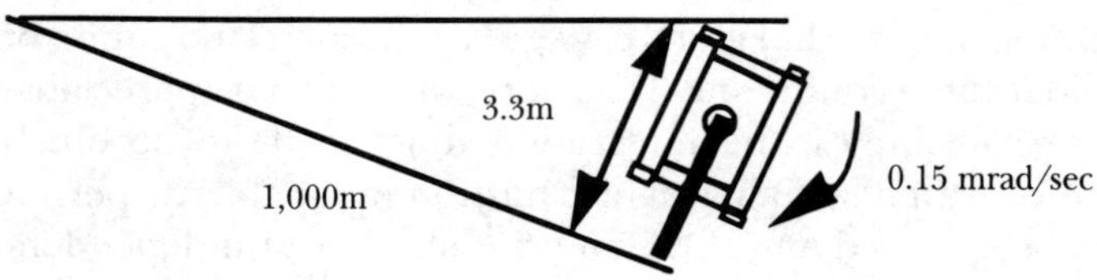

Table 5.5
Decorrelation Time for Rigid Targets at 35 and 95 GHz Using Target Movement Parameters From Table 5.4

Frequency (GHz)	*Wavelength λ (cm)*	*Tank Decorrelation Time t_d (sec)*	*Aircraft Decorrelation Time t_d (sec)*
35	0.86	8.7	0.005
95	0.32	3.2	0.002

the scope of this chapter. On the other hand, the relatively small decorrelation times afforded by aircraft target movement offer the real possibility of obtaining multiple independent samples with integration times of less than 1 sec.

5.5.5 Frequency Decorrelation

Multiple independent samples from frequency decorrelation require an RF bandwidth larger than the clutter or target decorrelation bandwidth Δf_F. Two definitions of decorrelation bandwidth are frequently encountered:

1. Width of the power spectrum at which its amplitude is one-half the maximum value, also called the half-power bandwidth;

2. Frequency separation at which the correlation coefficient of backscatter from two different frequencies goes through its first zero.

Half-Power Bandwidth

With this formulation, the decorrelation bandwidth is derived from the distribution of scattering cross-sections $\sigma(z)$ associated with scattering elements located at a distance z from the radar receiver along the direction of propagation. The delay function $P(t)$ [16–18] (i.e., the distribution in depth of the scattering elements) for the received energy field is given by

$$P(t) = \sigma(z/c) \tag{5.62}$$

where c is the speed of light.

The Fourier transform F of the delay function is defined as the transfer function $F(\omega)$; that is,

$$F(\omega) = F[P(t)] \tag{5.63}$$

The resulting power spectrum $W(\omega)$ is

$$W(\omega) = F(\omega)F^*(\omega) \tag{5.64}$$

where $F^*(\omega)$ represents the complex conjugate of $F(\omega)$. The half-power bandwidth $\Delta\omega$ of the power spectrum defines the decorrelation bandwidth $\Delta\omega_F$ from the relation

$$\frac{W(\Delta\omega_F)}{W(0)} = \frac{1}{2} \tag{5.65}$$

When the scattering cross-section is distributed exponentially in depth, the delay function takes the form

$$P(t) = \exp(-\alpha t) \tag{5.66}$$

The $1/e$ depth τ_0 of the delay function is

$$\tau_0 = \frac{1}{\alpha} \tag{5.67}$$

The Fourier transform $F(\omega)$ of $P(t)$ becomes

$$F(\omega) = F[\exp(-\alpha t)] = (\alpha + j\omega)^{-1} \tag{5.68}$$

producing a power spectrum

$$W(\omega) = F(\omega)F^*(\omega) = (\alpha^2 + \omega^2)^{-1} \tag{5.69}$$

Half-power bandwidth is found by substituting (5.69) into (5.65) as

$$\frac{W(\Delta\omega_F)}{W(0)} = \frac{1}{2} = \frac{\alpha^2}{\alpha^2 + (\Delta\omega_F)^2} \tag{5.70}$$

and solving for $\Delta\omega_F$, giving

$$\Delta\omega_F = \alpha = \frac{1}{\tau_0} \tag{5.71}$$

If the $1/e$ penetration depth D of the electromagnetic wave is

$$D = c\tau_0 \text{ meters} \tag{5.72}$$

the half-power bandwidth of the power spectrum is

$$\Delta f_F = \Delta\omega_F/2\pi = \frac{1}{2\pi\tau_0} = \frac{c}{2\pi D} = 0.16\,\frac{c}{D} \tag{5.73}$$

Decorrelation bandwidth defined in terms of the half-power bandwidth can be calculated for other scatterer distributions as well, taking the form

$$\Delta f_F = Kc/D \tag{5.74}$$

where K and D are functions of the particular distribution of the scatterers along the direction of propagation [17]. This expression applies to both clutter and target frequency decorrelation when the appropriate scatterer distribution for the object of interest is inserted. The values of K for several scatterer depth distributions along the direction of propagation are given in the second column of Table 5.6 by the coefficient. The characteristic length D is defined in the third column [18].

Decorrelation bandwidths corresponding to these distributions are shown in Figure 5.11 as a function of the object size. The exponential scatterer distribution gives the smallest decorrelation bandwidth, while the triangular gives the largest.

Table 5.6
Descriptive Parameters for Scatterer Depth Distributions Used in Half-Power Bandwidth Definition of Decorrelation Bandwidth [18]

Depth Distribution Function	*Half-Power Bandwidth (Hz)*	*Characteristic Length D*
• Exponential	0.16 c/D	$1/e$ penetration depth
• Gaussian	0.37 c/D	2σ scatterer distribution
• Uniform	0.44 c/D	Width of distribution
• Triangular	0.64 c/D	Width of triangle base
• Two delta functions	0.25 c/D	Separation between scatterer pair

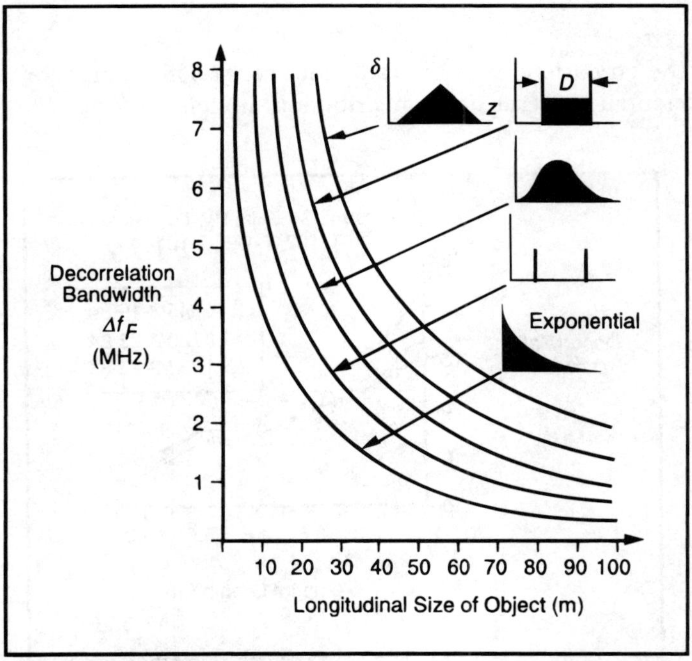

Figure 5.11 Decorrelation bandwidth as a function of object size and scatterer distribution. (*Source:* [17]. Reprinted with permission.)

The decorrelation bandwidth measured in the 35-GHz snow backscatter data shown in Figure 5.12 [19] fits the exponential scatterer model when D is set equal to the footprint length. The measured decorrelation bandwidth was set equal to the transmitted bandwidth at which the backscattered signal decreased to $1/e$ of its narrowband value. The data were obtained using a 35-GHz pulse radar (pulse length equal to 100 ns and pulse repetition frequency equal to 1.25 kHz) with an antenna whose two-way 3-dB beamwidth was 1.4 degrees. The transmitted frequency was changed on every other pulse in 5-MHz increments over the coherent bandwidth of 640 MHz. The transmitted and received polarizations were each vertical (VV). The footprint length was varied by changing the depression angle from 13 (largest footprint) to 35 (smallest footprint) degrees.

The 94-GHz snow decorrelation bandwidth measurements of Figure 5.13 are better approximated by the Gaussian distribution function when the footprint is 6.9m or less, and D is set equal to the footprint length. Hence the snow decorrelation bandwidth is larger at 95 GHz than at 35 GHz. The 95-GHz radar operating characteristics are the same as those of the 35-GHz radar, except for its noncoherent operation and smaller transmitted frequency bandwidth of 256 MHz.

The decorrelation bandwidth for targets is described by Barton [6] as

$$\Delta f_F = \frac{c}{2\pi L} \tag{5.75}$$

where L is the target length in the radial direction. This expression is equivalent to the exponential scatterer depth distribution model.

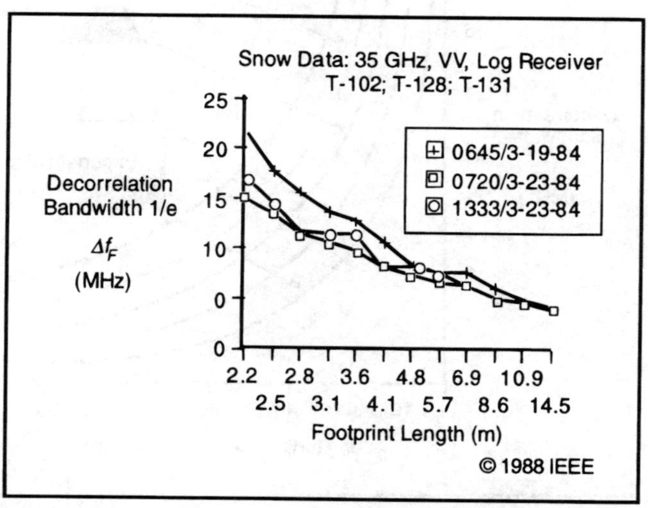

Figure 5.12 Decorrelation bandwidth of snow backscatter at 35 GHz appear to fit an exponential scatterer distribution. (*Source:* [19]. Reprinted with permission.)

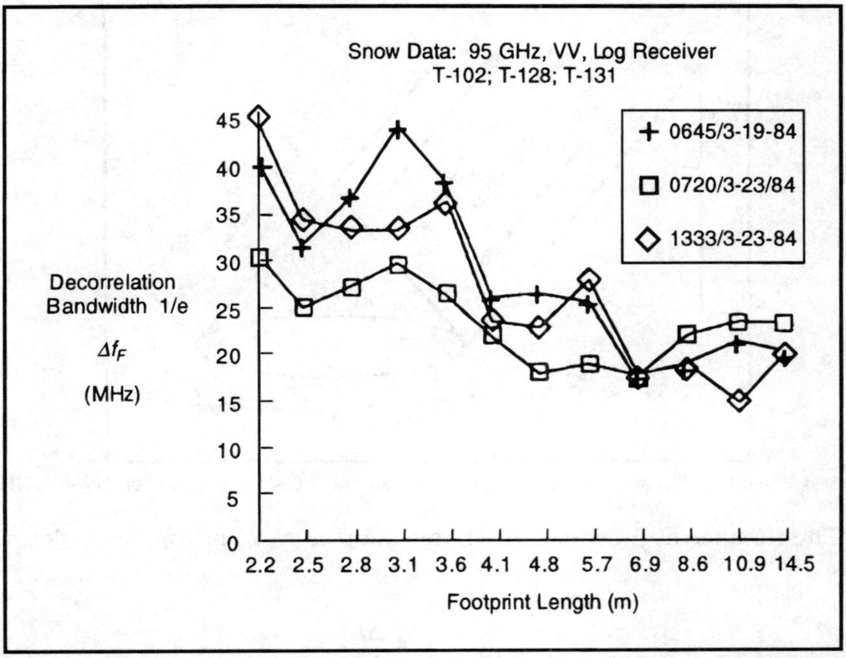

Figure 5.13 Decorrelation bandwidth of snow backscatter at 94 GHz appears to be approximated by a Gaussian scatterer distribution. (*Source:* [14]. Reprinted with permission.)

First Zero of Correlation Coefficient

When the decorrelation bandwidth is set equal to the frequency separation at which the correlation coefficient computed from backscatter measured at two different frequencies goes through its first zero, the decorrelation frequency is equal to [20–22]

$$\Delta f_F = \frac{c}{2D} \tag{5.76}$$

when the scatterers are uniformly distributed. The geometry that applies to this definition is shown in Figure 5.14. In comparison with (5.76), the decorrelation frequency calculated from the half-power bandwidth definition is $0.44c/D$ for uniformly distributed scatterers. The difference is approximately 12%, which is probably not significant for calculating the effects from this type of phenomenon.

A more generalized derivation of the relation between the backscattered signals at two frequencies and the first zero of the correlation coefficient is given by Weissman with respect to measuring ocean wave heights [23]. Measuring the correlation properties of the backscattered signal at different frequency separations gives information about the longitudinal distribution of scattering centers in a target or clutter [14].

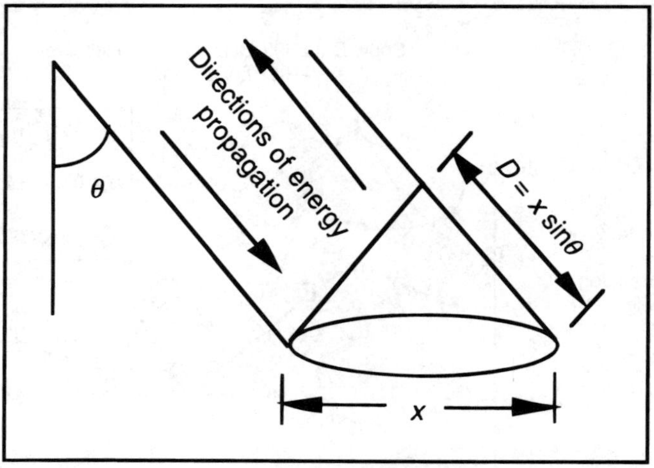

Figure 5.14 Geometry for decorrelation frequency as defined by first zero of the correlation coefficient.

The number of independent clutter samples N_F from frequency decorrelation is

$$N_F = 1 + \frac{\Delta F}{\Delta f_F} \tag{5.77}$$

where

ΔF = RF modulation bandwidth.

Effect of Frequency Decorrelation on the Reduction of the Standard Deviation of the Backscattered Clutter Signal

Lane [24] has measured the standard deviation of clutter areas consisting of barley, corn, grass, coniferous trees, deciduous trees, and a plowed field. Because the 256-MHz IF bandwidth of his radar was large compared with the effective pulse bandwidth of 5 MHz, approximately $N = 25$ independent samples ($B_{IF}/2B_P = 256/10 = 25.6$) from each type of clutter were averaged. This reduced the standard deviation of the amplitude return by a factor of $N^{1/2} = 5$. As an example of the effect of a wide-bandwidth receiver, compare the standard deviation of pine trees and deciduous trees reported by Currie and others [15] of 6.5 and 6.4, respectively, at 95 GHz using a narrow-bandwidth (20 MHz) receiver (see Table 4.9) with the Lane measurements (averaged over comparable depression angles) of 1.8 and 2.0, respectively.

5.5.6 Combining Samples From Both Temporal and Frequency Decorrelation

The expected number of independent samples N obtained from the combination of temporal and frequency decorrelation is given by Lind [12] as

$$N = N_T N_F \left[1 - \left(1 - \frac{1}{N_F} \right)^{n_p/N_T} \right] \tag{5.78}$$

where n_p = number of pulses received per integration time t_i. In this expression, frequency decorrelation is applied to the groups of pulses that are temporally independent.

Generally, the number of independent samples is less than the number of pulses or FMCW upsweep and downsweep cycles that are transmitted during the target or clutter illumination time or signal integration time. With frequency agility, the number of independent samples may be less than the number of transmitted pulses, especially if the target or clutter has a restricted depth or the RF bandwidth is limited. Several transmitted pulses can then occur so close to each other in frequency that their frequency spacing is insufficient for decorrelation. The number of independent samples from temporal decorrelation is limited by the width of the target, target and clutter movement, and motion of the radar platform.

When $N_F >> 1$, (5.78) can be simplified to

$$N = N_T N_F [1 - e^{-n_p/N_T N_F}] \tag{5.79}$$

This equation is illustrated in Figure 5.15 for $N_T = 10$ and $N_F = 10$.

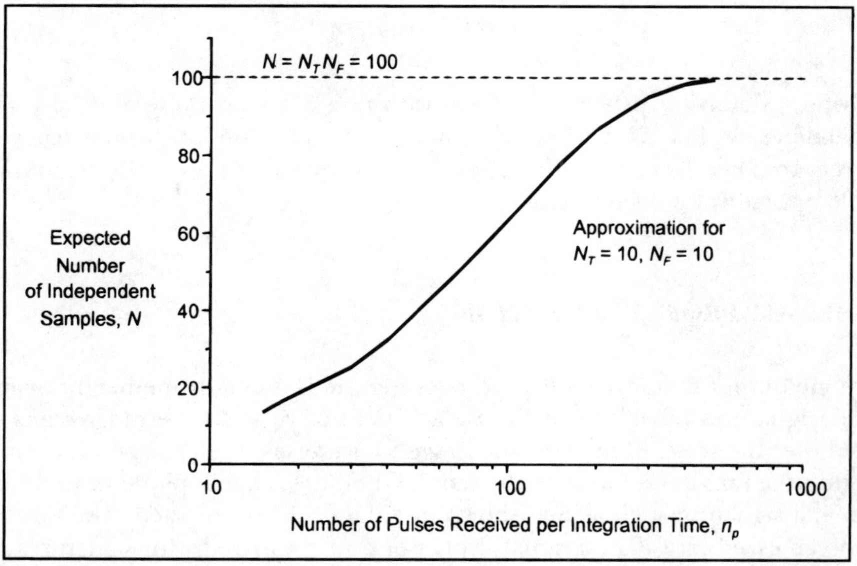

Figure 5.15 Number of independent clutter samples from the combination of temporal and frequency decorrelation assuming $N_F >> 1$.

Several conclusions may be drawn from (5.79):

1. For large n_p, the number of independent samples approaches the product of N_T and N_F.
2. When $N_F = 10$, the number of independent samples is approximately equal to the product of N_T and N_F when $n_p/N_T \geq 25$.
3. When $N_F = 100$, the number of independent samples is approximately equal to the product of N_T and N_F when $n_p/N_T \geq 250$.
4. When $N_F = 1$, the number of independent samples is equal to N_T.

5.6 RANGE TO TARGET

The ability of an FMCW radar to supply target range and detection data is also evaluated from the radar range equation. From (5.16),

$$S/N = \frac{(0.02)(10^{2.5})(10^{2.5})(8.6 \times 10^{-3})^2 \sigma_T}{(4\pi)^3 (50)^4 (1)(1.38 \times 10^{-23})(290)(5 \times 10^6)(10^{0.9})} \qquad (5.80)$$

where the radar performance parameters from Table 5.2 have been used. For a conservative estimate of radar cross-section equal to 1m^2, the signal-to-noise ratio becomes

$$S/N = 19 \text{ dB} \qquad (5.81)$$

implying a detection probability of 0.9 with one detected pulse and false alarm probabilities as low as 10^{-8} as shown by the detection probability curves in Figure 5.16. Therefore, target detection does not appear to pose an insurmountable problem for this low-flying radar.

5.7 HEIGHT PROFILE OPERATION

A unique output that evolves from processing the IF signal is the height or range profile signatures of objects in the radar's field of view. As the radar scans or is moved over the scene of interest, the range to scatterers on the targets is measured to within the range resolution of the radar. When the range is plotted as a function of time, a signature such as that shown in Figure 5.17 is obtained. The waveform can be analyzed in a digital signal processor and compared with signatures from known targets and other objects using the classification techniques discussed in the previous chapter.

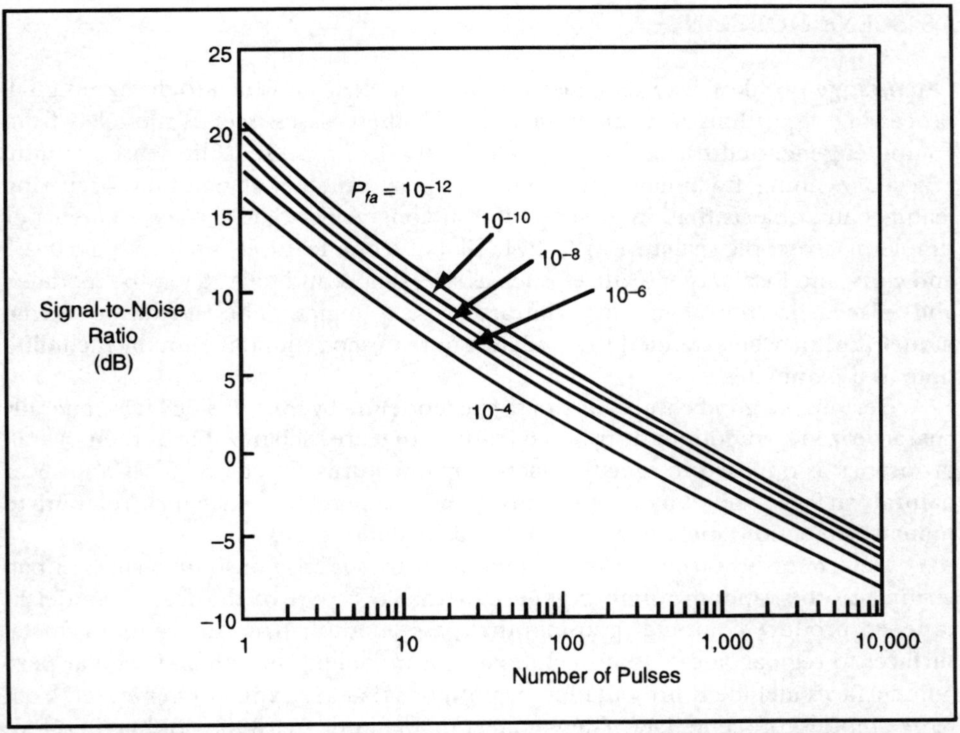

Figure 5.16 Signal-to-noise ratio as a function of number of samples noncoherently integrated to achieve a detection probability of 0.9 with a Swerling 3 target. (*Source:* [7]. Reprinted with permission.)

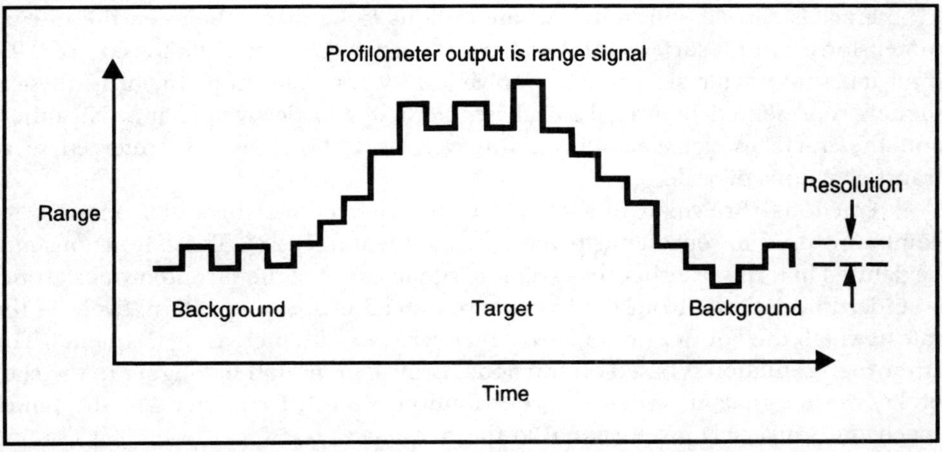

Figure 5.17 Range profile signature of target features relative to ground features.

5.8 SCENE MODELING

Often range profile sensor data may not be available from which to develop signal-processing algorithms in a timely manner. In these cases, scenes modeled from computer-generated fractal imagery may be used to represent the sensor output. Fractal, meaning fractional dimension, imagery can model naturally occurring features and objects that, in general, do not contain periodic surface elements or simple macroscopic structures [25–27]. Global scene features, such as hills, trees, and cliffs, and local target features, such as size, shape, and height, may be modeled and placed in the scene at arbitrary aspect angles. The stochastic fractal model [26] may be extended to simulate a time-varying phenomenon by including time as a parameter.

Terrain is a good choice for stochastic modeling by fractals since it is generally characterized by randomly distributed features that are recognizable by their overall properties as opposed to specific macroscopic features. Fractal surfaces appear as natural surfaces and thus capture all of the shape-related structure relevant to human perception, including textured and shaded regions.

Deterministic surfaces can be generated by adding constraints such as patterning to the random number generator that is a part of the fractal model in order to produce computer graphics that pass smoothly from the random fractal surfaces to regular ones. Other techniques for modeling smooth and regular periodic surfaces include representation by polygons that are explicitly generated from large amounts of scene data. These conceptually simple periodic objects therefore require a large amount of data storage. On the other hand, a complex and nondeterministic object may be effectively modeled with the fractal technique to significantly reduce the computer database requirements. A scene may, in general, contain both deterministic and stochastic features.

Fractals allow a continuum of dimensions to be placed between the integer dimensions. Thus a surface can be characterized with a fractal dimension of 2.05, implying smooth terrain, or 2.9, implying very rough terrain. In most physical models represented by fractals, each segment is statistically similar to all others and the fractal is statistically invariant over large distances, also referred to as transformations of scale.

Fractional Brownian functions that model fractional Brownian motion are commonly used to represent stochastic terrain features [28]. These functions may be defined in terms of either the variation with distance of the second-order statistics (standard deviations) from the fractally produced imagery, or alternatively, as the rate at which the Fourier power spectrum decreases with increasing frequency. The latter representation is based on fractional Brownian motion belonging to the class of $1/f$ noise signals in which the contribution of each frequency f to the power spectrum is inversely proportional to the frequency.

Real surfaces are not fractal at all scales since there are smallest and largest components that comprise their shape (e.g., grain size and region size). These largest and smallest components define the limits between which the surface can be described with a single fractal function. When the projected size of a pixel becomes comparable to either of these limits, the fractal approximation may fail. If the pixel size is large with respect to the largest shape components, Nyquist sampling behavior occurs (i.e., the surface appears to become smoother as the pixel size is increased). When pixel size is less than the smallest shape components, texture edges are observed (i.e., inhomogeneities in the shape structure appear). When the power spectrum representation is used, a surface generated with fractional Brownian motion possesses macroscopic features up to the order of magnitude corresponding to the lowest possible frequencies in the Fourier spectrum of the surface. The surface also contains arbitrarily small surface detail corresponding to the higher frequencies in the Fourier spectrum [27].

A three-dimensional model of a surface can be created using pixels derived from a fractional Brownian function. The shape or roughness of a pixel B_i can be written in terms of that of a neighboring pixel B_{i-1} as

$$B_i = \frac{RV}{|\Delta x|^{3-d}} + B_{i-1} \tag{5.82}$$

where

RV = Gaussian distributed random variable;
Δx = distance between pixels;
d = dimension of the surface modeled by the fractal.

The fractal model fails at boundaries between homogeneous regions when the fractal dimension is less than the topological dimension. This property can be used to identify edges as their topology is no longer represented by the fractal dimension corresponding to the region on the other side of the edge.

Mandelbrot [25] illustrated the effect of d on the perceived texture of fractal imagery. Increasing values of d cause images to become corrugated, resembling high Earth-like mountains. In order to obtain smooth valley regions and rolling terrain that are more typical of many real areas represented by a fractional Brownian function, the value of d is maintained throughout the image, but the vertical scale factor is varied in different portions of the modeled area. For example, reducing the vertical scale factor causes mountains to become rolling terrain. By making the vertical scale factor a function of altitude, mesa and canyon effects can be created.

Figure 5.18 shows three fractally generated areas that used (5.82) to model the output of a height profilometer, such as from the IF signal of an FMCW radar. Region 1 on the left represents a forest; region 2 in the middle, an unpaved road with a tank shown to the right of the center of this region (the turret and gun of the tank have been modeled to within the resolution of the sensor); and region 3 to the right, a cliff with rough terrain such as a forest. By using the sensor resolution as defined by the antenna, transmitted waveform, and signal processing, a transfer function can be derived to translate the features in the modeled scene into IF signals. Then, rule-based signal-processing algorithms that measure signal ampli-

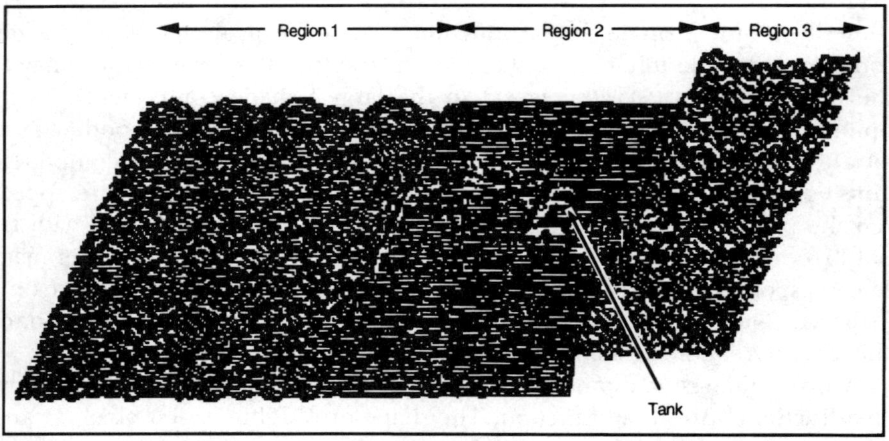

Figure 5.18 Fractally generated scene used to develop signal-processing algorithms for range profile data. Region 1 represents a forest, region 2 an unpaved road with a tank shown to the right of the center of this region, and region 3 a cliff with rough terrain such as a forest.

tudes, target lengths, heights, widths, and rates of change of target dimensions with respect to scan time can be developed to detect and classify the objects of interest. When real data are available, these trial algorithms can be applied and further enhanced to improve performance.

All phenomena are not modeled by fractals based on Gaussian random variables. For example, economics models that track price changes over time use the logarithm of the price as the random variable. Hyperbolic distributions are used to model ocean islands, wet and dry lake areas, and radar target and clutter returns [29].

The hyperbolic random variable u is described by

$$P(U) = \text{Prob}(U > u) = f\{u^{-\alpha}\} \tag{5.83}$$

where α = order of the hyperbolic distribution, related to the fractal dimension.

Hyperbolic fractal-generated imagery, corresponding to known rain rates, were used by Bogush [29] to represent the radar characteristics of rain-bearing clouds. For the particular altitude, latitude, and longitude that correspond to the interception of the radar beam with the cloud model, one notes how the fractal slice produced by the intersection of the antenna beam and the fractal sheet cloud model has been characterized in terms of rain rate. The corresponding absorption and backscatter are interpolated from tables or graphs, or their values are calculated from appropriate formulas and computer simulations, such as those given in Chapter 3.

5.9 SUMMARY

Since the intermediate frequency signal in an FMCW radar is proportional to range, its measurement can provide range to the background, range to the target, and range profile imagery of objects in the radar's field of view. The design and operation of such a radar must take into account range measurement errors produced by the finite extent of the antenna footprint on the ground and by mixing of noise, RF, and LO frequencies that occur in the receiver. The performance of the radar is improved when the signal-to-noise and signal-to-clutter ratios are increased through diversity and integration gains. Diversity gain decreases the signal fluctuation loss, while integration gain decreases the magnitude of the interference with respect to the signal of interest. These techniques result in an increase in the number of independent samples received by the radar and, hence, increase the effective signal-to-interference ratios. By using the height profile images available from the IF signal, an additional set of features can be developed to help differentiate targets from clutter and other nontarget objects or structures. Fractal images were proposed as substitutes for real range profilometer data in order to begin the development of signal-processing algorithms that could later be applied to profilometer data.

References

[1] Skolnik, M. I., *Introduction to Radar Systems*, New York, NY: McGraw-Hill, 1962.

[2] Goldwasser, R., et al., *IEEE Microwave and Millimeter-Wave Monolithic Circuits Symposium*, June 1988, pp. 55–58.

[3] Raffaelli, L., and E. Stewart, "A Standard Monolithic Transmitter for 38 GHz PCN Applications," *Microwave Journal*, Oct. 1992, pp. 24–30.

[4] McMillan, R. W., "Chapter 8: MMW Solid-State Sources," in *Principles and Applications of Millimeter-Wave Radar*, N. C. Currie and C. E. Brown, (Eds.), Norwood, MA: Artech House, 1987.

[5] Scheer, J. A., "Chapter 10: MMW Radar Receivers," in *Principles and Applications of Millimeter-Wave Radar*, N. C. Currie and C. E. Brown, (Eds.), Norwood, MA: Artech House, 1987.

[6] Barton, D. K., *Modern Radar System Analysis*, Norwood, MA: Artech House, 1988.

[7] Skolnik, M. I., *Radar Handbook*, New York, NY: McGraw-Hill, 1970.

[8] Long, M., *Radar Reflectivity of Land and Sea*, Norwood, MA: Artech House, 1983.

[9] Moore, R. K., "Chapter 25: Ground Echo," in *Radar Handbook*, M. I. Skolnik, (Ed.), New York, NY: McGraw-Hill, 1970.

[10] Elachi, C., *Spaceborne Radar Remote Sensing: Applications and Techniques*, New York, NY: IEEE Press, 1988.

[11] Barton, D. K., *Radar Systems Analysis*, Norwood, MA: Artech House, 1976.

[12] Lind, G., "Frequency Agility Radar Range Calculation Using Number of Independent Pulses," *IEEE Trans. on Aerospace and Electronic Systems*, Vol. AES-12, No. 6, Nov. 1976, pp. 811–815.

[13] Trebits, R. N., R. D. Hayes, and L. C. Bomar, "mm-Wave Reflectivity of Land and Sea," *Microwave Journal*, Aug. 1978, p. 49.

[14] Currie, N. C., R. D. Hayes, and R. N. Trebits, *Millimeter-Wave Radar Clutter*, Norwood, MA: Artech House, 1992.

[15] Currie, N. C., F. B. Dyer, and R. D. Hayes, "Radar Land Clutter Measurements at Frequencies of 9.5, 16, 35, and 95 GHz," Report A-1485-003, AD A012 709, Engineering Experiment Station, GA Institute of Technology, April 2, 1975.

[16] Gjessing, D. T., J. Hjelmstad, and T. Lund, "A Multifrequency Adaptive Radar for Detection and Identification of Objects: Results on Preliminary Experiments on Aircraft Against a Sea-Clutter Background," *IEEE Trans. on Ant. and Prop.*, Vol. AP-30, No. 3, May 1982, pp. 351–365.

[17] Gjessing, D. T., *Target Adaptive Matched Illumination Radar: Principles and Applications*, London, U.K.: Peter Peregrinus Ltd., 1986.

[18] Gjessing, D. T., "Target Adaptive Radar, Propagation Medium Adaptive Communication Systems, Present Day Methodology and Future Perspectives: A Review," *AGARD Conf. Proc. 284: Propagation Effects in Space/Earth Paths*, AD-A090 850, 1980.
[19] Currie, N. C., et al., "Millimeter-Wave Measurements and Analysis of Snow-Covered Ground," *IEEE Trans. on Geosci. and Rem. Sensing*, Vol. 26, No. 3, May 1988, pp. 307–318.
[20] Birkemeier, W. P. and N. D. Wallace, "Radar Tracking Accuracy Improvement by Means of Pulse-to-Pulse Frequency Modulation," *AIEE Trans. Comm. Electronics*, Vol. 81, Jan. 1983, pp. 571–575.
[21] Ulaby, F. T., and M. C. Dobson, *Handbook of Radar Scattering Statistics from Terrain*, Norwood, MA: Artech House, 1988.
[22] Nathanson, F. E., *Radar Design Principles*, New York, NY: McGraw-Hill, 1969.
[23] Weissman, D. E., "Two Frequency Radar Interferometry Applied to the Measurement of Ocean Wave Height," *IEEE Trans. on Ant. and Prop.*, Vol. AP-21, No. 5, Sept. 1973, pp. 649–656.
[24] Lane, T. L., "95 GHz Clutter Backscatter TABLIS V Data Base Summary," Final Technical Report on Boeing Airplane Company Contract No. BB3041, Georgia Institute of Technology, Atlanta, GA, Dec. 1982.
[25] Mandelbrot, B. B., *The Fractal Geometry of Nature*, San Francisco, CA: W. H. Freeman, 1982.
[26] Fournier, A., D. Fussell, and L. Carpenter, "Computer Rendering of Stochastic Models," *Comm. of the ACM*, Vol. 25, No. 6, June 1982, pp. 371–384.
[27] Pentland, A. P., "Fractal-Based Description of Natural Scenes," *IEEE PAMI-6*, No. 6, Nov. 1984, pp. 661–674.
[28] Mandelbrot, B. B., and J. W. Van Ness, "Fractional Brownian Motions, Fractional Noises and Application," *SIAM Review*, Vol. 10, No. 4, Oct. 1968, pp. 422–437.
[29] Bogush, Jr., A. J., *Radar and the Atmosphere*, Norwood, MA: Artech House, 1989.

CHAPTER 6

Passive Millimeter-Wave Sensors

Passive millimeter-wave (MMW) sensors receive energy that is emitted or transmitted from other sources. In space exploration projects, they are used as receivers for collecting radiation and signals from distant stars and galaxies. Radiometers operating in the microwave and millimeter-wave spectrum are also placed on weather satellites to provide atmospheric temperature and water vapor profiles. These data find use in weather forecasting for civilian and military applications. In fire-and-forget weapons, they are used as part of a suite of sensors that detect targets and help guide a missile or warhead to the designated vulnerable area on the target. Passive sensors also find use as electronic support measures (ESM) where they locate and track the source of the energy that is radiated from a potential target. In the ESM application, the energy may actually be transmitted by some other sensor such as a radar, whereas in most radiometry applications the received energy is emitted by the target, background, or atmosphere.

The chapter begins by describing Planck's radiation law that governs black-body emission at all frequencies. It is used to derive the Rayleigh-Jeans law that is applicable at microwave and millimeter-wave frequencies where the emitted energy is directly proportional to the absolute temperature of the object. Applications are reviewed that utilize frequencies at which atmospheric absorption is high and window frequencies at which absorption is low. Radiative transfer theory is discussed to understand the contributions of the atmosphere and ground-based materials to the passive signature of observed objects. Expressions are derived that relate reduction in radiometer sensitivity to lossy sensor components and impedance mismatches within the receiver. The operation and military and civilian applications are described for the total power, Dicke, and noise injection radiometers. Expressions are provided to calculate the signal-to-noise and signal-to-clutter ratios used to compute the detection probability of targets from radiometer data. These are shown to be a function of the target-to-background contrast temperature, antenna beamfill factor, minimum detectable temperature of the radiometer, and the standard deviation of the background temperature. Expressions are also given to calculate the predetection bandwidth in terms of the 3-dB bandwidth of the intermediate frequency filters, and the integration time in terms of the 3-dB bandwidth of the postdetection or video filters. The predetection bandwidth and integration time are the major factors that control the minimum detectable temperature of the sensor. The accuracy to which passive sensors can track objects is discussed. The passive signatures of several high-clutter backgrounds, including water and snow, are described in terms of their emissivities using modeled and experimentally obtained data. The chapter concludes with a comparison of the active and passive signatures of snow.

6.1 PLANCK RADIATION LAW

Blackbody objects (i.e., perfect emitters of energy) above absolute zero emit energy E per unit volume and per unit frequency at all wavelengths according to the Planck radiation law

$$E = \frac{8\pi hf^3}{c^3} \frac{1}{\exp\left(\frac{hf}{k_B T}\right) - 1} \frac{\text{J}}{\text{m}^3\text{Hz}} \tag{6.1}$$

where

h = Planck's constant = 6.6256×10^{-34} J sec;
k_B = Boltzmann's constant = 1.380662×10^{-23} J/K;
f = frequency at which radiation is measured in Hz;
c = speed of light = 3×10^8 m/sec;
T = temperature of radiating object in K.

Using the exponential series expansion for the exponential term in the denominator allows (6.1) to be rewritten as

$$E = \frac{8\pi hf^3}{c^3} \frac{1}{\frac{hf}{k_B T} + \left(\frac{hf}{k_B T}\right)^2 + \cdots} \frac{\text{J}}{\text{m}^3\text{Hz}} \tag{6.2}$$

When only the linear term is retained, as when the operating frequency f is less than $k_B T/h$ ($\approx 6 \times 10^{12}$ Hz), the Planck radiation law reduces to the Rayleigh-Jeans law given by

$$E = \frac{8\pi f^2 k_B T}{c^3} \frac{\text{J}}{\text{m}^3\text{Hz}} \tag{6.3}$$

Here, temperature is directly proportional to the energy of the radiating object, making calibration of the radiometer simpler. With perfect emitters, or blackbodies, the physical temperature of the object (T) is equal to the brightness temperature T_B that is detected by a radiometer. However, the surfaces of real objects do not normally radiate as blackbodies (i.e., they are not 100% efficient in emitting the energy forecast by the Planck radiation law). To account for this nonideal emission, a multiplicative emissivity factor is added to represent the amount of energy that is radiated by the object, now referred to as a graybody. The emissivity is equal to the ratio of T_B to T.

Radiometers used to detect objects measure the apparent temperature difference between the object and the background. As discussed in Section 6.2, apparent temperature includes the energy emitted by a graybody target, effects of antenna gain and sidelobes, energy added by antenna self-emission caused by losses, and noise reflected back into the receiver by mismatches in impedance between sensor components. In its simplest form, a radiometer receiver is similar to a pulse- or frequency-modulated continuous wave radar, with the exclusion of the transmitting components. It contains a low-loss narrow-beamwidth antenna whose sidelobes are minimized to restrict the energy that enters the receiver from areas other than the main beam. Signal processing components include a low-noise-figure mixer, wide-bandwidth intermediate frequency amplifier, a second detector (also called the video detector), and a low-pass video amplifier that provides integration of the signal. In military applications, millimeter-wave radiometers are used to detect metal targets such as a motorized tank, an armored personnel carrier, a fuel storage tank, or a ship. Their operating wavelengths make them suitable for detecting targets (e.g., the beamwidths are appropriate to the target size and the frequency can be chosen at an atmospheric window where emitted energy received from surface-based objects is maximized and emitted energy from the atmosphere is minimized) and for packaging within the envelopes of precision guided munitions. In the millimeter-wave spectrum, atmospheric windows exist at 35, 94, 140, 225, and 350 GHz. Because military targets are predominantly composed of metal, their emissivity is small and they reflect the downward-emitted sky radiation. Since the sky radiation is generally less than the radiation from the Earth's background, a contrast exists between the energy sensed by the radiometer from the target and background, making target detection possible. The detected signal can be analyzed for target features such as an intensity greater than a clutter-adaptive threshold, a pulse width corresponding to the target size, or ratios of the above discriminants.

In weather-forecasting applications, radiometers operating in the microwave and millimeter-wave spectrum are used to gauge atmospheric constituents such as oxygen and water vapor. These measurements allow temperature and humidity atmospheric profiles, respectively, to be calculated. Here, the operating frequency may be at the center or tails of absorption bands, which for oxygen are at 60 and 118 GHz and for water vapor are at 22, 183, and 320 GHz. When absorption at the higher frequencies is exploited, the quadratic term in the denominator of (6.2) is retained to minimize the error that would otherwise occur in converting the measured energy into temperature profiles. The magnitude of the error introduced when the quadratic term is neglected is given in Table 6.1 as a function of frequency.

6.2 RADIATIVE TRANSFER THEORY

Radiative transfer theory describes the contribution of cosmic, galactic, atmospheric, and ground-based emission sources to the passive signature of objects in the sensor's field of view. For the military target application, this theory provides a method for computing the energy measured by the radiometer when its antenna footprint intercepts the target as well as the background. This information is used to calculate the target-to-background contrast temperature that determines the detection probability of the target based on the radiometer data.

In Figure 6.1, a radiometer is shown flying in a missile or gun-fired round at a height h above the ground and is pointed toward the ground. The energy detected by the radiometer receiver is described by radiative transfer theory [1]. The received signal contains a combination of energy components emitted by the target, ground, and atmosphere. In the figure, T_c represents the sum of the cosmic brightness temperature and the galactic brightness temperature. The cosmic temperature is independent of frequency and zenith angle and is equal to 2.735K. Its origin is attributed to the background radiation produced when the universe was originally formed. The galactic temperature is due to radiation from the Milky Way galaxy and is a function of the viewing direction and frequency. Above about 10 GHz, the galactic contribution may be neglected in comparison with the downward emission from the atmosphere.

The value T_D represents the total downward atmospheric emission, including the variation of temperature $T(z)$ and absorption coefficient with height, and is equal to

Table 6.1
Effect of Quadratic Correction Term on Value of Emitted Energy Calculated From Planck Radiation Law

f (GHz)	hf/k_BT	$(hf/k_BT)^2$	% Change in E
2	0.00032	1.0243E-07	0.03200483
6	0.0009601	9.2188E-07	0.09601449
22	0.0035205	1.2394E-05	0.35205314
60	0.0096014	9.2188E-05	0.96014493
118	0.0188829	0.00035656	1.88828502
183	0.0292844	0.00085758	2.92844203
320	0.0512077	0.00262223	5.12077295

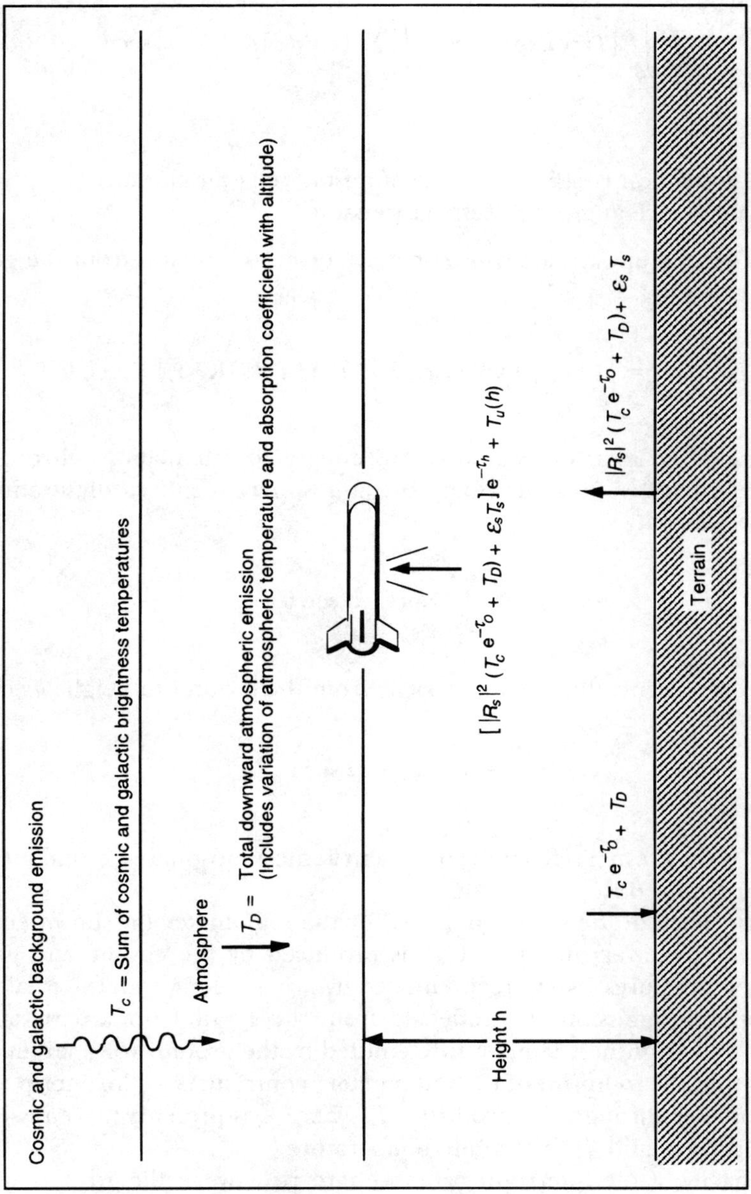

Figure 6.1 Radiative transfer in an Earth-looking radiometer sensor.

$$T_D = \int_0^\infty \{T(z) \exp[-\sec\theta \int_0^z K_\alpha(z')\, dz'] K_\alpha(z)\}\, dz \sec\theta \tag{6.4}$$

where

$K_\alpha(z)$ = absorption coefficient of the atmosphere at an altitude z;
θ = incidence angle with respect to nadir.

The value T_U is the upward atmospheric emission in the region from the ground to height h given by

$$T_U(h) = = \int_0^h \{T(z) \exp[-\sec\theta \int_0^z K_\alpha(z')\, dz'] K_\alpha(z)\}\, dz \sec\theta \tag{6.5}$$

The value τ_0 is the total one-way opacity (integrated attenuation) through the atmosphere. When $\theta < 70$ degrees and the atmosphere is spherically stratified, τ_0 may be expressed as

$$\tau_0 = \int_0^\infty K_\alpha(z)\, dz \sec\theta \tag{6.6}$$

The value τ_h represents the one-way opacity from the ground to height h equal to

$$\tau_h = \int_0^h K_\alpha(z)\, dz \sec\theta \tag{6.7}$$

The value R_s is the Fresnel reflection coefficient at the atmosphere-ground interface illustrated in Figure 6.1.

The first term in the energy detected by the radiometer (as shown entering the radiometer receiver in Figure 6.1) is produced by the cosmic and galactic brightness temperatures. They are attenuated by τ_0 and, along with the total downward atmospheric emission, are reflected from the ground upward toward the radiometer. The brightness temperature emitted by the ground (or a target if one is present within the footprint of the radiometer) contributes to the energy sensed by the radiometer through the product $\epsilon_s T_s$, where ϵ_s represents the emissivity of the ground surface and T_s its absolute temperature.

The opacity of the intervening atmosphere τ_h between the ground and the radiometer at height h attenuates (1) the cosmic, galactic, and downwelling atmospheric temperatures, and (2) the ground and target brightness temperatures before they are measured by the radiometer. To these energy sources is added the upward emission $T_U(h)$ due to atmospheric absorption phenomena that exist between the ground and height h. The total energy E received by the radiometer at a height h above the ground surface is found by combining the above energy sources as

$$E = [|R_s|^2(T_c e^{-\tau_0} + T_D) + \epsilon_s T_s]e^{-\tau_h} + T_U(h) \tag{6.8}$$

Two simplifications to the general radiative transfer equation of (6.8) can be made when the radiometer is deployed at low altitudes. First the cosmic, galactic, and downwelling atmospheric emission terms can be combined into one term called the sky radiometric temperature denoted by T_{sky}. The value T_{sky} is still a function of atmospheric water content, cloud cover, and radiometer operating frequency. A summary of the downwelling atmospheric temperature and atmospheric attenuation is given in Table 6.2 for a zenith-looking radiometer under clear air and 11 types of cloud conditions at S-band, X-band, and Ka-band frequencies [2]. As discussed in Chapter 3, both the downwelling atmospheric temperature and atmospheric attenuation increase with increasing frequency and increasing water content of the clouds. The cosmic and galactic temperatures are not included in the atmospheric temperature shown in the table. The second simplification that occurs when a radiometer is deployed at low altitudes (e.g., as part of a suite of sensors in a surface-to-surface missile) comes from neglecting the small upwelling atmospheric contribution.

6.2.1 Temperature Designations

Three temperature designations are often used to describe the energy detected by a radiometer: brightness temperature, antenna temperature, and equivalent system input noise temperature. A fourth, apparent temperature, is also defined in conjunction with the system input noise temperature.

Brightness Temperature

Brightness temperature, as introduced in the preceding section, includes the energy emitted by the target or ground surface and the energy emitted by the atmosphere that is reflected by the surface. As depicted in Figure 6.2, the brightness temperature can be written as

$$T_B(\theta, \phi) = T_{BS} + |R_s|^2 T_{sky} \tag{6.9}$$

where

$T_{BS} = \epsilon_s T_s =$ ground surface component of brightness temperature; (6.10)

$\epsilon_s = 1 - |R_s|^2 =$ emissivity of the surface;

$T_s =$ physical temperature of the surface;

$R_s =$ plane-wave Fresnel reflection coefficient for air-surface interface;

Table 6.2

Downwelling Atmospheric Temperature T_D and Atmospheric Attenuation A for a Zenith-Looking Radiometer as a Function of Cloud Type and Cloud Height for S-Band, X-Band, and *Ka*-Band Frequencies [2]

	Lower Cloud				*Upper Cloud*					*S-Band (2.3 GHz) Zenith*		*X-Band (8.5 GHz) Zenith*		*Ka-Band (32 GHz) Zenith*	
Case	*Density g/m³*	*Base km*	*Top km*	*Thickness km*	*Density g/m³*	*Base km*	*Top km*	*Thickness km*	*Remarks*	T_D *(K)*	*A (dB)*	T_D *(K)*	*A (dB)*	T_D *(K)*	*A (dB)*
1	—	—	—	—	—	—	—	—	Clear air	2.15	0.035	2.78	0.045	14.29	0.228
2	0.2	1.0	1.2	0.2	—	—	—	—	Light thin clouds	2.16	0.036	2.90	0.047	15.92	0.255
3	—	—	—	—	0.2	3.0	3.2	0.2		2.16	0.036	2.94	0.048	16.51	0.266
4	0.5	1.0	1.5	0.5	—	—	—	—		2.20	0.036	3.55	0.057	24.56	0.397
5	—	—	—	—	0.5	3.0	3.5	0.5		2.22	0.037	3.83	0.062	28.14	0.468
6	0.5	1.0	2.0	1.0	—	—	—	—	Medium clouds	2.27	0.037	4.38	0.070	35.22	0.581
7	—	—	—	—	0.5	3.0	4.0	1.0		2.31	0.038	4.96	0.081	42.25	0.731
8	0.5	1.0	2.0	1.0	0.5	3.0	4.0	1.0	Heavy clouds	2.43	0.040	6.55	0.105	61.00	1.083
9	0.7	1.0	2.0	1.0	0.7	3.0	4.0	1.0		2.54	0.042	8.04	0.130	77.16	1.425
10	1.0	1.0	2.0	1.0	1.0	3.0	4.0	1.0		2.70	0.044	10.27	0.166	99.05	1.939
11	1.0	1.0	2.5	1.5	1.0	3.5	5.0	1.5	Very heavy clouds	3.06	0.050	14.89	0.245	137.50	3.060
12	1.0	1.0	3.0	2.0	1.0	4.0	6.0	2.0		3.47	0.057	20.20	0.340	171.38	4.407

Notes: Cases 2–12 are clear air and clouds combined; antenna located at sea level, heights are measured from ground level; no cosmic background or ground contribution is included in downwelling temperature T_D; attenuation A is measured along a vertical path from ground to 30-km altitude.

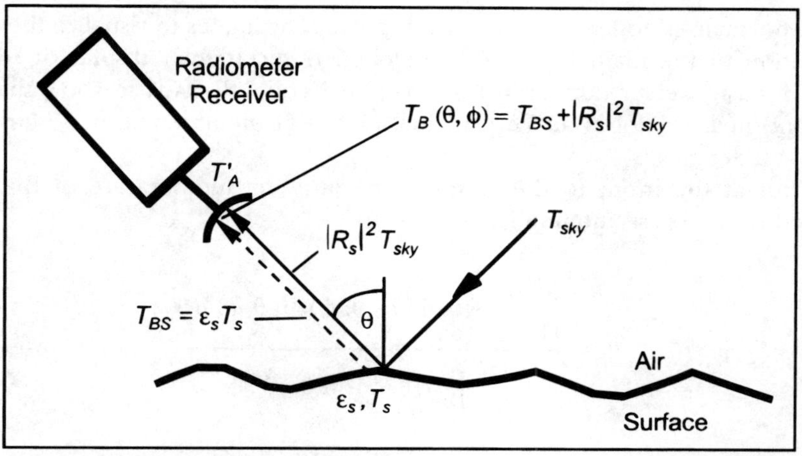

Figure 6.2 Components of the measured brightness temperature at low-altitude radiometer operation.

$$T_{sky} = T_c e^{-\tau_0} + T_D(\theta) \tag{6.11}$$

= cosmic and galactic brightness temperatures plus temperature due to downwelling emission in the specular direction;

ϕ = azimuth angle between 0 and 2π rad.

Although suppressed in (6.9), brightness temperature is also a function of antenna polarization, as discussed in Section 6.8. The relation between emissivity and reflectivity shows that objects with high emissivity reflect little of the energy that impinges upon them and vice versa. Therefore, the primary signature component from a highly reflective, low-emissivity metal target is due to reflection of the downwelling atmospheric energy that reaches the metal surface.

Upon substituting (6.10) for T_{BS} and expressing emissivity in terms of reflectivity, the brightness temperature in (6.9) becomes

$$T_B(\theta, \phi) = (1 - |R_s|^2)\, T_s + |R_s|^2 T_{sky} \tag{6.12}$$

Antenna Temperature

The second temperature designation, antenna temperature T_A, normalizes the brightness temperature with respect to the radiometer's antenna radiation pattern $f(\theta, \phi)$ over 4π steradians. In a passive sensor application such as a radiometer, the radiation pattern shows the received power distribution or gain of the antenna over the 4π steradian solid angle in watts per steradian with respect to the power density of a lossless isotropic antenna. Graphs of received power are usually normalized to unity and shown in polar coordinates to visualize the spatial distribution of the main beam and sidelobes. A rectangular display of received power intensity versus degrees off axis is also frequently used to show the main beam and near sidelobes in the principal planes of the antenna, $\theta = 90$ degrees and $\phi = 0$ degrees.

Thus at the input to the receiver, the antenna temperature of the scene observed by a lossless antenna is

$$T_A = \frac{\int_0^{2\pi} \int_0^{\pi} f(\theta, \phi)\, T_B(\theta, \phi) \sin\theta \, d\theta \, d\phi}{\int_0^{2\pi} \int_0^{\pi} f(\theta, \phi) \sin\theta \, d\theta \, d\phi} \tag{6.13}$$

Here, $f(\theta, \phi)$ is the normalized radiation pattern of a perfect plane-polarized antenna and $T_B(\theta, \phi)$ is the brightness temperature of the scene observed by a radiometer antenna with the same polarization [3]. The effect of the antenna radiation pattern on the observed antenna temperature may be approximated by using the mainlobe beam efficiency η_B. Beam efficiency is defined as the fraction of the antenna radiation pattern solid angle that is enclosed by the mainlobe. When the target of interest occupies the full ground area enclosed by the mainlobe, the antenna temperature is written as

$$T_A = \eta_B T_{B_{tgt}} + (1 - \eta_B)\, T_{B_{bkd}} \tag{6.14}$$

where

$T_{B_{tgt}}$ = brightness temperature of the target;
$T_{B_{bkd}}$ = brightness temperature of the background.

Equivalent System Input Noise Temperature

Figure 6.3 illustrates the third temperature designation, equivalent system input noise temperature T_{SYS}. It accounts for self-emission and, where present, reflection of energy by impedance mismatches between radiometer hardware components. When the radiometer components are perfectly matched such that internal reflection coefficients between the components are zero, the equivalent system input noise temperature is due to three sources: (1) emission by materials in the scene under observation, (2) antenna self-emission, and (3) noise generated by the transmission line and receiver (referred to the antenna terminals). When these sources are accounted for, T_{SYS} consists of the sum of two terms. The first, T'_A, sometimes referred to as *apparent temperature,* includes the antenna temperature produced by the scene and the noise temperature due to antenna self-emission. The second,

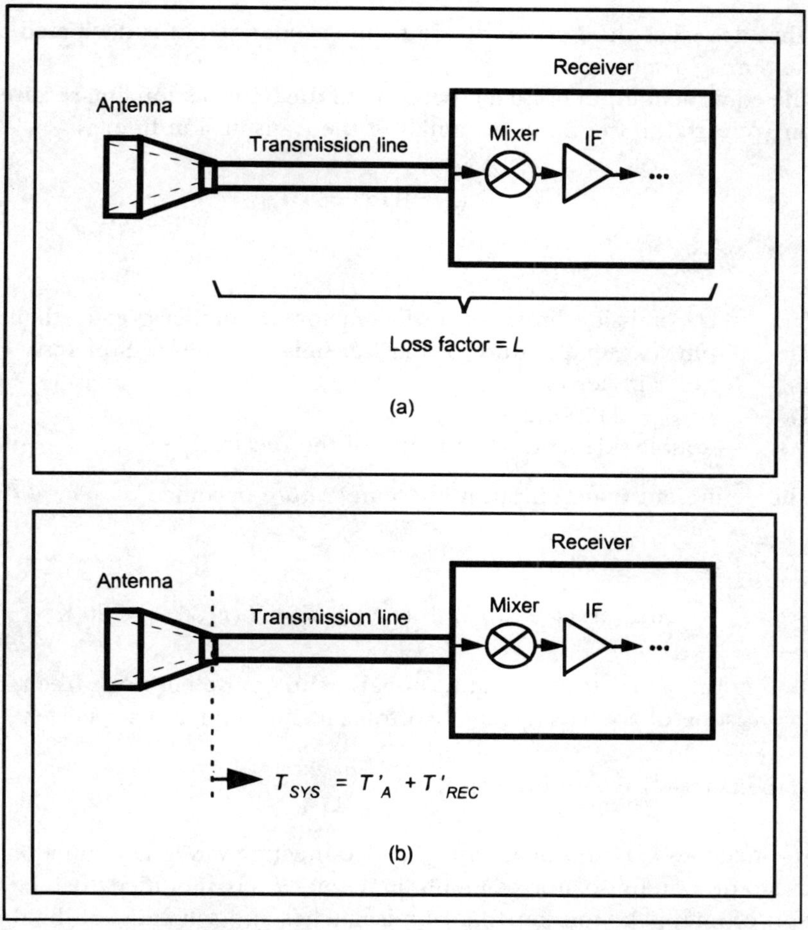

Figure 6.3 Equivalent system input noise power as contributed by the scene under observation, antenna self-emission, and noise generated by the transmission line-receiver combination: (a) antenna with radiation efficiency η_l and physical temperature T_p connected to a transmission line-receiver combination and (b) equivalent noise-free antenna and transmission line-receiver combination.

T'_{REC}, accounts for the equivalent input noise power of the transmission line-receiver combination.

If the antenna has a radiation efficiency η_l (ratio of power received at the interface of the antenna and transmission line in Figure 6.3 to the incident power at the antenna aperture) and physical temperature T_p, T'_A can be written as

$$T'_A = \eta_l T_A + (1 - \eta_l)\, T_p \tag{6.15}$$

The radiation efficiency is equal to $1/L_{ant}$, where L_{ant} (≥ 1) is the antenna loss factor that describes the portion of the incident energy that is dissipated as heat in the antenna structure.

The equivalent input noise temperature of the transmission line-receiver combination (referred to the input terminals of the transmission line) is

$$T'_{REC} = (L - 1)\,T_L + LT_{REC} \tag{6.16}$$

where

L = transmission line-receiver loss factor (a number greater than 1);
T_L = physical temperature of the transmission line-receiver lossy components in kelvin;
T_{REC} = $(F_{DSB} - 1)290\text{K}$;
F_{DSB} = double-sideband noise figure of the receiver.

Thus, the equivalent system input noise temperature becomes

$$\begin{aligned} T_{SYS} &= T'_A + T'_{REC} \\ &= \eta_l T_A + (1 - \eta_l)\,T_p + (L - 1)\,T_p + L(F_{DSB} - 1)290\text{ K} \end{aligned} \tag{6.17}$$

representing the signal at the input terminals of an equivalent noise-free receiving system consisting of a perfectly matched transmission line and a receiver.

6.2.2 Impedance-Mismatch Effects

When mismatches exist at the antenna feed-connecting waveguide junction and at the waveguide-switch/isolator assembly junction, T'_A is modified to include the reflection of energy by the impedance mismatches and self-emission by the feed, waveguide, and isolator. This is illustrated in Figure 6.4. The values T_f, T_w, and T_i represent the physical temperatures in kelvin of the antenna feed, the waveguide from the feed to the ferrite switch, and the ferrite switch and isolator assembly,

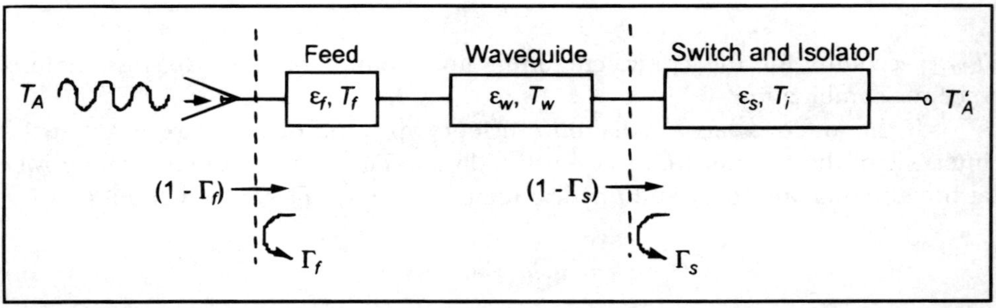

Figure 6.4 Transmission of energy through a radiometer having impedance mismatches at component junctions.

respectively. The values Π_f, Π_w, and Π_i are the power transmission coefficients through these same components, and Γ_f and Γ_s are the power reflection coefficients of the feed and ferrite switch, respectively. Multiple reflections are neglected under the assumption that internal impedance mismatches are small [4,5].

Ruf and others [4] show that the first effect of the impedance mismatches is incident energy attenuation from a value of T_A at the antenna to a reduced value of

$$T_A(1 - \Gamma_f)\,\Pi_f\Pi_w(1 - \Gamma_s)\Pi_s \tag{6.18}$$

at the output of the switch and isolator assembly.

The second set of contributions comes from thermal self-emission from the feed, waveguide, and switch and isolator components shown in Figure 6.4. Thermal self-emission from the feed itself has a direct and a reflected path to the output of the switch and isolator assembly. If the relative phase length between these two paths is denoted by ϕ_f, the emission by the feed becomes

$$T_f(1 - \Pi_f)\,(1 - \Gamma_s)\Pi_w\Pi_s|1 + (\Gamma_f\Pi_f)^{1/2}e^{j\phi_f}|^2 \tag{6.19}$$

The factor $T_f(1 - \Pi_f)$ represents the brightness temperature of the feed, where $(1 - \Pi_f)$ is the emissivity of the feed. Since the feed losses are distributed over the length of the component and not concentrated at a single point, each differential part of the loss has its own relative phase length associated with it. If the feed is an appreciable number of half-wavelengths long, then these phase lengths will average to zero. Thus, there will not be any net constructive or destructive interference between the direct and the reflected emission from the feed to the output. In this case, the power from the two paths can be added directly, giving

$$T_f(1 - \Pi_f)\,(1 - \Gamma_s)\Pi_w\Pi_s(1 + \Gamma_f\Pi_f) \tag{6.20}$$

Similarly, the thermal emission by the waveguide adds power from direct and reflected paths as

$$T_w(1 - \Pi_w)\,(1 - \Gamma_s)\Pi_s(1 + \Gamma_f\Pi_f^2\Pi_w) \tag{6.21}$$

The thermal emission by the switch and isolator assembly has three paths to the output. The direct path from the switch to the output contributes

$$T_i(1 - \Pi_s) \tag{6.22}$$

The other two paths are formed by reflection of the switch and isolator assembly brightness temperature at the junction with the waveguide and the junction with the antenna. The brightness temperature, which reflects off the switch mismatch at the waveguide, is equal to the physical temperature T_i of the isolator shown in

Figure 6.5. The nonreciprocal two-port isolator is constructed by terminating one port of a circulator with a matched load. The signal from port 1 (connected to the waveguide in this case) reaches port 2 (the isolator output), while the output from port 2 is directed to port 3, where it is absorbed. Thus as emission from the switch and isolator assembly travels toward the antenna, the physical temperature of the isolator is seen by the waveguide. Therefore, the brightness temperature from thermal emission reflected by the switch-waveguide and antenna-feed impedance mismatches is equal to the physical temperature of the isolator. Hence, the component of this emission that reflects off the switch-waveguide mismatch back into the switch reaches the output as

$$T_i \Gamma_s \Pi_s. \tag{6.23}$$

The component that reflects off the antenna-feed mismatch reaches the output as

$$T_i(1 - \Gamma_s)^2 \Gamma_f \Pi_f^2 \Pi_w^2 \Pi_s \tag{6.24}$$

The phase relationship between the emission described by (6.22) and by either of (6.23) or (6.24) averages out to zero in radiometers where the physical size of the switch and isolator exceed an appreciable number of half wavelengths. However, the exact phase relationship between the emission described by (6.23) and (6.24)

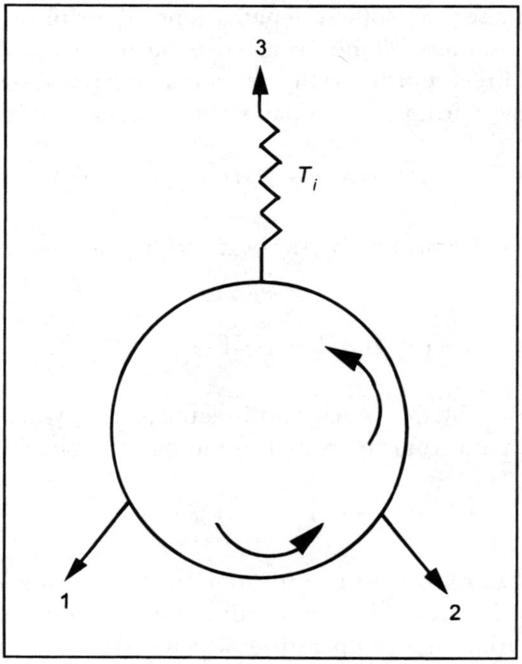

Figure 6.5 Isolator construction based on a circulator with a matched load at port 3.

must be considered since both can be referenced to a single plane, usually selected as the input port of the switch assembly. If ϕ denotes the relative phase between the contributions from (6.23) and (6.24), then their combined contribution is

$$T_i \Pi_s |\Gamma_s^{1/2} + (1 - \Gamma_s)\Gamma_f^{1/2}\Pi_f \Pi_w e^{j\phi}|^2 \qquad (6.25)$$

The phase angle is multiplied by unity since the internally reflected signals in the antenna signal path leave the system through the antenna and are not reintroduced into the receiver. A factor of two would be needed to account for the reintroduction of multiply-reflected signals into the output, if the phase angle associated with the emission of signals in a path with an internal calibration source were considered.

The value T'_A is thus equal to the sum of (6.18), (6.20), (6.21), (6.22), and (6.25) as given by

$$\begin{aligned} T'_A = {} & T_A(1 - \Gamma_f)(1 - \Gamma_s)\Pi_f \Pi_w \Pi_s + T_f(1 - \Pi_f)(1 - \Gamma_s)\Pi_w \Pi_s(1 + \Gamma_f \Pi_f) \\ & + T_w(1 - \Pi_w)(1 - \Gamma_s)\Pi_s(1 + \Gamma_f \Pi_f^2 \Pi_w) \\ & + T_i[(1 - \Pi_s) + \Pi_s |\Gamma_s^{1/2} + (1 - \Gamma_s)\Gamma_f^{1/2}\Pi_f \Pi_w e^{j\phi}|^2] \end{aligned} \qquad (6.26)$$

6.2.3 Significance of System Input Noise Temperature

The equivalent system input noise temperature T_{SYS} is representative of all the energy that a radiometer senses. Energy contributed from antenna pattern sidelobes and backlobes, ohmic losses, and impedance mismatches must be minimized to achieve the maximum sensitivity of a radiometer to measure the external energy entering through the mainlobe of the antenna. In a radar application, T_{SYS} is the system noise that competes with the desired radar backscattered signal. In a radiometer application, T_{SYS} is the desired signal that contains information about the emission characteristics of the scene under observation.

A radiometer is designed to measure T'_A as accurately and precisely as required for a particular application. The accuracy is controlled by the gain stability of the intermediate frequency (IF) amplifiers and the stability of any reference temperature calibration sources in the receiver. The precision of the measurement is dictated by the minimum detectable temperature, as discussed in the following sections.

6.3 TOTAL POWER RADIOMETER

Although many radiometer hardware configurations exist, such as total power, Dicke, and noise injection, the total power type, illustrated in Figure 6.6, finds the greatest use in military target detection applications because of its relatively low cost and compact configuration. Energy entering through the antenna is down-converted in the mixer, amplified and filtered in the IF section, square law detected,

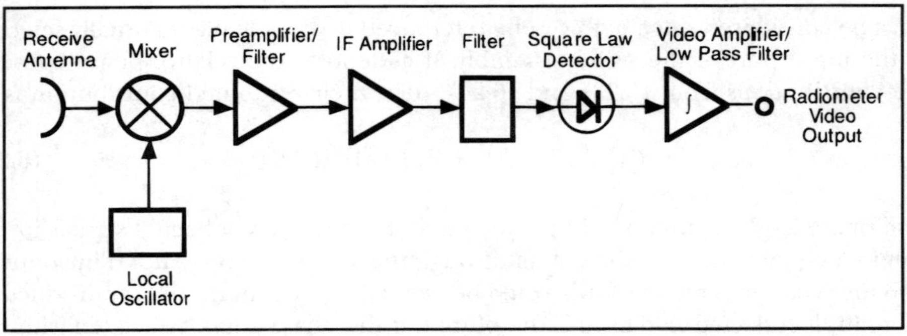

Figure 6.6 Total power radiometer.

and then integrated in the video section to yield a signal that is processed with the target detection algorithms.

The major weakness of the total power radiometer is that fluctuations in the internal gain of amplifiers used in the receiver can mask as changes in the input energy, ordinarily produced by variations in the brightness temperature of the scene in the radiometer ground footprint. The total power radiometer cannot separate the variations in the video signal caused by the gain fluctuations from the real changes in the energy emitted by the scene. Modern amplifier design and the relatively short operating time of radiometers in military weapons applications, however, make the use of the total power radiometer viable.

6.4 DICKE RADIOMETER

Dicke radiometers often find use when signal integration times are relatively large and gain fluctuations are of concern, such as in retrieval of atmospheric temperature and water vapor profiles. These radiometers, illustrated in Figure 6.7, compare the antenna temperature generated by a scene with an internal reference temperature [1,6]. An internal switch, called the reference or Dicke switch, is inserted into the waveguide in between the antenna and the input to the receiver as close to the antenna as possible. The switch alternates the inputs from the scene and the constant temperature reference into the receiver. A synchronous demodulator located between the square law detector and the low-pass filter alternately switches the detected signal envelope representing the scene and reference temperature into one of two unity gain amplifiers with opposite polarity. One amplifier receives the antenna temperature signal and the other receives the reference temperature signal.

The switching rate of the Dicke switch and the synchronous demodulator is chosen to maintain a constant system gain over a period of one switching cycle. Since the sampling theorem states that the sampling frequency must be at least twice the highest frequency of interest, the switching frequency f_s is selected such that

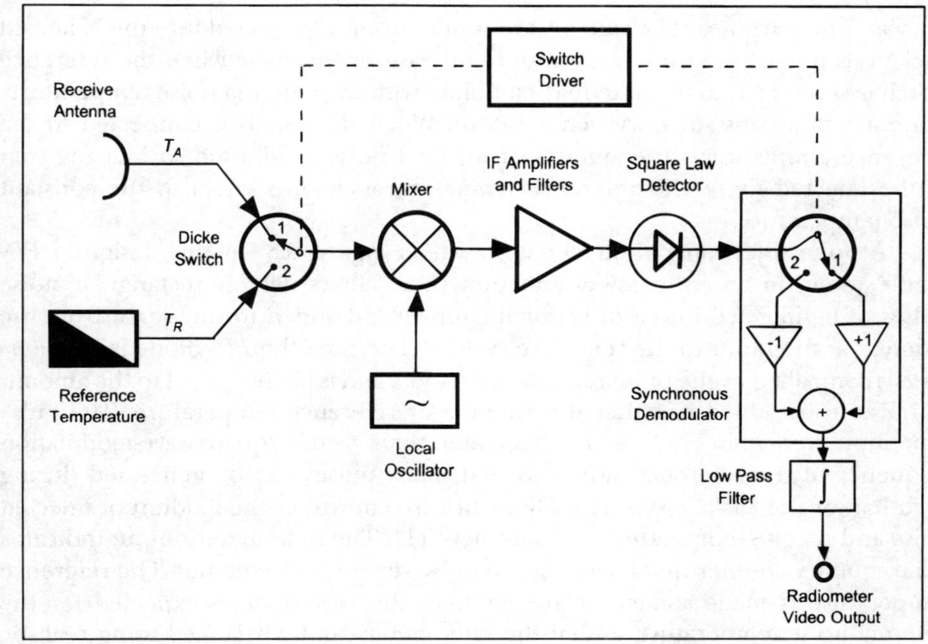

Figure 6.7 Dicke radiometer.

$$f_s \geq 2B_{3dB} \tag{6.27}$$

where B_{3dB} = 3-dB bandwidth of the low-pass filter = bandwidth of the fluctuating component of the input signal that is retained in the output voltage.

The sensitivity of the Dicke radiometer is maximized when the reference temperature is equal to the expected antenna temperature including the effects of losses. In this balanced configuration, any changes in the signals from the antenna and reference temperatures due to gain fluctuations cancel, maximizing the sensitivity of the radiometer to the fluctuating component of the input signal. Since the Dicke radiometer dwells on the scene of interest for only half the total available integration time, the minimum detectable temperature of the balanced Dicke radiometer is twice that of a total power radiometer (i.e., it is a factor of two less sensitive).

6.5 NOISE INJECTION RADIOMETER

Other radiometer configurations use noise injection to achieve the balanced Dicke condition over a range of antenna temperatures at the expense of sensitivity or minimum detectable temperature. Applications for this type of radiometer include accurate measurement of emissivities and, hence, brightness temperatures of objects for scientific purposes. The variable amount of noise T'_A injected into the balanced Dicke receiver (during the portion of the measurement cycle when the reference switch is connected to the antenna) combines with the antenna noise temperature, summing to a constant noise temperature. When the switch is connected to the reference temperature, the amount of injected noise is adjusted so that the sum of the injected noise and reference temperature signal is equal to the constant noise temperature.

A noise injection radiometer is shown in Figure 6.8. In this design, a *PIN* diode switch, in its *on* or low-attenuation state, allows narrow rectangular noise pulses to be injected into a directional coupler and added to the signal from the antenna at the input to the reference switch.[1] The time the *PIN* diode is in the *on* state is controlled by the pulse repetition period t_r and is proportional to the amount of noise to be added to either the antenna or reference temperature. The pulse repetition frequency, $1/t_r$, is much greater than f_s, the square-wave-modulation frequency of the reference switch so that many pulses can be generated during each half of the Dicke cycle. This allows precise control of the amount of injected noise and precise temperature measurement [1]. The radiometer output indicator is a frequency counter that measures the pulse repetition frequency. The reference temperature is made somewhat greater than the largest value expected for the antenna noise temperature so that the radiometer can be balanced using realistic values of t_r. Noise diodes are readily available with excess noise power ratios large enough to balance a wide range of antenna temperatures at the lower millimeter-wave frequencies. As the operating frequency is increased, however, state-of-the-art limitations in available noise diode power may not allow as great a range of antenna temperatures to be balanced. The minimum detectable temperature of a noise injection radiometer is greater than that of the Dicke radiometer, in proportion to the amount the reference temperature in the noise injection radiometer exceeds the reference temperature in the balanced Dicke radiometer that does not use noise injection.

6.6 RADIOMETER SIGNAL-TO-NOISE AND SIGNAL-TO-CLUTTER RATIO

As with a radar sensor, the energy detected and processed by a passive radiometer sensor can be used to locate targets of interest. In order to calculate the detection probability of an object using the radiometer signal, the signal-to-noise or signal-to-clutter ratio, whichever is limiting, must be known. The signal-to-noise plus clutter ratio S/NC is expressed as

$$S/NC = \frac{(\Delta T_B)(BFF)}{\sqrt{MDT^2 + \sigma^2_{bkd}}} \tag{6.28}$$

1. A *PIN* diode is a semiconductor device consisting of a layer of high resistivity intrinsic material between highly doped *p*- and *n*-materials.

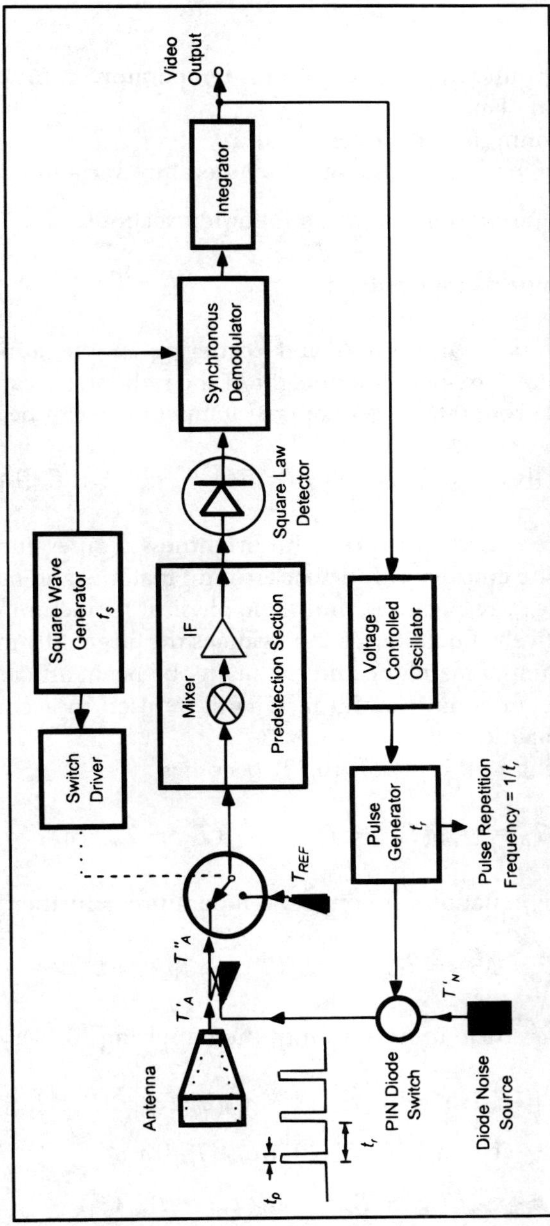

Figure 6.8 Balanced Dicke radiometer with pulsed noise injection.

where

ΔT_B = background-to-target brightness temperature contrast;
BFF = antenna beamfill factor;
MDT = minimum detectable temperature;
σ_{bkd} = one-sigma (1σ) background temperature variation.

The factors in (6.28) are discussed in the following sections.

6.6.1 Brightness Temperature Contrast

At low-altitude operation, absorption and scattering in the atmospheric layer between the ground and the operating height of the radiometer can be neglected. In this case, the background-to-target contrast temperature can be written as

$$\Delta T_B = \{[\epsilon_{bkd} T_{bkd} + |R_{bkd}|^2 T_{sky}] - [\epsilon_{tgt} T_{tgt} + |R_{tgt}|^2 T_{sky}]\}\eta_l \tag{6.29}$$

where (6.12) has been used to express the brightness temperature of the background and target. The emissivity of the background material and target have been denoted by ϵ_{bkg} and ϵ_{tgt}, respectively, and their physical temperatures in kelvin by T_{bkg} and T_{tgt}, respectively. Equation (6 29) assumes the target fills the ground-projected area of the antenna mainlobe and the mainlobe beam efficiency is unity. In the next section, the antenna beamfill factor will be used to account for targets that do not fill the mainlobe.

Upon substituting $\epsilon = 1 - |R|^2$, (6.29) becomes

$$\Delta T_B = [\epsilon_{bkd}(T_{bkd} - T_{sky}) - \epsilon_{tgt}(T_{tgt} - T_{sky})]\eta_l \tag{6.30}$$

When $T_{bkd} = T_{tgt}$, the equation for contrast temperature is further simplified to

$$\Delta T_B = (\epsilon_{bkd} - \epsilon_{tgt})(T_{bkd} - T_{sky})\eta_l \tag{6.31}$$

If the mainlobe beam efficiency is not unity, then applying (6.14) yields

$$\Delta T_B = \{[\epsilon_{bkd} T_{bkd} + |R_{bkd}|^2 T_{sky}] - \eta_B[\epsilon_{tgt} T_{tgt} + |R_{tgt}|^2 T_{sky}] - (1 - \eta_B)[\epsilon_{bkd} T_{bkd} + |R_{bkd}|^2 T_{sky}]\}\eta_l \tag{6.32}$$

$$= \eta_B[\epsilon_{bkd}(T_{bkd} - T_{sky}) - \epsilon_{tgt}(T_{tgt} - T_{sky})]\,\eta_l \tag{6.33}$$

If $T_{bkg} = T_{tgt}$, (6.33) becomes

$$\Delta T_B = \{\eta_B(\epsilon_{bkd} - \epsilon_{tgt})(T_{bkd} - T_{sky}) \tag{6.34}$$

Comparing (6.31) and (6.34), we see that $\eta_B \neq 1$ decreases the background-to-target contrast temperature by a multiplicative factor equal to η_B. It also increases the equivalent system input noise temperature through its effect on T_A as shown by (6.14).

The contrast temperature between a metal object and an emissive background as seen by a downward-looking radiometer may be calculated using (6.34). If the emissivities of the background and reflective metal object are 0.8 and 0.15, respectively; temperature of the background and target are 290K; sky temperature is 35K (a typical value for 35-GHz systems); antenna radiation efficiency is 75% and antenna mainlobe beam efficiency is 90%; then

$$\Delta T_B = (0.90)(0.8 - 0.15)(290\text{K} - 35\text{K})(0.75) = 112\text{K} \tag{6.35}$$

The contrast temperature in (6.35) is large enough to be detected by a radiometer designed for a fire-and-forget weapon application. Of course, as the water vapor content of the atmosphere increases, T_{sky} increases and the contrast decreases. Covering the metal with absorbent materials will also reduce the contrast temperature by causing the object to appear radiometrically warmer.

6.6.2 Antenna Beamfill Factor

The beamfill factor accounts for the reduced background-to-target contrast temperature measured when the target does not completely fill the ground-projected area of the antenna mainlobe. It can be modeled using one of two approaches. In the first technique, whose geometry is illustrated in Figure 6.9, the beamfill factor is approximated by an expression that incorporates an *antenna beamshape factor* as

$$BFF = 1 - e^{-\alpha_B A_T / A_B} \tag{6.36}$$

where

α_B = antenna beamshape factor;

A_T = target area in m^2;

A_B = ground-projected footprint area in m^2

$$= \pi R^2 \theta_a \theta_e / (4 \cos\theta); \tag{6.37}$$

R = operating range of the radiometer in m;

θ_a, θ_e = azimuth, elevation one-way 3-dB antenna beamwidths in radians;

θ = incidence angle of the radiometer with respect to nadir.

The antenna beamshape factor, which is a function of the aperture taper, has been modeled for three types of antenna aperture illuminations, as shown in Table 6.3 [7].

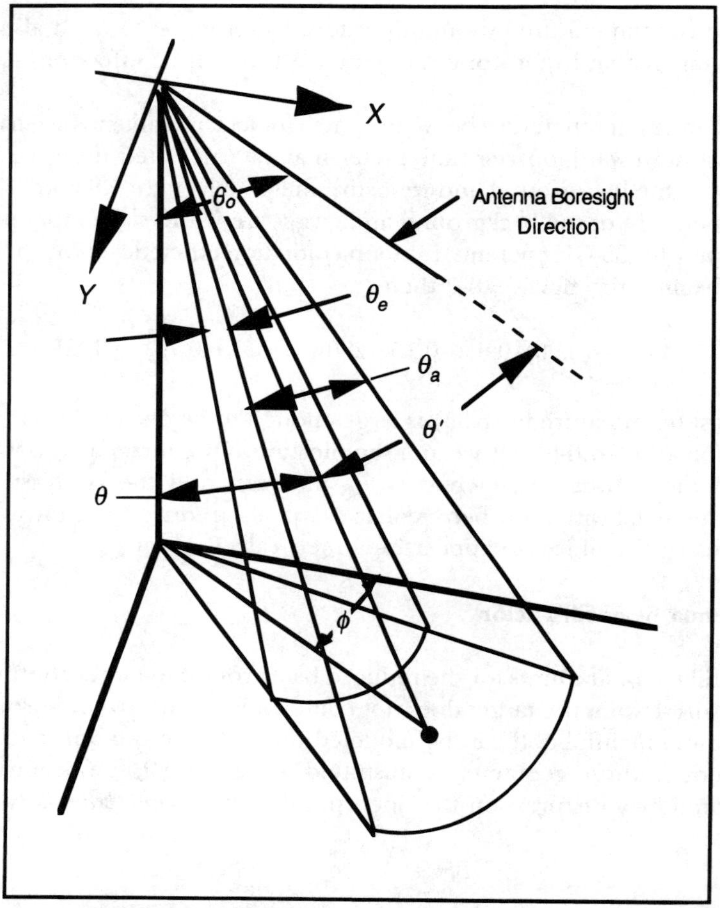

Figure 6.9 Beamfill factor-defining geometry.

Table 6.3
Antenna Beam Factor for Three Aperture Illumination Tapers [7]

Antenna Aperture Illumination Taper	*Beamshape Factor* α_B
Parabolic	0.503
10 dB	0.580
Uniform	0.658

With $\theta_a = \theta_e = 6$ degrees $= 0.105$ rad (a value typical of a *Ka*-band antenna constrained by missile diameter), $\theta = 0$ degrees (nadir looking), $R = 50$ m, $A_T = 18$ m^2, and $\alpha_B = 0.66$, (6.36) gives a value for the beamfill factor of 0.44.

If the value of the antenna beamshape factor is not known, it may be approximated using the following measurement process:

1. Measure antenna pattern noting one-way 3-dB beamwidth and sidelobe levels.
2. Calibrate radiometer receiver in the laboratory to account for emission from ohmic losses and reflected receiver noise from impedance mismatches [8–10].
3. Point the radiometer at zenith and measure the sky temperature.
4. Measure the thermometric temperature (that recorded by a thermometer) of the ground in the vicinity of the measurement site and its emissivity. Alternatively, the ground emissivity may be noted from measurements made by others of similar surfaces.
5. Place a large metal plate with an unoxidized (highly reflective) surface characterized by a reflectivity of at least

$$|R_s|^2 \approx 0.9$$

   over the measurement site.
6. Position the radiometer at a known height above the metal plate for the incidence angle of interest and record the measured antenna temperature.
7. Calculate the beamfill factor from the geometry of the measurement configuration and the antenna parameters, such as 3-dB beamwidth and sidelobe levels.
8. Repeat steps 6 and 7 for other heights that vary the beamfill factor while keeping the incidence angle fixed.
9. Confirm 100% beamfill factor at the height where the antenna temperature corresponds to the sky temperature measured in step 3.
10. Confirm 25%, 50%, and so forth beamfill factors for the antenna temperatures that correspond to scenes consisting of (1/4 · sky temperature + 3/4 · ground thermometric temperature), (1/2 · sky temperature + 1/2 · ground thermometric temperature), etc.
11. Plot the beamfill factor as a function of height.
12. Use the above values of beamfill factor and height to fit the beamfill factor to an expression of the form

$$1 - \exp(-\alpha_B A_T / A_B)$$

and then solve for the antenna beamshape factor α_B as

$$\alpha_B = -\frac{A_B}{A_T} \ln(1 - BFF) \tag{6.38}$$

In this relation, A_B is a function of range R, incidence angle θ, and the antenna's azimuth and elevation one-way beamwidths θ_a and θ_e, respectively.

The second method of finding the beamfill factor is by using modeled or actual mainlobe antenna pattern data to compute the portion of the mainlobe that intercepts the target.

In step 1 of this three-step process, the one-way gain of the antenna G_R in the radiometer receiver is written as [11]

$$G_R(\theta, \phi) = G_0 \exp[-a(\theta'/\theta_b)^2] \tag{6.39}$$

when a Gaussian model is used for the mainlobe. Here

G_0 = antenna peak gain;
a = 2.7726;
θ_b = one-way 3-dB antenna beamwidth;
θ' = $\cos^{-1}(\sin\theta \sin\theta_0 \cos\phi + \cos\theta \cos\theta_0)$;
θ_0 = boresight angle measured from nadir;
θ = angle of incidence;
ϕ = azimuth angle measured from x-y projection of the boresight direction.

In step 2, (6.39) is used to compute the area of the mainlobe that is over the target from knowledge of the range, incidence angle, and target location using the geometrical relations developed from Figure 6.9. If the actual antenna pattern data are known, they may be used in place of the modeled beam. The beamfill factor is found in step 3 as

$$BFF = \frac{\text{Mainlobe area over target}}{\text{Total mainlobe area}} \tag{6.40}$$

6.6.3 Minimum Detectable Temperature

When gain fluctuations can be neglected, the minimum detectable temperature for a total power radiometer is given by

$$MDT = \frac{T_{SYS}}{\sqrt{B_{IF} t_i}} \tag{6.41}$$

where

T_{SYS} = equivalent system input noise temperature in K;
B_{IF} = predetection intermediate frequency bandwitdh in Hz;
t_i = integration time of video amplifier stage in sec.

The predetection IF bandwidth, also called the reception bandwidth, is the width of a rectangular filter having the same maximum gain as the real one and that produces the same statistical error in output power measurement. It is to be contrasted with the noise bandwidth of an ideal rectangular filter that has the same maximum gain as the real one, and produces the same output power when the input noise is white [12].

When gain fluctuations are significant, the minimum detectable temperature for the total power radiometer is given by

$$MDT = T_{SYS}\sqrt{1/(B_{IF}t_i) + (\Delta G/G)^2} \tag{6.42}$$

where

G = IF amplifier gain;
ΔG = gain fluctuation.

By contrast, the minimum detectable temperature for a balanced Dicke radiometer having a reference temperature equal to the antenna temperature is

$$MDT = \frac{2T_{SYS}}{\sqrt{(B_{IF}t_i)}} \tag{6.43}$$

or twice that of a total power radiometer with negligible fluctuations in gain.

The minimum detectable temperature of the noise injection radiometer is

$$MDT = \frac{2(T_0 + T'_{REC})}{\sqrt{B_{IF}t_i}} \tag{6.44}$$

where

T_0 = reference temperature > maximum expected value of T'_A.

Radiometers are designed with large predetection bandwidths to maximize the sensitivity (i.e., achieve a small minimum detectable temperature). The predetection bandwidth B_{IF} is determined by the type of filter used in the IF stage. The relation of B_{IF} to the 3-dB bandwidth ΔB of the IF filter is shown in Table 6.4 [6].

The integration time t_i is usually set equal to the dwell time of the radiometer on the target to maximize signal-to-noise or signal-to-clutter ratio. The dwell time, or time-on-target, is a function of the forward velocity, spin rate, sensor aperture, squint angle, target size, and orientation of the radiometer. The values of these parameters are a function of the radiometer application.

A general expression for the integration time t_i in terms of the 3-dB bandwidth of the video filter is

$$t_i = \frac{1}{K_F B_{3dB}} \tag{6.45}$$

where B_{3dB} represents the 3-dB bandwidth of the video filter and K_F is a function of the filter implementation as described in Table 6.5. The table also shows the relation of the video filter noise bandwidth B_N to the 3-dB bandwidth of the filter. The output of the video filter produces a dc voltage representing the average value of T_{SYS} and an ac component whose root mean square (rms) value represents the statistical uncertainty associated with the measurement of T_{SYS}.

An ideal integrator is not the same as an ideal low-pass filter. An ideal integrator is described by a rectangular function in the time domain, while an ideal low-pass filter is described by a rectangular passband in the frequency domain. In terms of filtering out fluctuations in the ac component of the video signal, however, the two

Table 6.4
Predetection IF Bandwidth Compared to Half-Power Bandwidth for Several IF Filter Designs [6]

Filter Type	*G(f)*	$B_{IF}/\Delta B$
n cascaded single tuned stages	$\{1 + [\Delta\omega/(\pi\Delta B_s)]^2\}^{-n}$	
$n = 1$		3.14
$n = 2$		1.96
$n = 3$		1.76
$n = 5$		1.62
$n = \infty$ (Gaussian)	$2^{-[\Delta\omega/(\pi\Delta B)]^2}$	1.50
m cascaded $2n$-pole Butterworth filters	$\{1 + [\Delta\omega/(\pi\Delta B_s)]^{2n}\}^{-m}$	
$m = 1, n = 2$		1.48
$m = 1, n = 3$		1.26
$m = 1, n = \infty$		1.00
$m = 2, n = 2$		1.30

Note: $G(f)$ = IF filter transfer function; $\Delta\omega = \omega - \omega_0$, where ω_0 = center frequency in radians; ΔB_s = 3-dB bandwidth of a single filter section; ΔB = 3-dB bandwidth of the composite filter.

Table 6.5
Integration Time Filter Constant K_F and Noise Bandwidth for Several Types of Video Filters

Filter Type	K_F	B_N/B_{3dB}
Single RC	3.14	1.57
Two RC filters in cascade	2.44	1.22
Three RC filters in cascade	2.31	1.55
Box car filter	2.26	1.13
Five-pole Butterworth	2.02	1.01
Five-pole Bessel	2.08	1.04
Gaussian	2.12	1.06
Rectangular (ideal)	2.00	1.00

are equivalent if the integration time of the ideal integrator is equal to the reciprocal of twice the bandwidth of the ideal low-pass filter.

The minimum detectable temperature of a total power radiometer given by (6.41) can be deduced by physical reasoning [6]. The IF signal corresponding to the equivalent system input noise temperature is correlated over a time period of the order of $1/B_{IF}$ seconds. Hence, there are B_{IF} independent signal samples per second that contribute to the detector output. These samples are averaged in the postdetection integrator over the integration time t_i producing $B_{IF} \cdot t_i$ independent samples. Therefore, the rms deviation of any sample is $1/\sqrt{B_{IF} t_i}$ times the average dc output. The IF bandwidth is limited by the usable bandwidth of the amplifiers and by interfering signals. The integration time is limited by the desire to confine the dwell time to the time the target fills the 3-dB beamwidth of the radiometer antenna. This may require t_i to be no more than the time required for a point source to traverse the half-power points of the antenna beam. A long integration time may also cause a significant displacement in the position from which the energy is sensed.

To calculate the minimum detectable temperature typically found in a total power radiometer designed for a fire-and-forget weapon application, assume the physical temperature of the lossy components is 290K, the double-sideband noise figure is 6 dB (a factor of 4), and the ohmic loss is equal to the RF component losses of 2 dB (a factor of 1.6). Using the numerical values for ΔT_B and *BFF* computed above,

$$T'_A = (\Delta T_B)(BFF) = (112\text{K})(0.44) = 49\text{K} \qquad (6.46)$$

Substituting the design parameters of the preceding paragraph into (6.17) gives

$$T_{SYS} = 1{,}687\text{K} \qquad (6.47)$$

For $B_{IF} = 500$ MHz and $t_i = 3$ msec,

$$MDT = 1{,}687 \sqrt{\frac{1}{500 \times 10^6 \times 3 \times 10^{-3}}} = 1.4\text{K} \tag{6.48}$$

As a rule of thumb, a radiometer is used to detect minimum target-to-background contrast temperature differences that are several times greater than the *MDT* in order to increase the probability that the detected signal is not a random fluctuation.

6.6.4 Signal-to-Noise Plus Clutter Ratio Calculation

The signal-to-noise plus clutter ratio can now be found using the calculated values for the factors in (6.28). Thus,

$$S/NC = \frac{(\Delta T_B)(BFF)}{\sqrt{MDT^2 + \sigma_{bkd}^2}} = \frac{(112)(0.44)}{\sqrt{(1.4)^2 + (3)^2}} = 15 = 12.0 \text{ dB} \tag{6.49}$$

where 3K represents a moderate value for the 1σ fluctuation in the background-clutter signature. In this example, the radiometer is more clutter-limited than noise limited due to the particular values for the integration time and clutter fluctuation that were postulated. The reverse situation, where a radiometer operates in noise-limited conditions, is also frequently true.

6.7 ANGLE-TRACKING ERROR

A MMW radiometer can track objects or locate geological features such as rivers, forests, and crop areas similarly to a radar as was described in Section 4.11. Since a radiometer only receives energy emitted by objects, the one-way antenna beam-width is used in the expression for angle tracking error σ_p. Thus,

$$\sigma_p = \frac{0.49\theta_p}{\sqrt{(S/N)_p}} \tag{6.50}$$

for $S/N >> 1$ and

$$\sigma_p = \frac{0.58\theta_p}{\sqrt{(S/N)_p}} \tag{6.51}$$

for $S/N << 1$,

where

θ_p = radiometer (one-way) 3-dB antenna beamwidth;
$(S/N)_p$ = radiometer signal-to-noise ratio.

Glint, which can degrade radar tracking performance, does not affect radiometer tracking accuracy as individual scatterers do not dominate the passive signature. Thus whereas a radar can be used in early phases of a mission to guide a missile to a target, a radiometer can be used in the end-game missile scenario to improve the accuracy of the missile impact on the target. In this type of application, the radar transmitter is turned off when the missile is within several hundred feet of the target, allowing the radar to operate as a radiometer and guide the missile to its final aimpoint. This technique avoids the aimpoint wander that could be caused by target scintillation had the radar been transmitting until target impact. However, since the one-way beamwidth in (6.50) and (6.51) is greater than the two-way beamwidth used in radar tracking, the 1σ passive angle-tracking error due to the finite antenna beamwidth is greater for radiometers than for radars. The passive target location error (in linear distance measure computed as the range multiplied by the 1σ angle error) may be tolerable, though, at short range and with adequate S/N ratios as found in end-game scenarios and ESM applications employing multiple passive sensors to locate the energy radiators.

Equation (6.50) was used to plot the angle-tracking error shown in Figure 6.10. The upper curve displays σ_p as a function of S/N when the antenna beamwidth θ_p is held constant at 6 degrees. The linear curve gives σ_p as a function of antenna beamwidth when S/N is fixed at 13 dB.

6.8 BACKGROUND-CLUTTER SIGNATURES

The passive signature of the background is dependent on its emissivity and absolute temperature. Since emissivity is a function of incidence angle, frequency, roughness, polarization, conductivity, and water content, so is the passive background signature. The emissivity for horizontally polarized $[\epsilon_H(\theta)]$ and vertically polarized $[\epsilon_V(\theta)]$ emission at an air-dielectric interface is written in terms of the relative complex dielectric constant of the emissive background material as

$$\epsilon_H(\theta) = 1 - \left| \frac{\cos\theta - \sqrt{e_r - \sin^2\theta}}{\cos\theta + \sqrt{e_r - \sin^2\theta}} \right|^2 \tag{6.52}$$

and

$$\epsilon_V(\theta) = 1 - \left| \frac{e_r\cos\theta - \sqrt{e_r - \sin^2\theta}}{e_r\cos\theta + \sqrt{e_r - \sin^2\theta}} \right|^2 \tag{6.53}$$

where the second term in each equation is the square of the Fresnel reflection

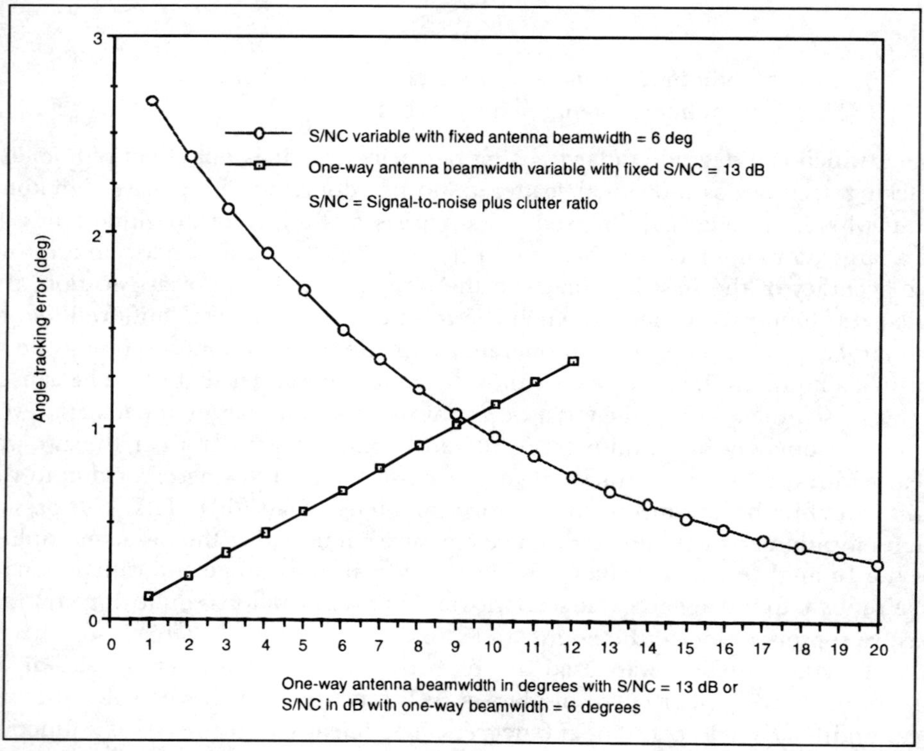

Figure 6.10 One-sigma tracking error for a radiometer sensor as a function of antenna beamwidth and signal-to-noise ratio.

coefficient (also referred to as reflectivity), e_r is the relative complex dielectric constant of the emissive surface with respect to the free-space permittivity e_0, and θ is the angle of incidence at which the downward atmospheric emission and background emissive and reflective energies are measured with respect to a line normal to the surface [1]. The emissive material is assumed to be nonmagnetic; hence, the relative magnetic permeability was set equal to unity in (6.52) and (6.53).

At normal incidence, the expression for emissivity simplifies to

$$\epsilon(\theta = 0) = 1 - \left|\frac{1 - \sqrt{e_r}}{1 + \sqrt{e_r}}\right|^2 \tag{6.54}$$

showing that the horizontally and vertically polarized emissivities are equal at nadir.

The signatures of emissive materials that are composed of polar molecules (i.e., those where the positive and negative centers of charge are separated such as in water) can be analyzed by expressing the relative complex dielectric constant in terms of its real and imaginary parts as

$$e_r = e'_r - je''_r \tag{6.55}$$

where e'_r is the real part and e''_r is the imaginary part. The Debye formulation of the dielectric constant that accounts for the relaxation time[2] of the polar molecule is then used to write the relative complex dielectric constant as [13,14]

$$e_r = e_\infty + \frac{e_s - e_\infty}{1 + (j\omega\tau)^{1-\alpha_\tau}} - j\frac{\sigma_c}{\omega e_0} \tag{6.56}$$

where

ω $2\pi f$ = radian frequency where f is in Hz;
e_∞ = dielectric constant at infinite frequency;
e_s = static value of the dielectric constant;
τ = relaxation time in sec;
σ_c = ionic conductivity in mhos/m;
α_τ = empirical parameter that describes the distribution of relaxation times;
e_0 = 8.854×10^{-12} = permittivity of free space in F/m.

6.8.1 Water Signature

The simplicity of (6.56) for calculating the dielectric constant of surfaces that contain water is deceptive because e_s, τ, and σ_c are all functions of the water temperature and salinity. The polar structure of the water molecule gives rise to the frequency dependence of the relatively large real and moderate imaginary components of its dielectric constant.

The emissivity of water has been modeled at several microwave and MMW frequencies by applying regression equations to experimental data to find the specific dependence of e_s, τ, and σ on temperature and salinity in the Debye model [15–17]. The emissivity at 37 GHz is shown in Figure 6.11(a) as a function of incidence angle and water surface temperature for rainwater and freshwater lakes and ponds where the ionic conductivity is assumed zero. Figure 6.11(b) applies to open ocean with a salinity of 34‰ (parts per thousand). The emissivity shows very little dependence on salinity at 37 GHz. At lower frequencies, such as those in the L , S, and C microwave bands, emissivity is a stronger function of salinity due to the larger contribution of conductivity to the imaginary part of the dielectric constant. The strong polarization dependence of emissivity is due to the polar nature of the water molecule.

---

2. The relaxation time is defined as the time required for the positive and negative moments of the molecule to revert to a random distribution after removal of the applied electric field [13]. The polarity of the water molecule is responsible for the dispersion in the dielectric constant, where dispersion is the variation of the dielectric constant with frequency.

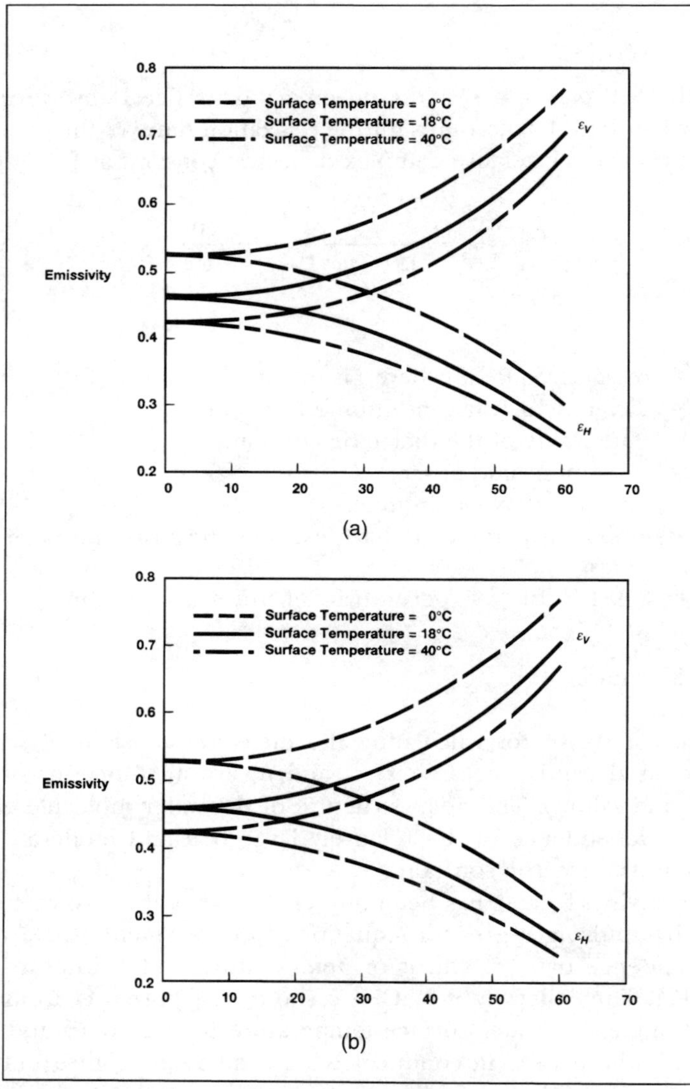

Figure 6.11 Modeled emissivity of water at 37 GHz. e_V and e_H represent the vertically- and horizontally-polarized emissivities: (a) Salinity = 0 $^0/_{00}$ (ionic conductivity = 0 mho/m) and (b) Salinity = 34 $^0/_{00}$ (open ocean).

The Klein-Swift model [17] was used to generate Figure 6.11 from the model's polynomial expressions for static dielectric constant, relaxation time, and conductivity of water as a function of temperature and salinity with $\alpha_T = 0$ and $e_\infty = 4.9$. The equations for the static dielectric constant component of the model are

$$e_s(T, S) = e_s(T)\, a(S, T) \tag{6.57}$$

where T is the temperature in °C, S is salinity in parts per thousand,

$$e_s(T) = 87.134 - 1.949 \times 10^{-1}T - 1.276 \times 10^{-2}T^2 + 2.491 \times 10^{-4}T^3 \quad (6.58)$$

and

$$a(S, T) = 1.000 + 1.613 \times 10^{-5}ST - 3.656 \times 10^{-3}S + 3.210 \times 10^{-5}S^2 - 4.232 \times 10^{-7}S^3 \quad (6.59)$$

The model for relaxation time is

$$\tau(T, S) = \tau(T, 0)\, b(S, T) \quad (6.60)$$

where

$$\tau(T, 0) = 1.768 \times 10^{-11} - 6.086 \times 10^{-13}T + 1.104 \times 10^{-14}T^2 - 8.111 \times 10^{-17}T^3 \quad (6.61)$$

and

$$b(S, T) = 1.000 + 2.282 \times 10^{-5}ST - 7.638 \times 10^{-4}S - 7.760 \times 10^{-6}S^2 + 1.105 \times 10^{-8}S^3 \quad (6.62)$$

The ionic conductivity of sea water is given by

$$\sigma_c(T, S) = \sigma_c(25, S)\exp(-\Delta\beta) \quad (6.63)$$

where

$$\Delta = 25 - T \quad (6.64)$$

$$\beta = 2.033 \times 10^{-2} + 1.266 \times 10^{-4}\Delta + 2.464 \times 10^{-6}\Delta^2 - S(1.849 \times 10^{-5} - 2.551 \times 10^{-7}\Delta + 2.551 \times 10^{-8}\Delta^2) \quad (6.65)$$

and

$$\sigma_c(25, S) = S(0.182521 - 1.46192 \times 10^{-3}S + 2.09324 \times 10^{-5}S^2 - 1.28205 \times 10^{-7}S^3) \quad (6.66)$$

Reviews of the Klein-Swift model for sea surface emissivity show that the uncertainty in the modeled brightness temperature increases with frequency above 10 GHz, to a level corresponding to ≈2K at 37 GHz, and that the error varies with sea surface temperature [18,19]. At 19 and 22 GHz, the error is less than 1K, while at 37 GHz the error is ≈2K when the surface temperature is 275K and is negligible when the surface is 295K. Below X-band, the model is designed to give undisturbed smooth sea surface brightness temperature accuracies of 0.3K or better.

6.8.2 Boulder, Gravel, Ground, and Vegetation Signatures

Figure 6.12 illustrates the effects of surface roughness on brightness temperature [20]. The measurements were made at 42 GHz at a quarry where limestone was sized and stockpiled prior to shipment. Five sizes of limestone were studied: (1) blocks and boulders ranging from 30 cm to a few meters in diameter, (2) cobbles ranging from 10 to 15 cm in diameter, (3) cobbles ranging from 5 to 10 cm in diameter, (4) gravel ranging from 2 to 5 cm in diameter, and (5) gravel of less than 2 cm in diameter. The data were recorded as the radiometer was scanned in elevation over the different stockpiles. The brightness temperature from the diffuse scatterers shows a weak dependence on incidence angle for angles up to approximately 35 degrees. The boulder scans do not exhibit appreciable sensitivity to polarization. The more abrupt changes in the brightness temperature observed from the large boulders was due to scattering of energy by sharp changes in the topography of the boulders. As illustrated by the relatively high brightness temperatures, these rock fragments have high emissivities. The lower right graph in Figure 6.12 shows the difference between the vertically and horizontally polarized brightness temperatures and, thus, emphasizes the polarization dependence of the brightness temperature on the size of the limestone fragments.

The measured brightness temperatures of asphalt, concrete, plowed ground, a road, sand, grass, bushes, and water at 35 GHz are presented in Figure 6.13 [21]. The brightness temperatures of these surfaces exhibit polarization dependence with incidence angle similar to that displayed by the smaller rock fragments. All but water have high emissivities. Water shows the greatest change of brightness temperature with incidence angle and polarization because it is a polar molecule.

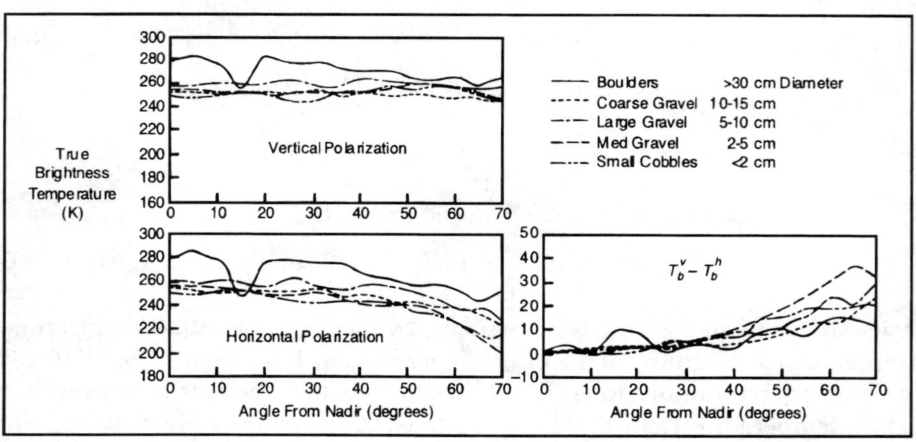

Figure 6.12 Measured brightness temperatures of boulders, cobbles, and gravel at 42 GHz [20].

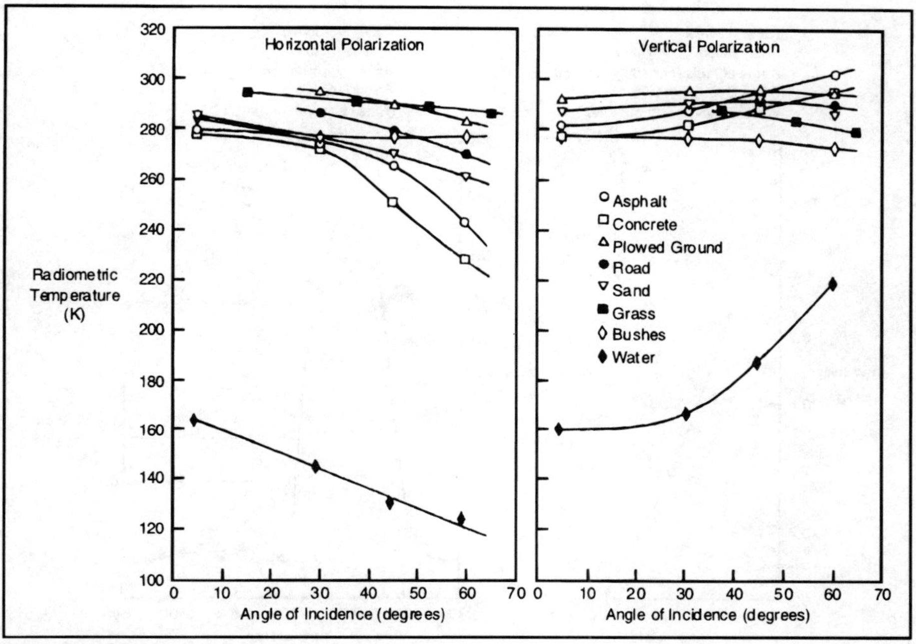

Figure 6.13 Measured background temperatures at 35 GHz [21].

6.8.3 Snow Signature

The measured horizontally-polarized emissivity of snow at 10.7, 37, and 94 GHz is depicted in Figure 6.14 for dry and wet snow [22]. These ground-based measurements were made using radiometers secured to truck-mounted booms. The radiometer specifications are given in Table 6.6. As discussed with respect to the snow radar cross-section data presented in Chapter 4, these passive snow data represent *typical* signatures that may vary from location to location depending on surface roughness, snow depth, water content, and grain size.

As shown on the left in Figure 6.14, the emissivity of dry and wet snow decrease with increasing frequency. However, the emissivity of dry snow is less than that of wet snow, causing the measured brightness temperature of dry snow to be lower also. Since metal targets have low emissivity and derive the majority of their signature from reflected atmospheric emission, they generally appear colder than the background. Therefore, the background-to-target contrast temperature is lower when a metal target is on dry snow than on wet snow, making it more difficult to detect the manmade target on dry snow.

The curves on the right in Figure 6.14 illustrate the diurnal change in the apparent temperature of snow over a 24-hour period at 10.7 and 37 GHz. The diurnal change is due to the variation in the water content of the snow and the change in the physical temperature of the snowpack that occurs during the transition from warm day to cool night. When the snow is dry, characteristic of data

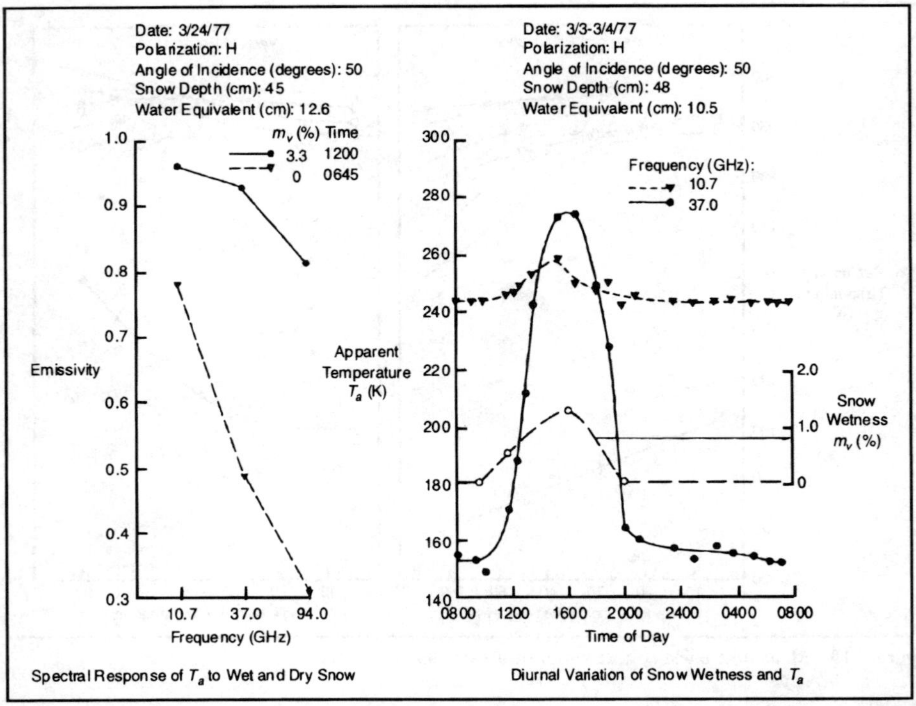

Figure 6.14 Measured emissivity and apparent temperature of dry and wet snow versus frequency and time of day [22].

Table 6.6
Specifications of Radiometers Used for 10.7-, 37-, and 94-GHz Measurements of Apparent Temperature of Snow [22]

Frequency (GHz)	*Radiometer Type*	*Polarization*	*Bandwidth (MHz)*	*Sensitivity (K)*	*Accuracy (K)*
10.69	Dicke	H	200	0.2	1
37	Dicke	H and V	300	0.5	±1
94	Total Power	H	730	3.5	$\pm[0.05(300 - T_s) + 6]$*

* T_s = surface temperature of emissive material.

gathered before 10 a.m. and after 8 p.m., the apparent temperature is lower than when the snow is wet (liquid water volume $m_v > 0\%$) as in the noon to 8 p.m. period. The diurnal change in apparent temperature is greater at 37 GHz because the shorter wavelength energy is attributable to emission from the snow surface where a greater amount of change in the water content of the snow occurs. The small change in apparent temperature after 8 p.m. is attributed to the decrease in the temperature of the surface layer of the snowpack. At the longer 10.7-GHz wavelength, the detected energy is due to emission from deeper regions in the snow that are not as sensitive to an increase in free water from the sun load. The constant apparent temperature observed between midnight and approximately 10 a.m. is attributed to the constant temperature of the deeper layers of the snowpack and the underlying ground.

Similar changes in apparent temperature appear in Figure 6.15 for wet and dry snow, this time plotted as a function of incidence angle [22]. The early morning measurements of apparent temperature, made when the snow surface was −14.6°C, show lower values at all incidence angles as compared to values obtained when the snow surface temperature was 0°C in the early afternoon. The approximately 5 to 15K difference in apparent temperature between the dry and wet snow at 10.7 GHz is due to the difference in the thermometric temperature of the snowpack during the dry and wet snow conditions. The larger change in apparent temperature between dry and wet snow at 37 GHz is due to the effects of snow wetness that (1) decrease the received energy attributable to scattering in the snowpack volume, (2) smooth the snow surface, and (3) reduce the depth from which the emitted energy is detected.

Snow brightness temperature measurements made from space using the Scanning Multichannel Microwave Radiometer (SMMR) sensor are shown in Figure 6.16 [23]. SMMR contains passive radiometers that receive energy in horizontal and vertical polarizations at 6.6, 10.7, 18, 22, and 37 GHz. The scan of the 780-km

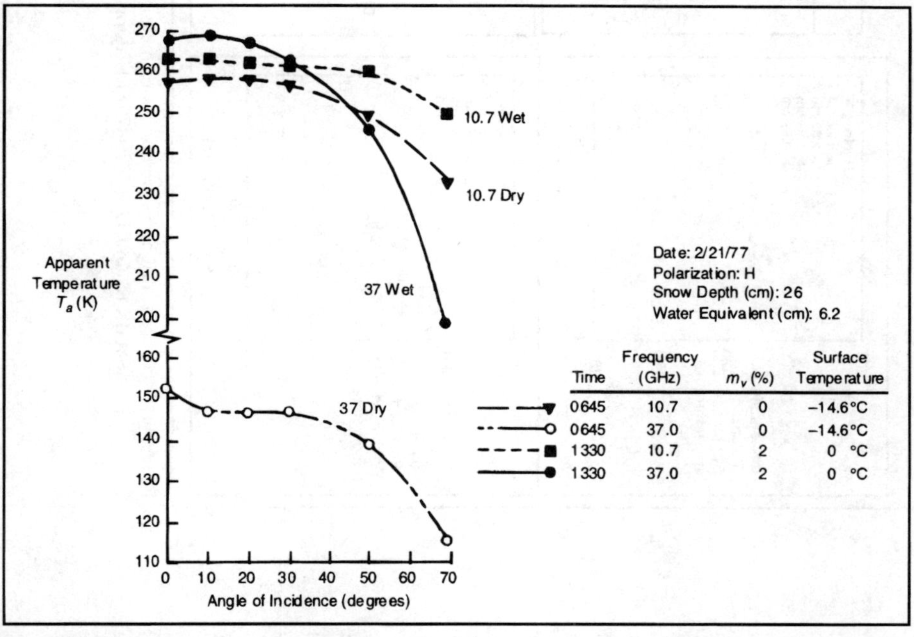

Figure 6.15 Response of apparent temperature of dry and wet snow to incidence angle as measured at 10.7 and 37 GHz [22].

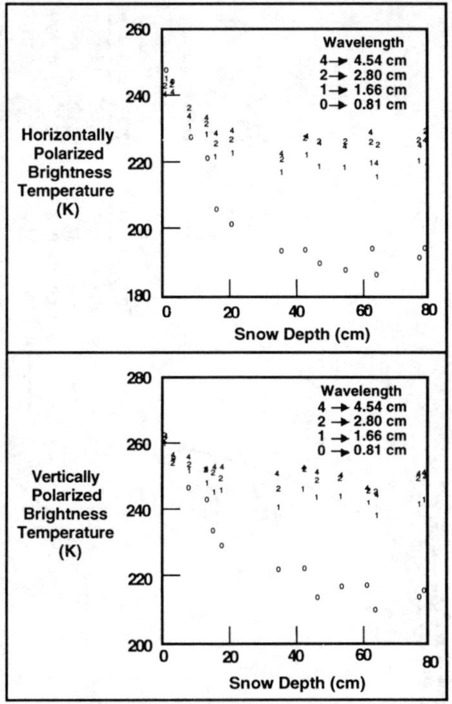

Brightness temperature differential, horizontal polarization, night pass to day pass for selected conditions for R02C08 grid point

Snow Surface Condition	CD	Snow Depth (cm)	Wavelength (cm) 0.81	1.66	2.80	4.54
Frozen Night, Frozen Day	60 60	78 78	+03	+03	+01	0
Frozen Night, Melted Day	70 72	77 66	-30	-13	-06	-03
Frozen Night, Melted Day	88 88	21 21	-27	-16	-07	-02
Frozen Night, Frozen Day	94 94	22 22	0	0	+03	0
Frozen Night, Melted Day	98 100	14 9	-20	-06	+05	+15
Melted Night, Melted Day	104 106	24 12	-01	+05	+12	+12

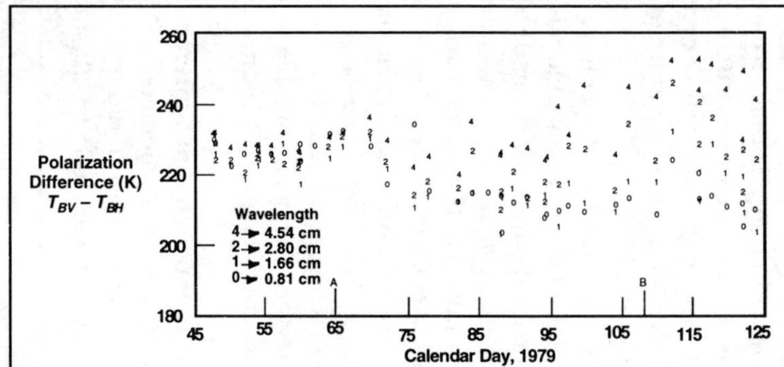

Figure 6.16 Diurnal brightness temperature measurements of snow at 6.6, 10.7, 18, and 37 GHz as made from space using the SMMR (*Source:* [23]. © 1987 IEEE. Reprinted with permission.)

wide swath is conical at a constant angle of 50 degrees with the Earth's surface at an orbital altitude of 955 km. The surface resolution elements are 148 × 95, 91 × 59, 55 × 41, 46 × 30, and 27 × 18 km, respectively, at the five frequencies.

Brightness temperature data at 6.6, 10.7, 18, and 37 GHz, corresponding to wavelengths of 4.54, 2.80, 1.66, and 0.81 cm, are given on the left of Figure 6.16 for horizontal and vertical polarization as a function of snow depth. These data were measured during the snow accumulation phase that occurred from calendar day (CD) 310 of 1978 to CD 60 of 1979, when maximum depth was observed. The major effect of the accumulating snow was to reduce the horizontally and vertically polarized 37-GHz brightness temperatures relative to other frequencies. The low emissivity of the dry snow at 37 GHz was attributed to scattering of the emitted energy. The difference in vertically and horizontally polarized measurements of brightness temperature, as the snow changed from dry to wet, is shown in the graph on the lower right. When the snow was dry, the polarization difference was a weak function of frequency remaining nearly constant between 21 and 27K. After the snow began to melt, the snow depth decreased and the polarization differences showed more dependence on frequency. The 37-GHz polarization difference decreased to less than 20K, while the 6.6-GHz value increased to 37K. The 37-GHz result was attributed to standing water in melt ponds lying on the snow surface that exhibited the strong polarization dependence characteristic of water.

The table on the upper right of the figure presents the difference between night and day measurements of horizontally polarized brightness temperatures. The frozen night and frozen day data show little brightness temperature difference. Most frozen night and melted day difference data show negative values, indicating larger daytime brightness temperatures corresponding to the wet snow. By CD 88, the melting phase was well underway as indicated by the decrease in snow depth from 78 to 21 cm. The sharp decrease in brightness temperature at 37 GHz was attributed to a significant decrease in scattering as the snow particles changed into liquid water. On CD 94, the snow was refrozen and exhibited night-to-day changes in brightness temperature that were similar to those of CD 60. On CDs 98 and 100, the frozen night and melted day data turn positive at the longer wavelengths as emission from the underlying ground is detected at the relatively small snow depth of 9 cm, and as the brightness temperature varies due to the difference in night and day surface temperatures. The latter explanations also apply to the CD 104 and 106 data. Here it is also more likely that liquid water was present in the snowpack as the minimum air temperature was 272K on CD 104.

The effects of surface temperature, snow particle radius, depth, wetness, and underlying surface characteristics on energy emission from snow are summarized in Table 6.7 [24]. To recapitulate, the apparent temperature of snow increases as snowpack water equivalent increases. With increased water equivalent, the apparent temperature is more sensitive to the surface properties of the snow, such as surface temperature and liquid water accumulation, and emission from deeper layers in the snow pack is not as readily detected. As the operating frequency of the radiometer decreases (i.e., longer wavelength energy is detected), the apparent temperature

Table 6.7
Effect of Snow Physical Characteristics on Energy Emission

Physical Characteristic	*Effect on Emission*
Physical temperature	• Increases linearly as temperature increases
Snow particle radius	• Decreases as radius increases
Depth	• Decreases as depth increases until saturation depth (a function of wavelength) is reached
Wetness	• Increases rapidly in presence of free water
Underlying surface	• Modifies emission for thin and dry snow layers as the emissivity of the underlying surface becomes dominant

becomes a stronger function of the bulk properties of the snowpack such as physical temperature and water equivalent. At higher frequencies, apparent temperature is influenced more by the surface properties of the snowpack. Emissivity, and hence apparent temperature, decrease as frequency increases.

6.8.4 Summary of Effects of Snow Wetness, Incidence Angle, and Frequency on the Active and Passive Signatures of Snow

Table 6.8 presents a summary of the influence of snow wetness, incidence angle, and observation frequency on the radar cross-section and apparent temperature signatures of snow. The effects on radar cross-section are from discussions presented in Chapter 4.

6.9 TARGET DETECTION IN NONMETAMORPHIC AND METAMORPHIC SNOW

Figures 6.17 and 6.18 show the results of a data collection experiment where a corner reflector target simulator was placed in snow and an integrated active and passive 35-GHz sensor was scanned over the scene while flying in a helicopter. The sensors shared the same antenna and most receiver components, so that their data were coregistered. The data processing illustrates the potential of a dual sensor and data fusion algorithm to detect a target in a high-clutter environment where a low-cost single passive or active sensor would typically experience difficulty in detecting the target with high probability and low false alarm.

The images in Figure 6.17 were processed from signature data taken in the afternoon when free water was present in the snow. Hence, the apparent temperature of the background was high and a large background-to-target contrast existed. When a threshold was applied to the raw passive data, much of the nontarget clutter was removed. Further processing resulted in additional clutter reduction by elimination of signals whose time duration did not correspond to the target geome-

Table 6.8
Radar Cross-Section and Apparent Temperature Dependence on Snow Wetness, Incidence Angle, and Frequency

Variable	*Radar Cross-Section*	*Apparent Temperature*
Wetness	• Decreases with increasing wetness	• Increases with increasing wetness
Incidence angle	• Weak sensitivity near nadir even when wetness varies • Ground plays major role at lower frequencies • As frequency increases, snowpack effects dominate, especially for wet snow	• Shows same diurnal behavior at angles between 0 degrees and 50 degrees
Frequency	*Parallel polarization* • Increases with frequency, faster for dry snow than for wet snow • Backscatter at 1 GHz dominated by ground contribution; snowpack effects are minor • As frequency increases, loss through snow increases, diminishing ground contribution while increasing snow backscatter contribution • Therefore, snowpack contribution is greater at higher frequencies particularly for wet snow	• Decreases with increasing frequency • Sensitivity to wetness is greater at higher frequencies
Frequency	*Cross-polarization* • Above 8 GHz the increase in backscatter levels off for wet and dry snow or, at most, slowly increases as frequency increases	• Not applicable

try and scan rate (pulse width discrimination). Similar processing of the active sensor data removed clutter and resulted in data that were correlated with the end product of the passive sensor processing. This noncoherent cross-sensor processing further eliminated false target detections in the final images.

When the early morning data were analyzed, as shown in Figure 6.18, both the active and passive sensor clutter were greater as predicted by the signature theory for metamorphic and dry snow. Although the threshold algorithm was effective in the passive sensor, it had limited usefulness in the active sensor. Pulse width discrimination was again effective for the passive data and of limited effectiveness for the active data. However, once again when the active and passive data were cross-correlated, the false targets were greatly reduced, enabling selection of the real target returns.

6.10 SUMMARY

Radiative transfer theory was introduced to explain how atmosphere and ground-based constituents give rise to passive signatures that are detected by radiometer

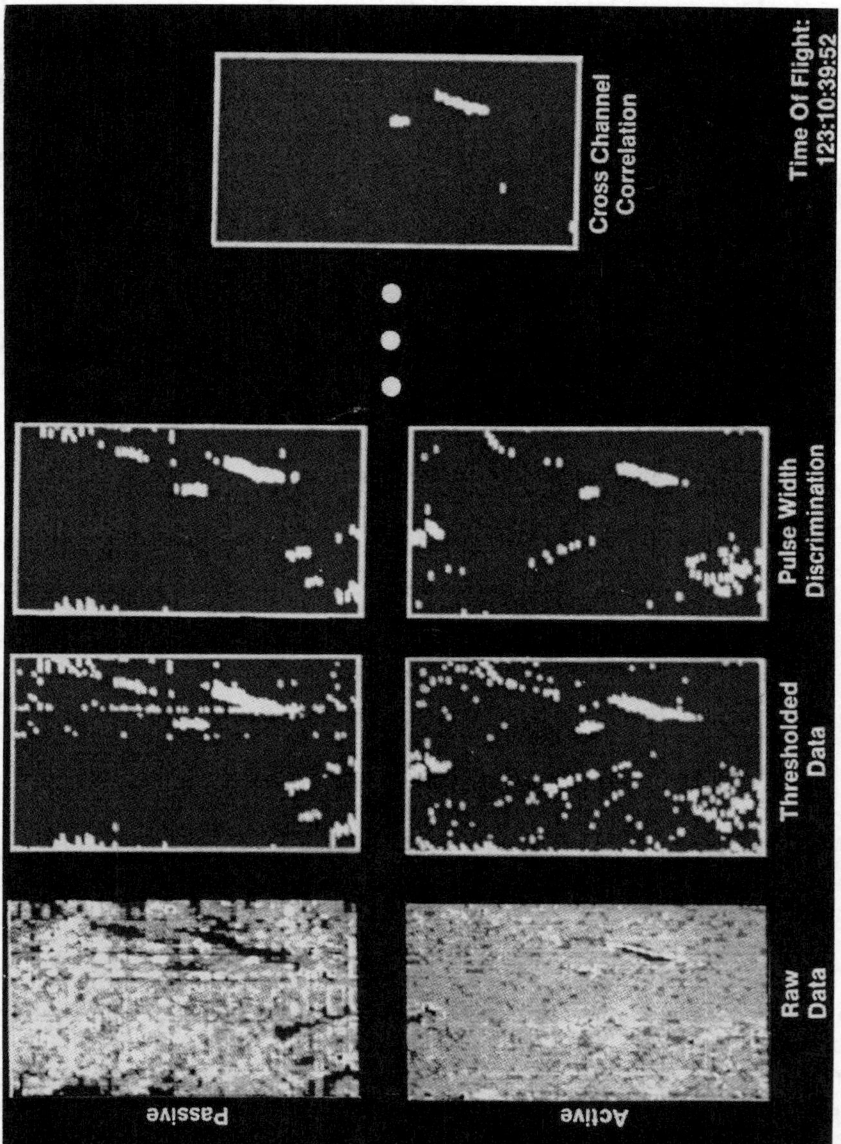

Figure 6.17 Target detection in wet (nonmetamorphic) snow using active and passive 35-GHz sensors (*Source:* Aerojet Corporation, Azusa, CA. Reprinted with permission.)

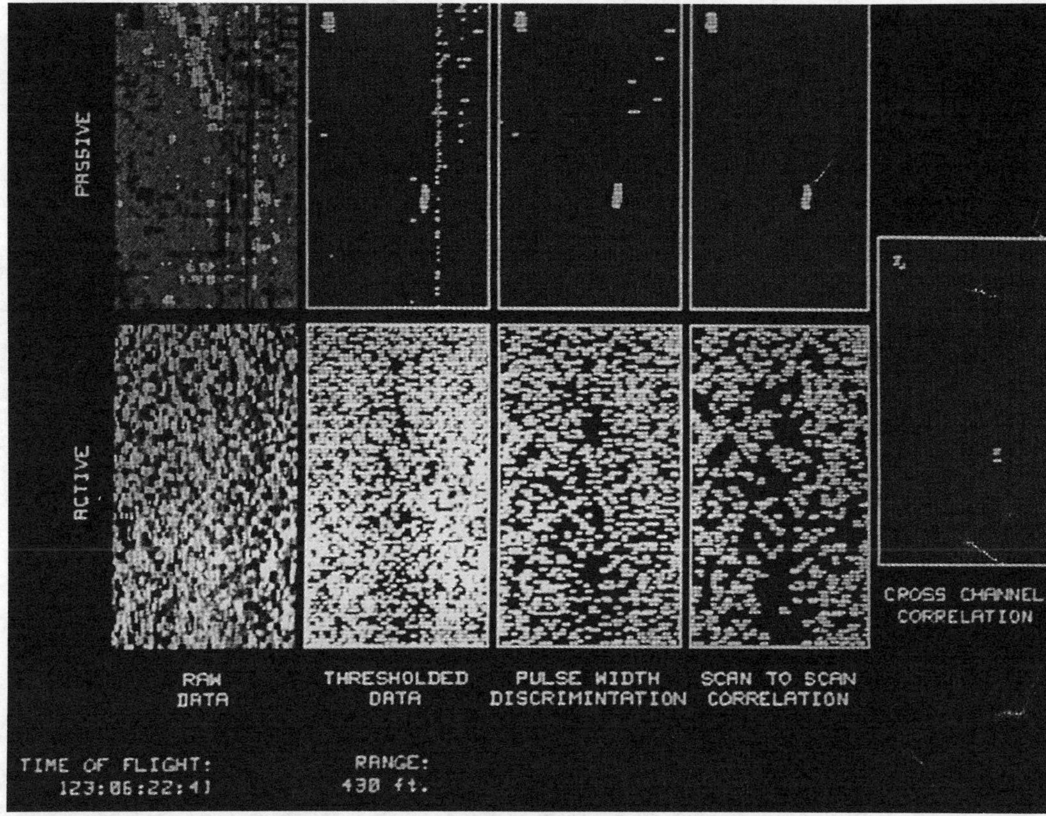

Figure 6.18 Target detection in dry (metamorphic) snow using active and passive 35-GHz sensors (*Source:* Aeroject Corporation, Azusa, CA. Reprinted with permission.)

sensors. In general, the passive signature has a component due to emission from the ground or target being detected and another due to energy emitted by the atmosphere that is reflected from the ground or target into the radiometer receiver. Ideally, a radiometer measures the brightness temperature of an object. However, real-world constraints such as antenna patterns that receive energy from areas other than the target of interest, lossy components in the receiver that emit energy of their own, and impedance mismatches that reflect receiver-generated noise back into the receiver all modify the signal measured by a radiometer. The modeling of these effects allows them to be eliminated (to within the ability to precisely measure or otherwise calculate the parameters needed for the model) from the equivalent system input noise temperature measurement in order to derive the true brightness temperature and emissivity of an object.

The total power radiometer finds use in military applications as an auxiliary sensor because of its relatively simple and low-cost hardware configuration, which allows it to locate predefined areas on a target suitable for munition or warhead impact. In applications where more accurate brightness temperature measurements are needed, Dicke and noise-injection radiometers are used. The minimum detectable temperature of a radiometer was shown to be inversely related to the predetection IF bandwidth and the integration time.

Most backgrounds that lack water show high emissivity over a large range of incidence angles. Water, because of its polar molecule, exhibits an emissivity having greater sensitivity to incidence angle and receiver polarization. Since snow may or may not contain free water, it has two signature states, one of low emissivity when it is dry and one of high emissivity when it is wet. In the low emissivity state the background brightness temperature is lower, reducing the target-to-background contrast temperature and possibly making metal target detection more difficult. The passive snow signature also decreases at higher frequencies. A summary of the dependence of the active (radar) and passive (radiometer) signatures on the free water content of the snow, incidence angle, frequency was presented. Both the active and passive signatures of the snow background make it more difficult to detect metal targets when the snow is dry: active because the backscatter increases and passive because the apparent temperature decreases.

References

[1] Ulaby, F. T., R. K. Moore, and A. K. Fung, *Microwave Remote Sensing: Active and Passive, Vol. I, Microwave Remote Sensing Fundamentals and Radiometry*, Norwood, MA: Artech House, 1981.

[2] Slobin, JPL Publication 81-46, *Microwave Noise Temperature and Attenuation of Clouds at Frequencies Below 50 GHz*, July 1981.

[3] Croswell, W. F., and M. C. Bailey, "Radiometer Antennas," Chapter 8 in *Antenna Applications Reference Guide*, R. C. Johnson, and H. Jasik, (Eds.), New York, NY: McGraw-Hill, 1987.

[4] Ruf, C. S., S. J. Keihm, and M. A. Janssen, "TOPEX/Poseidon Microwave Radiometer (TMR): I. Instrument Description and Antenna Temperature Calibration," *IEEE Trans. on Geoscience and Remote Sensing*, Vol. GE-33, No. 1, Jan. 1995, pp. 125–137.

[5] Callahan, P. S., *TOPEX Ground System Science Algorithm Specification*, Doc. D-7075, Rev. A, Jet Propulsion Laboratory, Pasadena, CA, April 1991.

[6] Tiuri, M. E., "Radio Astronomy Receivers," *IEEE Trans. on Military Electronics*, Vol. MIL-8, July-Oct. 1964, pp. 264–272.

[7] Roeder, R. S., R. E. Wilt, and M. W. Milsted, "Target Detection by Millimeter Wave Radiometry," ARPA Millimeter Wave Conference, Nov. 1977, pp. 140–150.

[8] Stelzried, C. T., "Temperature Calibration of Microwave Thermal Noise Sources," *IEEE Trans. on Microwave Theory and Techniques*, Jan. 1965, pp. 128–130.

[9] Singer A., R. Ulrich, and E. Naess, "Thermal Calibrators in Millimeter-Wave Radiometry A Word of Caution," *Proc. IEEE*, June 1997, pp. 1094–1096.

[10] Hardy, W. N., "Precision Temperature Reference for Microwave Radiometry," *IEEE Trans. on Microwave Theory and Techniques*, Mar. 1973, pp. 149–150.

[11] Ulaby, F. T., C. T. Allen, and A. K. Fung, "Method for Retrieving the True Backscatter Coefficient from Measurements with a Real Antenna," *IEEE Trans. on Geoscience and Remote Sensing*, Vol. GE-21, No. 3, July 1983, pp. 308–313.

[12] Evans, G., and C. W. McLeish, *RF Radiometer Handbook*, Artech House, Norwood, MA, 1977.

[13] Debye, P., *Polar Molecules*, New York, NY: Dover reprint, originally published by Reinhold Publishing Corp., 1929.

[14] Cole, K. S., and R. H. Cole, "Dispersion and Absorption in Dielectrics," *J. Chem. Phys.*, Vol. 9, 1941, pp. 341–351.

[15] Stogryn, A., "The Apparent Temperature of the Sea at Microwave Frequencies," *IEEE Trans. Ant. and Prop.*, Vol. AP-15, 1967, pp. 278–286.

[16] Weyl, P., "On the Change in Electrical Conductance of Seawater with Temperature," *Limnol. Oceanogr.*, Vol. 9, 1964, pp. 75–78.

[17] Klein, L. A., and C. T. Swift, "An Improved Model for the Dielectric Constant of Sea Water at Microwave Frequencies," *IEEE Trans. on Antennas and Propagation*, Vol. AP-25, No. 1, Jan. 1977, pp. 104–111. Also appears in *IEEE Journal of Oceanic Engineering*, Vol. OE-2, No. 1, Jan. 1977, pp. 104–111.

[18] Wentz, F. J., "Measurement of Oceanic Wind Vector Using Satellite Microwave Radiometers," *IEEE Trans. Geosci. and Remote Sensing*, Vol. 30, 1992, pp. 960–972.

[19] Keihm, S. J., M. A. Janssen, and C. S. Ruf, "TOPEX/Poseidon Microwave Radiometer (TMR): III. Wet Troposphere Range Correction Algorithm and Pre-Launch Error Budget," *IEEE Trans. on Geosci. and Remote Sensing*, Vol. 33, Jan. 1995, pp. 147–161.

[20] Edgerton, A. T., et al., *Microwave Emission Characteristics of Natural Materials and the Environment*, Report 9016R-8, Aerojet ElectroSystems Company, Prepared for Geography Branch, Earth Science Division, Office of Naval Research, Arlington VA, Reprinted Oct. 1973.

[21] Peake, W. H., and T. L. Oliver, *The Response of Terrestrial Surfaces at Microwave Frequencies*, Report AFAL-TR-70-301, ElectroScience Laboratory, Ohio State University, May 1971.

[22] Ulaby, F. T., and W. H. Stiles, *The Active and Passive Microwave Response to Snow Parameters*, Report RSL 340-2, Remote Sensing Laboratory, Univ. of Kansas Center for Research, Dec. 1978.

[23] McFarland, M. J., G. D. Wilke, and P. H. Harder, II, "Nimbus 7 SMMR Investigation of Snowpack Properties in the Northern Great Plains for the Winter of 1978–1979," *IEEE Trans. on Geoscience and Remote Sensing*, Vol. GE-25, No. 1, Jan. 1987, pp. 35–46.

[24] Rango, A., "Progress in Snow Hydrology Remote-Sensing Research," *IEEE Trans. on Geoscience and Remote Sensing*, Vol. GE-24, No. 1, Jan. 1986, pp. 47–53.

CHAPTER 7

Passive Infrared Sensors

Passive infrared (IR) sensors are used to detect energy emitted from targets, backgrounds, and the atmosphere in applications that include military surveillance, missile guidance and target detection, crop and forest management, and weather forecasting. One type of infrared sensor, the forward-looking infrared (FLIR), provides high-resolution real-time imagery of a scene that has been historically displayed on a monitor. Multiple detector arrays composed of pixels (picture elements) having a small instantaneous field of view are used in FLIRs in support of security and surveillance, target detection and classification, target aimpoint selection, and false target and countermeasure rejection. Signal processing that identifies features contained in the multiple pixel image is often used to automate the target recognition process. Another type of sensor, the infrared search and track (IRST), locates and tracks objects by capturing the energy emitted within the field of view of the sensor. IRST sensors are designed to operate with a small noise-equivalent temperature difference so that small target-to-background contrast temperatures can be detected. Therefore, their performance is enhanced with designs that incorporate a large instantaneous field of view. Some airborne IRSTs often scan a 360-degree hemisphere in azimuth and 50 to 90 degrees in elevation. Therefore, their total search area can be much larger than for FLIRs. The large number of scene pixels encountered by an IRST device requires computer-controlled data processing to separate the large number of false targets from the true targets. A third type of infrared sensor is the infrared radiometer. It combines the functional attributes of a FLIR and an IRST sensor to provide imagery of a scene and also operate with a small noise-equivalent temperature difference in order to better identify regions with small differences in temperature that may exist on an object. By using filters to channelize the received energy into narrow wavelength bands, weather phenomena such as the altitude of cloud top heights can be determined from radiometers onboard space-based satellites. Still other space-based hyperspectral radiometer sensors divide the visible and infrared spectrum into hundreds of 10-nm to 20-nm narrow spectral bands. These sensors employ signal processing to combine the narrow spectral band data to aid in identification of geologic, ocean, soil, vegetation, atmospheric, and snow/ice boundary features.

This chapter discusses the design of imaging and nonimaging passive infrared sensors and the figures of merit commonly used to characterize their performance. The Johnson criteria for object discrimination in terms of sensor resolution and signal-to-noise ratio (S/N) are reviewed and compared with the results of other studies. Methods are presented for calculating the detection probability of infrared sensors that operate in direct detection and heterodyne configurations to show the effects of different photon arrival statistics and noise generation mechanisms on the results. The chapter concludes by describing a general five-step process for analyzing multipixel imagery for automatic target recognition applications. The process is illustrated with specific algorithms for each of the five steps.

7.1 PASSIVE INFRARED SENSOR DESIGN ISSUES

The design of a passive IR sensor is constrained by the goals and scenario under which the sensor operates as described in Figure 7.1. Design goals address the use of imaging or nonimaging sensors; the amount of target discrimination to be performed (i.e., detection, orientation, classification, or identification); offline versus real-time signal processing; sensor operating life, size, weight, and maintenance requirements; and tracking or nontracking capability. Specification of the operating scenario includes the identification of the ambient environment, detectable object characteristics, background features, geometrical constraints, and the controlling dynamics as dictated by the velocity and acceleration of the sensor, scanning versus staring array design, and the motion of the detectable objects of interest [1].

Once the system constraints are understood, the detailed sensor design can proceed. Generally, the spectral band is chosen in a wavelength region where a strong target signature exists, although consideration must also be given to interfering background and atmospheric radiation and absorption that may be present. Infrared transmission bands that have relatively small atmospheric attenuation were shown in Figure 3.7. They include the short-wavelength infrared (SWIR) typically extending from 0.87 to 1.5 μm, mid-wavelength infrared (MWIR) from 3 to 5 μm, and long-wavelength infrared (LWIR) from 8 to ≥12 μm. Focal plane detector selection is usually based on the specific detectivity of the material (which is inversely proportional to noise-equivalent power), cooling temperature required to achieve a satisfactory signal-to-noise ratio, cost of the array material versus value of the target to be observed, and the number and size of the pixels in the array needed to meet the resolution, spatial frequency, and field of view requirements of the mission. Signal processing is driven by the level of target discrimination required and the available computing resources. As target discrimination increases from detection through identification, more pixels must intercept the target or the signal-to-noise ratio must be increased to keep the probability of correct discrimination the same. The system design process is an iterative one where the goals are compared

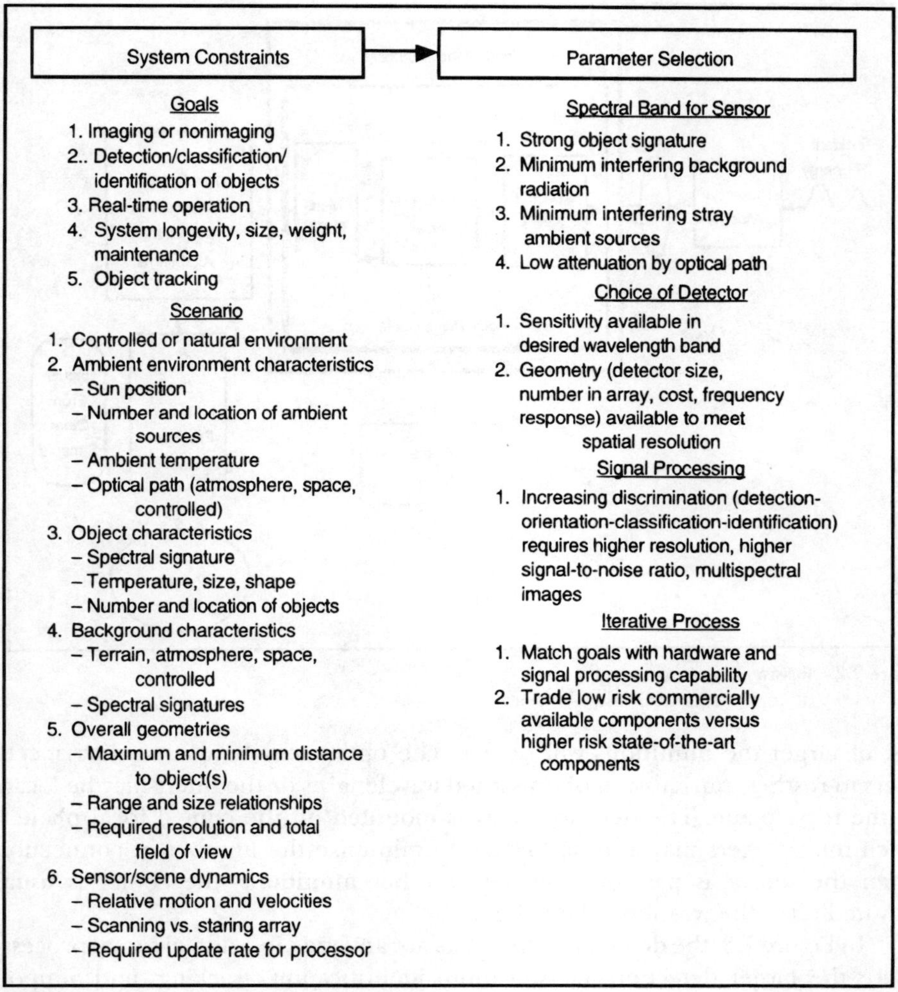

Figure 7.1 Passive IR sensor design issues.

with the performance of the hardware and signal processing and the sensor cost. If all goals cannot be satisfied with a given design, then either the goals are modified or the design is improved by modifying the optics, detector selection, signal processing, and scan pattern, and through the use of multiple sensor systems.

In the generalized IR sensor block diagram of Figure 7.2, incident IR energy is shown entering through a window on the left. The window material is selected to transmit the wavelengths of interest and withstand the heat, shock, and vibration that may be generated during delivery and operation of the sensor over the target area. The optics consist of one or more lenses that focus the energy onto the focal plane. In a fire-and-forget munitions application, the number of optical elements is partially determined by the impact on the cost of the sensor versus the type and

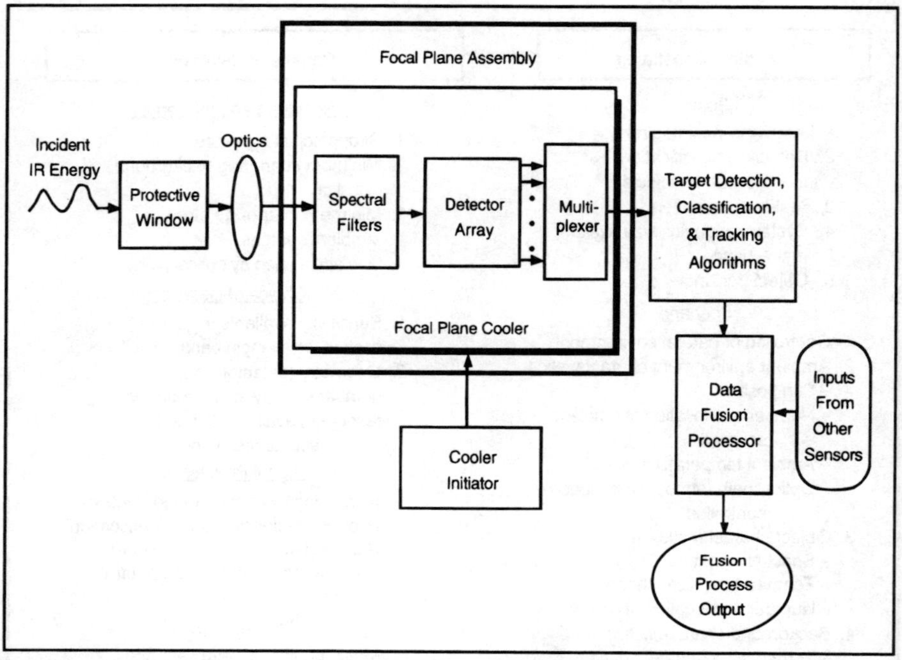

Figure 7.2 Passive IR transducer.

cost of target the munitions will defeat. The optics may be coated with spectral filters to restrict transmission of unwanted wavelengths or the filters may be located on the focal plane. The detector array is mounted on the cooled focal plane on which multiplexers may also be located to minimize the number of connections. When the sensor is part of a missile or other munitions, the cooler is usually activated after the weapon is launched.

In Figure 7.2, the detector output data are analyzed in a digital signal processor where the target detection, classification, identification, tracking, and aimpoint information are extracted. When used as part of a multiple-sensor system, processed IR sensor data may be combined with data from other sensors, as in a data fusion algorithm, to improve the detection probability, classification, or selection of the munitions impact point.

7.2 SCENE-IMAGING TECHNIQUES

A scene can be imaged in a number of ways. A single detector can be moved in a raster or rosette scan as illustrated in the left of Figure 7.3. A linear array of detectors can be scanned in a pushbroom or circular fashion as shown in the middle, and arrays of detectors can be used in a staring mode as illustrated on the right. Lower cost applications generally use linear arrays with smaller numbers of detectors.

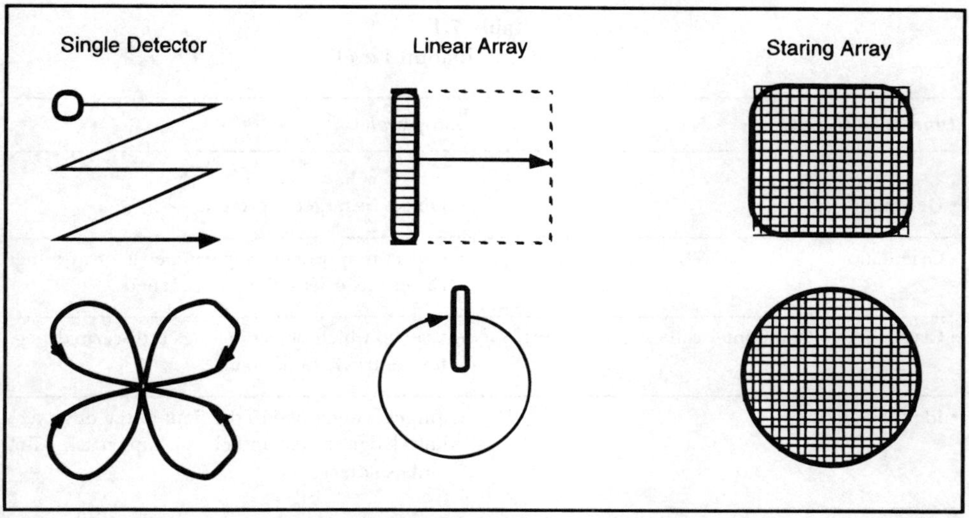

Figure 7.3 Developing an image from a single detector element, linear array, and a staring array.

They create two-dimensional images by incorporating the scanning motion of the munitions into the image-generation process. Surveillance and tracking applications requiring higher resolution, such as air-to-ground and air-to-air combat, use staring arrays to create high-resolution imagery.

7.3 OBJECT DISCRIMINATION LEVELS

Johnson [2] performed some of the first investigations to determine the number of pixels that must be resolved on a target in order for a human observer to know either that it is present (detection), its symmetry and orientation (orientation), the class to which it belongs (classification), or its description up to the limit of the observer's knowledge (identification). These discrimination levels and their interpretations are summarized in Table 7.1. The ability of the observer to interpret an image in these quantum steps is a function of the contrast and brightness of the object, properties of the specific intensifier device used to enhance the image, and the physiological response of the human eye-brain interface. Contrast is a function of the image signal-to-noise ratio, while brightness is determined by the luminance of the target.

In the experimental procedure used by Johnson, the ability of a human observer to discern a bar pattern of a given resolution was correlated with the discrimination level (detection, orientation, classification, or identification) for an object. The range was increased until the object was just barely detected, or its orientation determined, classified, or identified by the observer. Then a U.S. Air Force standard tri-bar chart, scaled to the image, was placed in the field of view. The standard tri-bar chart contains resolutions from 0.25 to 14.3 lines per mm.

Table 7.1
Object Discrimination Levels

Discrimination Level	*Interpretation*
• Detection	• Object or target is present
• Orientation	• Object is approximately symmetric or asymmetric and its orientation is discerned
• Classification (sometimes called recognition)	• Class to which object belongs is discerned (e.g., house, truck, tank, man)
• Identification	• Target is described to the limit of the observer's knowledge (e.g., motel, pickup truck, T-62 tank, soldier)

The number of resolved line pairs per foot (1 line pair equals 2 television lines or two pixels) that the observer needed to perform each level of discrimination was recorded. This value varied with the dimensions of the targets or objects. When the number of resolved line pairs was referenced to the minimum dimension of the target, as indicated on the left of Figure 7.4, the variability was reduced such that the maximum difference in the number of line pairs needed at all levels of discrimination among the test targets was not more than 25%. The resolution required for discrimination was therefore set equal to the number of dark and light line pairs subtended by the object's minimum dimension. Johnson's experiments were limited to broadside views of the targets. He also noted that the bar pattern must have the same contrast as the internal contrast of features in the target. While the level of discrimination of the targets themselves is range-dependent, the

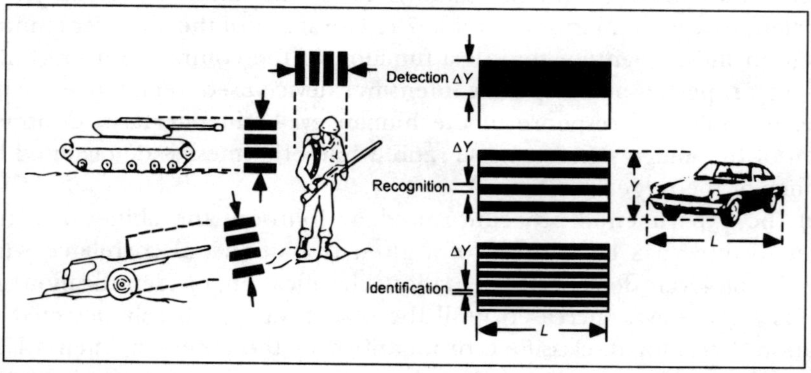

Figure 7.4 Correlation of a target object's minimum dimension with the spatial frequency of a bar pattern.

discrimination level represented by the given number of resolved line pairs is independent of distance. Higher target discrimination levels occur at closer ranges for a given sensor design and object-to-background contrast.

The results of Johnson's experiments are shown in Table 7.2. On the average, 2 pixels (1 line pair) are required to be resolved within an object's minimum dimension for detection, 2.8 pixels (1.4 line pairs) for orientation, 8.0 pixels (4.0 line pairs) for classification, and 12.8 pixels (6.4 line pairs) for identification. The increased number of line pairs needed for each higher discrimination level is illustrated on the right side of Figure 7.4. Johnson's original findings apply to first-generation FLIR sensors that incorporate one-dimensional detector arrays and a scene-scanning mechanism. Two-dimensional arrays require fewer pixels to be resolved over the target area, as will be discussed later in Section 7.6. The number of line pairs required for a particular discrimination level assumes that adequate image signal-to-noise ratio (image contrast) exists. Even when these conditions are met, only a 50% target discrimination probability is guaranteed by the numbers of line pairs in Table 7.2. To increase the probability, either sensor resolution (the number of pixels that subtends the minimum dimension of the target), or the signal-to-noise ratio has to be increased.

Johnson later extended his work to relate any value for the probability of discrimination to the number of line pairs that can be resolved across the minimum dimension of a target. He defined the target transform probability function (TTPF) that correlates the discrimination probability to the ratio of the actual number of line pairs resolved across the target's minimum dimension to the standard number given by the Johnson criteria in Table 7.2 [3–5]. A plot of the TTPF is shown in Figure 7.5 as the desired discrimination probability versus the ratio of the number of resolvable cycles across the minimum dimension of the target to the standard number.

Table 7.2
Johnson's Object Discrimination Criteria for One-Dimensional Arrays [2]

Target Broadside View	*Resolution per Minimum Dimension in Line Pairs*			
	Detection	*Orientation*	*Classification*	*Identification*
Truck	0.90	1.25	4.5	8.0
M-48 tank	0.75	1.20	3.5	7.0
Stalin tank	0.75	1.20	3.3	6.0
Centurion tank	0.75	1.20	3.5	6.0
Half track	1.00	1.50	4.0	5.0
Jeep	1.20	1.50	4.5	5.5
Command car	1.20	1.50	4.3	5.5
Soldier (standing)	1.50	1.80	3.8	8.0
105 howitzer	1.00	1.50	4.8	6.0
Average	1.0 ± 0.25	1.4 ± 0.35	4.0 ± 0.8	6.4 ± 1.5

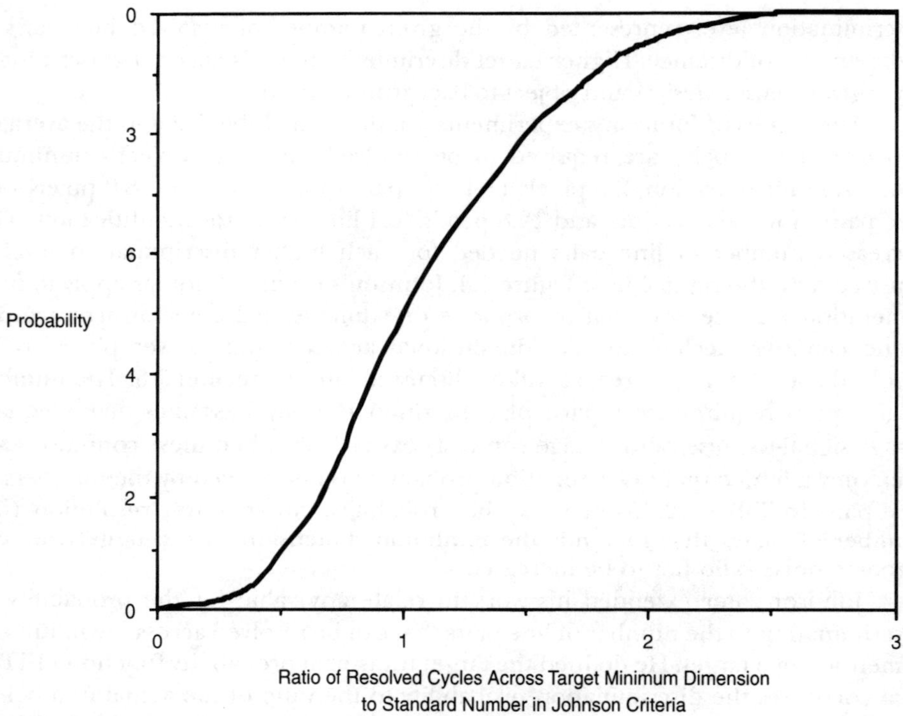

Figure 7.5 Target transform probability function.

Consider the following example of how to use Johnson's criteria to calculate the resolution required to detect and classify an M-48 tank. Table 7.2 specifies that 1.50 pixels (0.75 line pairs) are needed across the minimum (height) dimension for detection. If the height is 3.3 m, the resolution of each pixel must be 3.3 m/1.50 = 2.2 m. For classification, 7.0 pixels (3.5 line pairs) are required, implying a pixel resolution of 3.3 m/7 = 0.47 m.

Rosell and Willson [6] have confirmed Johnson's results for object detection, classification, and identification, and Schmieder and Weathersby [7] have confirmed the results for detection. Rosell and Willson superimposed transparencies of vehicles on transparencies of one of four background regions, namely (1) a uniform white background, (2) a road, (3) grass, and (4) grass among trees. The vehicle images used in the classification experiments were of a tank, a van truck, a halftrack with top-mounted radar antenna, and a tracked bulldozer with a derrick. The vehicle images used in the identification experiments were of M-47 Patton, M-48 Centurion, Panther, and Stalin tanks. The transparencies of the vehicles were made from photographs taken at a 45-degree depression angle, perpendicular to the vehicle's longitudinal axis, such that the sides and tops of the vehicles were imaged. The vehicle types and video signal-to-noise ratio were randomly varied and the probabilities of classification and identification were determined.

Table 7.3 shows that target classification becomes more challenging with added background clutter typical of a realistic environment. On the average, a 30% increase in image signal-to-noise ratio was needed before the observers could classify the targets in the clutter backgrounds as compared to the uniform white background with the same sensor resolution. In the uniform background, 1 pixel was required across the target's minimum dimension for *detection* at the 50% probability level when the image signal-to-noise ratio was 2.8. Eight pixels were needed across the target's minimum dimension for *classification* when the image signal-to-noise ratio was 2.5. Thirteen pixels were needed for *identification* when the image signal-to-noise ratio was 3.0. These results are in general agreement with Johnson's. In the clutter background, 2 pixels were required for detection with an image signal-to-noise ratio of 2.5. For classification in clutter, 8 pixels were required (as in the uniform background), but with an increased image signal-to-noise ratio of 3.4.

The results of the Rosell and Willson classification experiments were replotted by Gordon and others [8], as in Figure 7.6. This illustrates the probability of classification as a function of signal-to-noise ratio with a constant 4 line pairs (8 pixels) per minimum object dimension, and as a function of the number of line pairs with the signal-to-noise ratio fixed at 2.8. The dashed curve with the signal-to-noise ratio variable shows that the 50% probability of classification occurs with a relatively small signal-to-noise ratio of 2.8 and 4 line pairs resolved across the target's minimum dimension. The solid curve indicates that approximately 7.5 line pairs are required to achieve a 90% probability of classification at the same signal-to-noise ratio.

The Schmieder and Weathersby detection results for targets viewed in synthetically generated clutter are shown in Table 7.4. In this experiment, observers were placed in front of a television monitor and had unlimited viewing time to detect the targets. Tree, road, and grass clutter were synthetically generated in the 8 to 12-μm wavelength band. Three different-shaped targets were used: a Soviet T-10 tank, a U.S. Army jeep with canvas-enclosed cab, and a U.S. Army 2-1/2 ton truck with a canvas top. The signal-to-clutter ratio was varied by adjusting the target signal relative to a given set of clutter statistics in a given scene. Target and scene resolution were controlled using a two-dimensional Gaussian filter as a point spread function.

Table 7.3

Increase in Image Signal-to-Noise Ratio Required for Classification in Clutter as Compared With a Vehicle in a Uniform White Background

Background	*Increase in Image Signal-to-Noise Ratio*
• Road	15%
• Grass	24%
• Grass among trees	52%
• Clutter average	30%

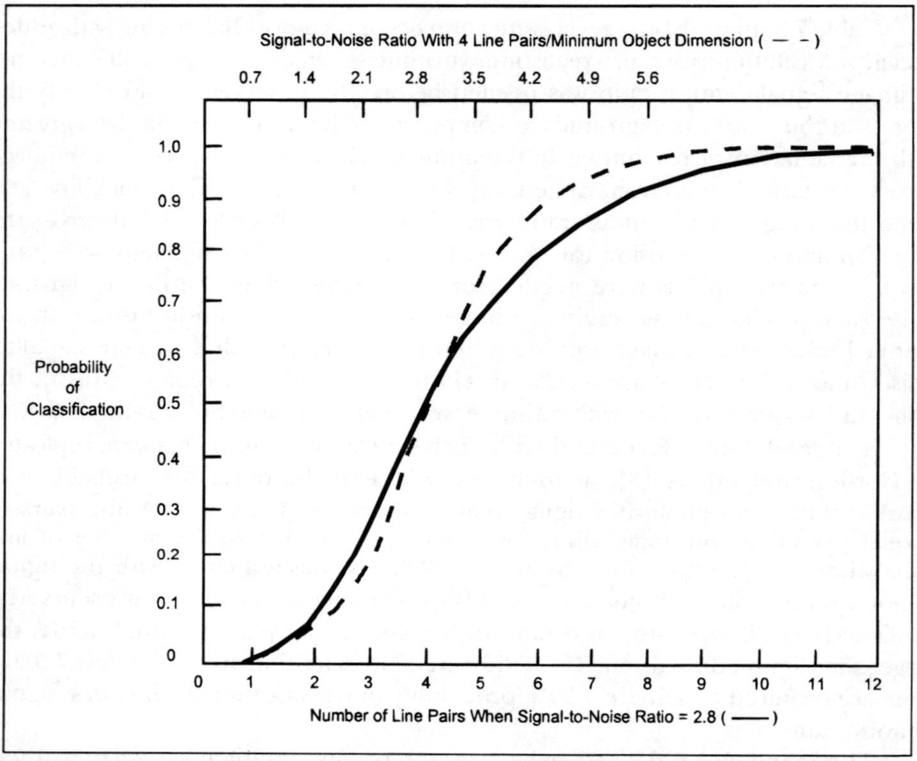

Figure 7.6 Probability of classification as a function of signal-to-noise ratio and resolution. (*Source:* [8]. Reprinted with permission.)

Table 7.4
Detection Criteria for Broad Categories of Clutter [7]

	Number of Line Pairs/Target		
Detection Probability	*Low Clutter* (*S*/*C* > 10)	Moderate Clutter (1 ≤ *S*/*C* ≤ 10)	High Clutter (*S*/*C* < 1)
1.0	1.7	2.8	—
0.95	1.0	1.9	—
0.90	0.9	1.7	7.0 (est.)
0.80	0.75	1.3	5.0
0.50	0.5	1.0	2.5
0.30	0.3	0.75	2.0
0.10	0.15	0.35	1.4
0.02	0.05	0.1	1.0
0.00	0.0	0.0	0.0

The signal-to-clutter ratio (S/C) in [7] is defined for positive contrast targets as

$$S/C = \frac{\text{Maximum target radiance} - \text{Mean background radiance}}{\text{rms clutter radiance}} \tag{7.1}$$

and for negative contrast targets as

$$S/C = \frac{|\text{Maximum target radiance} - \text{Mean background radiance}|}{\text{rms clutter radiance}} \tag{7.2}$$

The rms clutter radiance is computed as the average of the standard deviations σ_i of the radiance of the N clutter cells in the scene. Thus,

$$\text{rms clutter radiance} = \left(\sum_{i=1}^{N} \frac{\sigma_i^2}{N}\right)^{1/2} \tag{7.3}$$

The data in the table demonstrate that as the signal-to-clutter ratio increases, fewer line pairs need be resolved on the target to achieve a given detection probability. The Johnson results that require an average of one line pair to achieve a detection probability of 0.5 are confirmed for a broad range of signal-to-clutter ratios between 1 and 10.

7.4 ENERGY EMISSION BY BLACKBODY OBJECTS IN THE INFRARED SPECTRUM

Before discussing the performance measures used to characterize the operation of IR sensors, we will describe the emission of energy by objects that leads to their detection by the sensors. The emissive signature of objects is governed by the Planck radiation law. It describes the energy radiated at all wavelengths by blackbody (ideal emitter) objects with a temperature greater than absolute zero. In infrared applications, the emitted radiation may be calculated in terms of the spectral radiant excitance $M(\lambda, T)$ as a function of the wavelength at which the energy is measured and the temperature of the blackbody. In radiometric units of Watts per square centimeter of area and micrometer of wavelength, $M(\lambda, T)$ is a measure of the radiant flux or power density emitted by a surface and is given by [1]

$$M(\lambda, T) = \frac{2\pi hc^2}{\lambda^5 (e^{hc/\lambda k_B T} - 1)} \frac{\text{W}}{\text{cm}^2 \mu\text{m}} \tag{7.4}$$

where

λ = emitted wavelength in μm;
T = absolute temperature of the blackbody in K;

h = Planck's constant = $6.6256176 \times 10^{-34}$ W sec^2 or J sec;
c = speed of light = 3×10^{10} cm/sec;
k_B = Boltzmann's constant = 1.380662×10^{-23} J/K.

The total flux density over a wavelength interval λ_1 to λ_2 is obtained by integrating (7.4) over the interval as

$$\int_{\lambda_1}^{\lambda_2} M(T)\, d\lambda = \int_{\lambda_1}^{\lambda_2} \frac{2\pi hc^2}{\lambda^5(e^{hc/\lambda k_B T} - 1)}\, d\lambda\, \frac{\text{W}}{\text{cm}^2} \tag{7.5}$$

Substituting

$$c = \lambda f \tag{7.6}$$

and

$$d\lambda = \frac{-\lambda^2\, df}{c} \tag{7.7}$$

allows the emitted power density given by (7.4) to be rewritten as a function of frequency in hertz as

$$M(f, T) = \frac{2\pi hf^3}{c^2(e^{hf/k_B T} - 1)}\, \frac{\text{W}}{\text{cm}^2\text{Hz}} \tag{7.8}$$

The minus sign in (7.7) reflects the usual inverse relation between wavelength and frequency.

The excitance expressed by (7.4) is plotted in Figure 7.7 for several blackbody temperatures. Cooler objects such as those at room temperature maximize their emitted power in the far infrared as illustrated in the figure. As the temperature of the object increases, the emitted power peaks at shorter wavelengths.

For a given blackbody temperature, the wavelength at which maximum power is emitted is found by differentiating (7.4) with respect to wavelength. This operation leads to

$$e^{hc/\lambda_m k_B T}\left(1 - \frac{hc}{5\lambda_m k_B T}\right) = 1 \tag{7.9}$$

One solution of (7.9) is $hc/\lambda_m k_B T = 0$. But this implies that $\lambda_m T$ equals infinity and hence is of no practical interest. Returning to (7.9) and solving for its other solution leads to

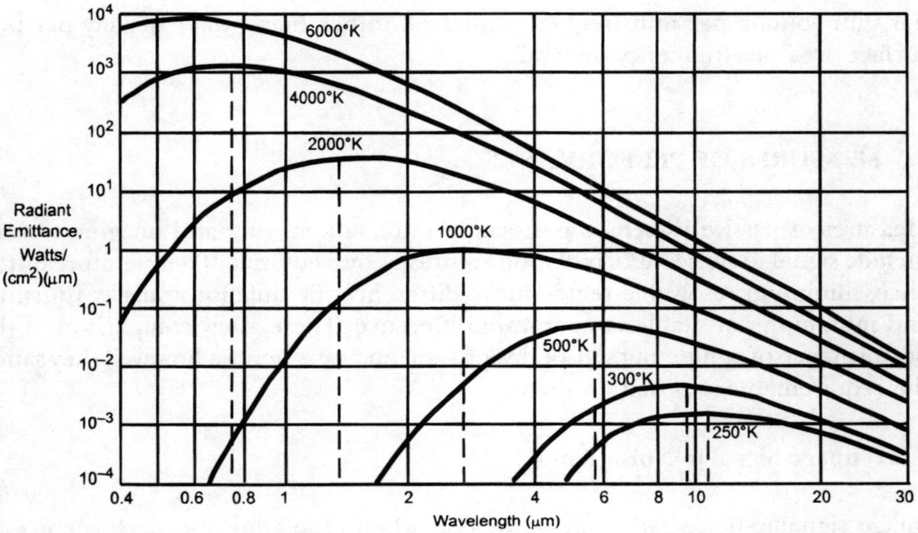

Figure 7.7 Blackbody emittance as given by Planck's radiation law. Vertical dashed lines mark wavelength of maximum energy emission for the indicated blackbody temperature.

$$\frac{hc}{k_B\lambda_m T} = 4.965 \tag{7.10}$$

Upon inserting the values for h, c, and k_B, we get Wein's displacement law for the wavelength at which the maximum power density occurs as

$$\lambda_m T = 2{,}897.8\ \mu\text{m K} \tag{7.11}$$

Thus signatures from hot exhaust gases from airplanes and ground vehicles peak in the near infrared, while thermal emission from sunlight-loaded surfaces and surfaces indirectly heated by hot internal combustion engines peak in the far infrared.

In the MMW spectrum where

$$hf << k_B T \tag{7.12}$$

the emitted power density is given by the Rayleigh-Jeans law as

$$M(f, T) = \frac{2\pi f^2 k_B T}{c^2} \frac{\text{W}}{\text{cm}^2\,\text{Hz}} \tag{7.13}$$

This expression was explored further in Chapter 6 in conjunction with the application of space-based microwave and millimeter-wave radiometers to retrieval of atmospheric data. There, the units of the emitted radiation were given in terms of energy per unit volume per unit frequency interval, rather than power density per unit surface area per frequency interval.

7.5 MEASURES OF PERFORMANCE

Measures of passive IR sensor performance for nonimaging and imaging sensors include signal-to-noise ratio, radiation contrast, noise-equivalent temperature difference, minimum resolvable temperature difference, modulation transfer function, and minimum detectable temperature difference. These allow comparison of the performance of competing sensor designs and an assessment of how well they satisfy the requirements imposed on them.

7.5.1 Image Signal-to-Noise Ratio

Image signal-to-noise ratio can be helpful when evaluating the performance of imaging sensors that use a human observer because it accounts for the eye response of the observer and the size of the visual display used to image the scene of interest relative to the area of the detected object. The image signal-to-noise ratio $(S/N)_I$ increases directly as the area of the detected image and inversely as the total area of the display as shown by [6]

$$(S/N)_I = [2t_i(\Delta F)(a/A)]^{1/2}(S/N)_V \tag{7.14}$$

where

t_i = integration time of the eye for a given display brightness in sec;

ΔF = video noise bandwidth in Hz;

a = area of the detected image;

A = total area of the display;

$(S/N)_V$ = video signal-to-noise ratio $= \Delta T/\text{NE}\Delta\text{T}$ (7.15)

ΔT = object-to-background contrast temperature;

NEΔT = noise-equivalent temperature difference.

7.5.2 Video Signal-to-Noise Ratio

Video signal-to-noise ratio, more commonly referred to as simply the signal-to-noise ratio S/N, is an important performance measure for search and track sensors. These devices are designed to locate regions of high thermal emission rather than to produce scene imagery. The S/N is expressed as [9]

$$S/N = \frac{K_e D_o^2}{4f}\left(\frac{\text{IFOV}}{\Delta F}\right)^{1/2} \int_{\lambda_L}^{\lambda_U} \tau_o(\lambda)\tau_a(\lambda)\epsilon(\lambda)\,\frac{\partial W_b(\lambda, T)}{\partial T}\,D_d^*(\lambda)\,d\lambda\,\Delta T \quad (7.16)$$

where

K_e	=	electrical efficiency (function of filter implementation, center frequency, and bandpass; telescope and detector spatial filtering; and scan velocity);
D_o	=	optical clear aperture;
f	=	focal length of optics;
IFOV	=	instantaneous field of view (solid angle subtended by one detector);
ΔF	=	electrical noise bandwidth;
$\tau_o(\lambda)$	=	optical spectral transmission;
$\tau_a(\lambda)$	=	atmospheric spectral transmission;
$\epsilon(\lambda)$	=	target emissivity;
$W_b(\lambda, T)$	=	blackbody spectral emittance;
T	=	absolute temperature of emitting body;
ΔT	=	target-to-background contrast temperature;
$D_d^*(\lambda)$	=	specific detector detectivity (equals system detectivity when detector noise >> amplifier noise);
λ_U	=	upper wavelength limit to spectral band of interest;
λ_L	=	lower wavelength limit to spectral band of interest.

The influence of sensor design parameters on S/N is shown by (7.16). It is seen that S/N increases directly as the square of aperture size, the square root of the instantaneous field of view, and linearly with specific detector detectivity. It decreases inversely with focal length and the square root of electrical noise bandwidth. Signal-to-noise ratio also increases as the target-to-background contrast temperature increases.

No one of the sensor design parameters can be altered without having a secondary impact on the other parameters. Parametric studies that vary design parameters must therefore account for their influence on the others. For example, a change in the IFOV may dictate a change in the scan mechanism, number of detectors, or focal length. If the focal length changes, then the detector subtense changes and potentially so does the cold shield efficiency that can impact the S/N. Cold shields reduce the background photon flux density or background-contributed noise by reducing the field of view and, hence, increase the value of D_d^*.

7.5.3 Noise-Equivalent Temperature Difference

Noise-equivalent temperature difference NEΔT is defined as the target-to-background contrast temperature of a large ($K_e = 1$) blackbody source [$\epsilon(\lambda) = 1$] that yields a $S/N = 1$. It is derived from (7.16) by solving for ΔT as

$$\text{NE}\Delta\text{T} = \frac{4f\sqrt{\Delta F}}{D_o^2\sqrt{\text{IFOV}}} \frac{1}{\int_{\lambda_L}^{\lambda_U} \tau_o(\lambda)\tau_a(\lambda) \frac{\partial W_b(\lambda, T)}{\partial T} D_d^*(\lambda)\, d\lambda} \tag{7.17}$$

The bandwidth ΔF is usually referenced to the noise bandwidth of a single RC low-pass filter. In this case,

$$\Delta F = \frac{\pi}{2} B_{3dB} = \frac{\pi}{2} \frac{1}{2\tau_d} \tag{7.18}$$

where τ_d = detector dwell time = time required for a detector to scan over a point. Noise bandwidths for other filter implementations are shown in Table 6.5.

Since NEΔT is proportional to the inverse of S/N, this performance measure is also used with search and track sensors. Thus the same design parameters that make S/N large also make NEΔT small. NEΔT is not constant at all background temperatures. It varies due to the changing value of the integral of the derivative of spectral emittance with respect to temperature that appears in the denominator. The spectral integral of the emittance derivative, shown in Table 7.5, is smaller at

Table 7.5
Spectral Integral of the Derivative of the Thermal Emittance at Several Background Temperatures T_{bkd} [10]

		$\frac{\partial W_b}{\partial T} = \int_{\lambda_1}^{\lambda_2} \frac{\partial W_b(\lambda\ T)}{\partial T} d\lambda\ \frac{\text{W}}{\text{cm}^2\text{K}}$ *for*			
λ_1 (μm)	λ_2 (μm)	T_{bkd} = 280K	T_{bkd} = 290K	T_{bkd} = 300K	T_{bkd} = 310K
3	5	1.1×10^{-5}	1.54×10^{-5}	2.1×10^{-5}	2.81×10^{-5}
3	5.5	2.01×10^{-5}	2.73×10^{-5}	3.62×10^{-5}	4.72×10^{-5}
3.5	5	1.06×10^{-5}	1.47×10^{-5}	2×10^{-5}	2.65×10^{-5}
3.5	5.5	1.97×10^{-5}	2.66×10^{-5}	3.52×10^{-5}	4.57×10^{-5}
4	5	9.18×10^{-6}	1.26×10^{-5}	1.69×10^{-5}	2.23×10^{-5}
4	5.5	1.83×10^{-5}	2.45×10^{-5}	3.22×10^{-5}	4.14×10^{-5}
8	10	8.47×10^{-5}	9.65×10^{-5}	1.09×10^{-4}	1.21×10^{-4}
8	12	1.58×10^{-4}	1.77×10^{-4}	1.97×10^{-4}	2.17×10^{-4}
8	14	2.15×10^{-4}	2.38×10^{-4}	2.62×10^{-4}	2.86×10^{-4}
10	12	7.34×10^{-5}	8.08×10^{-5}	8.81×10^{-5}	9.55×10^{-5}
10	14	1.3×10^{-4}	1.42×10^{-4}	1.53×10^{-4}	1.65×10^{-5}
12	14	5.67×10^{-5}	6.10×10^{-5}	6.52×10^{-5}	6.92×10^{-5}

colder temperatures, causing NEΔT to increase as the temperature of the background decreases. This may have negative implications for the detection of targets where ΔT is small and the background is cold. Furthermore, NEΔT has a stronger dependence on background temperature at mid-infrared wavelengths than at long infrared wavelengths because blackbody radiation from objects at or near ambient (300K) is centered at 10 μm, while the 3 to 5-μm band is located on the limb of the spectral distribution.

7.5.4 Auxiliary Relationships for Noise-Equivalent Temperature Difference

Noise-equivalent temperature difference can also be written in terms of the angular subtense of a single detector and the area of the optical clear aperture. This is accomplished by replacing IFOV and D_o in (7.17) by equivalent expressions to get

$$\mathrm{NE}\Delta\mathrm{T} = \frac{\pi}{A_o} \frac{(ab\Delta F)^{1/2}}{\alpha\beta} \frac{1}{\displaystyle\int_{\lambda_L}^{\lambda_U} \tau_o(\lambda)\tau_a(\lambda) \frac{\partial W_b(\lambda, T)}{\partial T} D_d^*(\lambda)\, d\lambda} \tag{7.19}$$

where

$$F/\# = f\text{-number} = f/D_o \tag{7.20}$$

a, b = detector dimensions;

α, β = detector angular subtenses;

= angular projections of detector through the optical system when the detector is located at the focal plane;

$$\mathrm{IFOV} = \alpha\beta = ab/f^2 \tag{7.21}$$

and

$$A_o = \text{area of optical clear aperture} = \pi D_o^2/4 \tag{7.22}$$

7.5.5 Application of S/N and NEΔT to IR Sensor Design

System detectivity is perhaps the most important parameter influencing sensor performance through S/N and sensor cost. System detectivity is a function of D_d^* and electronic noise.

Detector material selection, and hence D_d^*, is generally based on the spectral characteristics of the target and background, sensor bandwidth, cooling requirements, operating lifetime, and cost versus performance trades between detectors meeting the above requirements.

Signal-to-noise ratio is maximized (as may be required in an IRST sensor) by increasing the IFOV as shown by (7.16). However, if an imaging sensor such as a FLIR device is required, then a large IFOV leads to reduced resolution and, hence, reduced image quality. An example of a requirements flowdown for an IRST sensor is shown in Figure 7.8.

The intercept geometry between the sensor and the target controls the design. The tracker requirements affect the IFOV, while the guidance range is used to specify the field of regard (FOR) or total search area. The FOR, in turn, constrains the detector array dimensions and the optical clear aperture. The minimum irradiance of the target affects the selection of detector material and acquisition range. Although signal-to-noise ratio is increased by maximizing the aperture and minimizing the focal length, these parameters are constrained by the design envelope of the sensor and the cost goals. Maximizing optical transmission is eventually limited by technology and cost.

The role of wavelength selection in minimizing NEΔT, as in tracker applications, is seen by examining the derivative of the blackbody spectral emittance with respect to temperature. The spectral emittance and its derivative are illustrated in

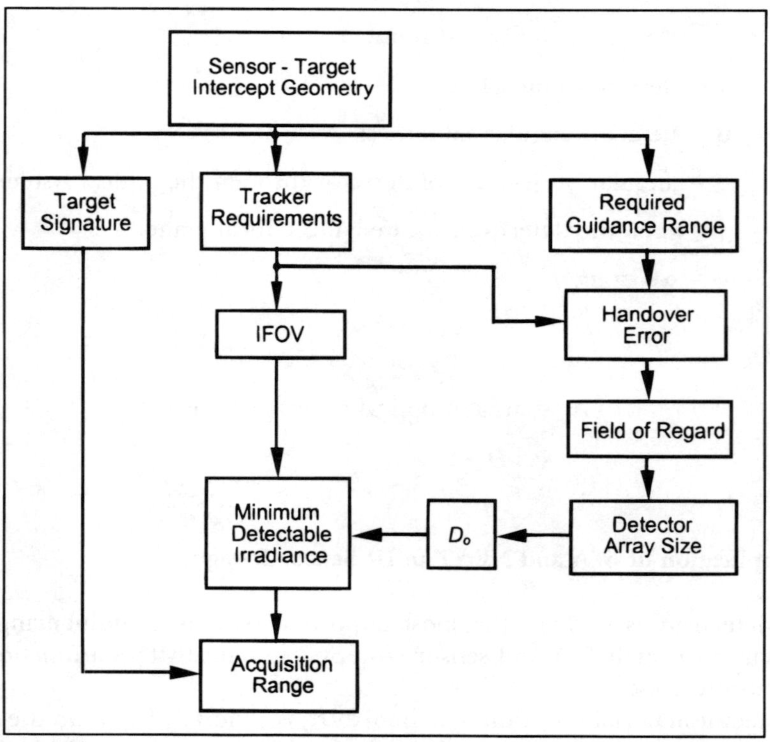

Figure 7.8 Requirements flowdown for an IRST sensor.

Figure 7.9 for three background temperatures [10]. In evaluating NEΔT, the derivative of the emittance W_b is integrated over the wavelength interval λ_1 to λ_2 in which the detector is operating as described by (7.19). Table 7.5 shows that the spectral integral of the emittance derivative is greater at longer wavelengths, minimizing NEΔT at long-wavelength sensor operation because of the inverse relation between NEΔT and the spectral integral. For example, a sensor operating between 3.5 and 5 μm has a spectral integral value of 2×10^{-5} when $T_{bkd} = 300$ K and a value ten times greater if it operates between 8 and 14 μm. Thus an infrared tracker that requires NEΔT to be minimized for optimal performance is designed to operate at long wavelengths.

7.5.6 Radiation Contrast

When the primary function of an IR sensor is collecting thermal images of a scene, the radiation contrast C_R may be the governing design parameter. Radiation contrast is defined as

$$C_R = \frac{W_t - W_b}{W_t + W_b} \tag{7.23}$$

where

W_t = emittance of the target object;
W_b = emittance of the background.

As indicated in Figure 7.10, the radiation contrast is maximized in the mid-infrared band at the background temperatures shown. Thus, other design parameters being equal, thermal imaging sensors operate more optimally at mid-infrared wavelengths than at long wavelenths.

7.5.7 Minimum Resolvable Temperature Difference

Noise-equivalent temperature difference is a measure of performance well-suited for a sensor that locates and tracks objects having high thermal emission. It is also valuable as a quality control parameter to ensure uniform sensitivity of individual detector elements in an array. However, it is not ideal for application to IR sensors that create multipixel images of a scene, such as FLIR sensors used in target classification and identification in background clutter. The minimum resolvable temperature difference (MRTD) is a better performance indicator in this application. MRTD includes effects of sharpness, graininess, contrast rendition, and interference by artifacts. It measures the S/N-limited thermal sensitivity as a function of spatial frequency (the rate of change of light-to-dark features with respect to a distance measure).

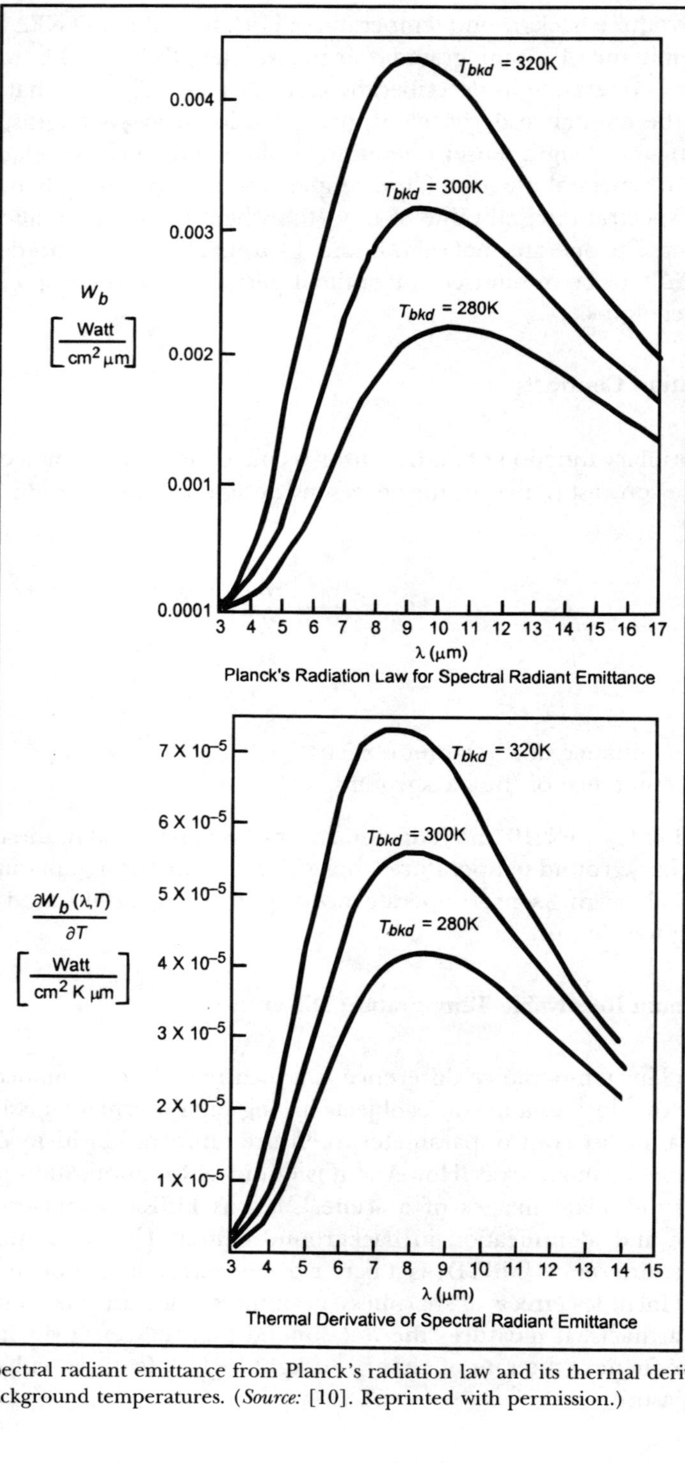

Figure 7.9 Spectral radiant emittance from Planck's radiation law and its thermal derivative at three background temperatures. (*Source:* [10]. Reprinted with permission.)

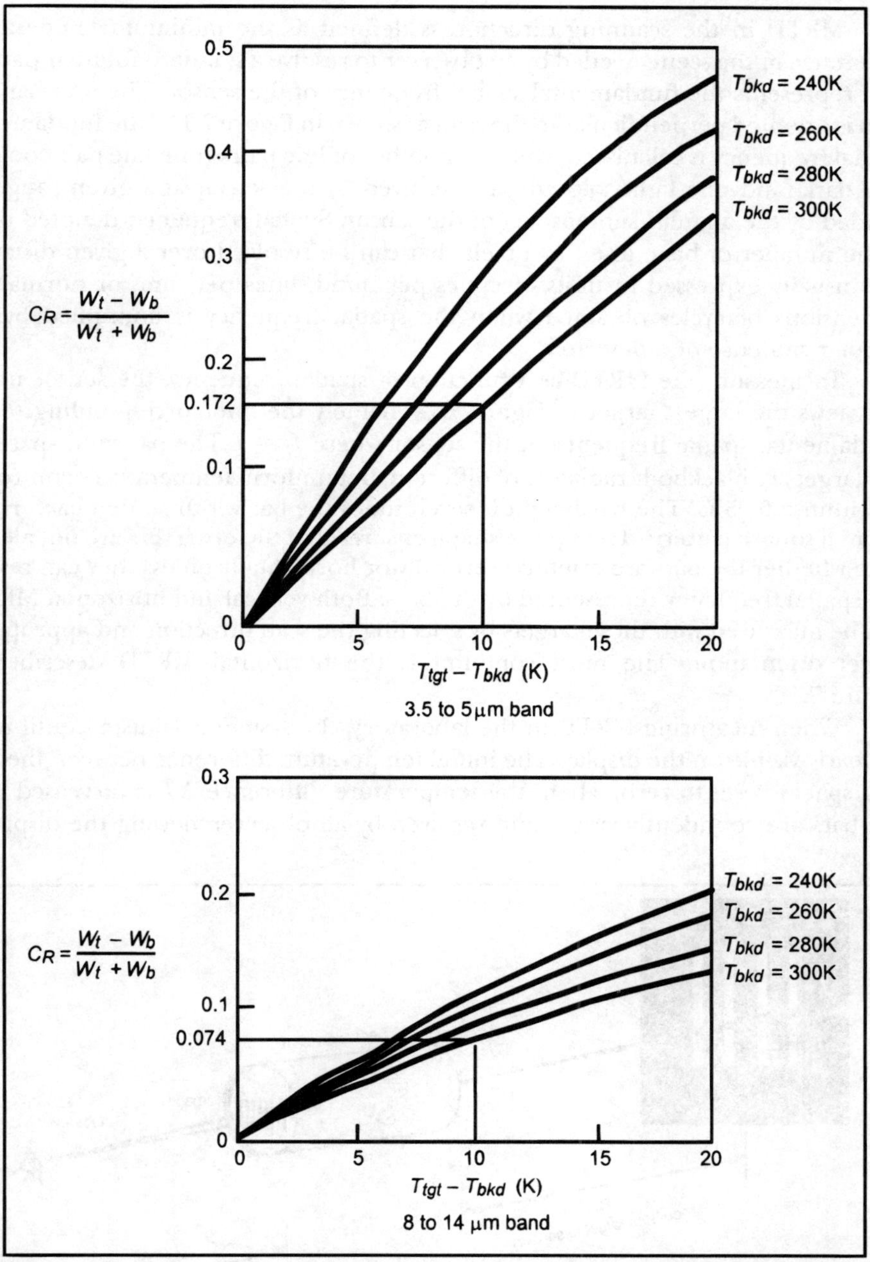

Figure 7.10 Radiation contrast at mid- and long-infrared wavelengths. (*Source:* [10]. Reprinted with permission.)

MRTD in the scanning direction is defined as the minimum temperature difference in the scene needed by an observer to resolve a standard four-bar pattern that represents the fundamental spatial frequency of the sensor. The four-bar pattern is oriented perpendicular to the scan as shown in Figure 7.11. The fundamental spatial frequency is calculated from the number of line pairs (one line pair contains one dark- and one light-colored bar) resolved by the sensor at a given range R_T divided by the angular subtense α_T of the sensor. Spatial frequency, denoted by f_T, is the number of bars, lines, or pixels that can be resolved over a given distance. It is usually expressed in units of cycles per mrad, lines per mm, or normalized dimensions of cycles obtained when the spatial frequency is multiplied by the angular subtense of a detector.

To measure the MRTD as a function of spatial frequency, the sensor under test views the largest target in Figure 7.12, namely the one corresponding to the fundamental spatial frequency of the sensor where $f_T = f$. The bars and spaces of the target are blackbody radiators of different, but uniform temperatures controlled to within ±0.05°C. The bar height is seven times the bar width so that each target forms a square pattern. The square shape ensures that the observers are not alerted as to whether the bars are oriented vertically or horizontally unless they can resolve the spatial frequency represented by the bars. Both vertical and horizontal MRTDs can be measured with these targets by selecting the scan direction and appropriate target orientation. The most common is the horizontal MRTD described in Figure 7.12.

When measuring MRTD in the laboratory, the system is adjusted until noise is clearly visible on the display. The initial temperature difference between the bars and spaces is set to zero. Then, the temperature difference ΔT is increased until the bars are confidently visible and resolved by an observer viewing the display at

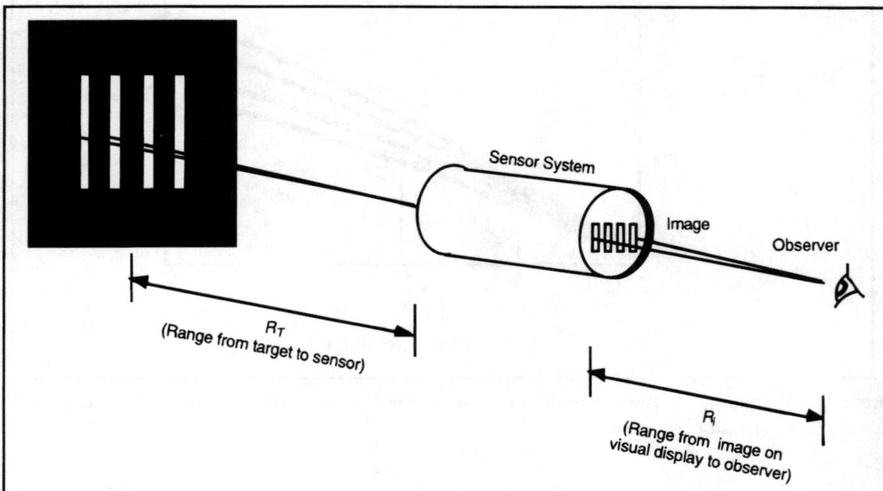

Figure 7.11 Minimum resolvable temperature difference measurement.

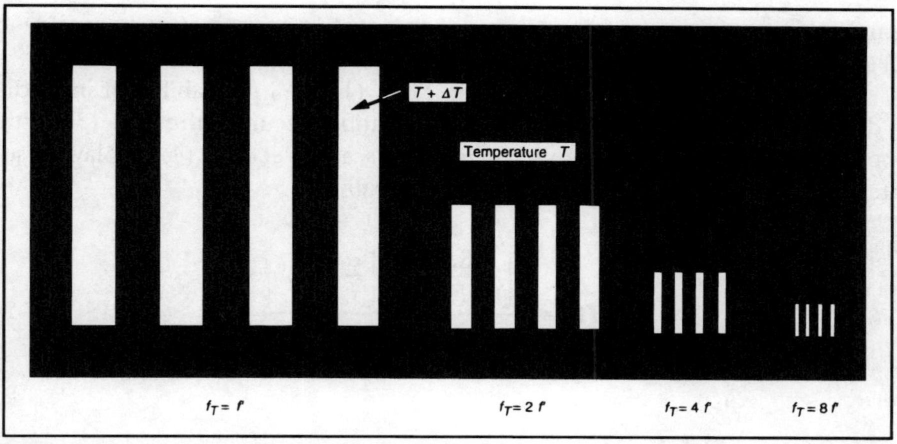

Figure 7.12 Four-bar targets of increasing spatial frequency [10].

a distance R_i. This ΔT is the MRTD at the fundamental spatial frequency of the sensor. The process of setting the initial temperature difference to zero and increasing it until the bars are resolved is repeated at the higher spatial frequencies, shown in Figure 7.12, to generate the MRTD curve represented by Figure 7.13. Curves of this type show that human resolution of higher spatial frequency information

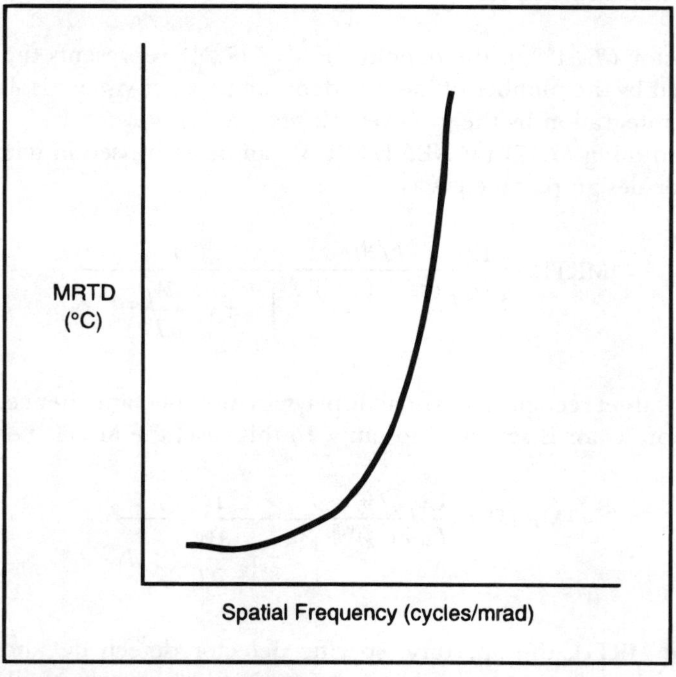

Figure 7.13 Example of minimum resolvable temperature difference curve.

requires larger contrast temperatures between the target features and the background.

When four assumptions are satisfied [10], (1) 90% probability of individual bar detection, (2) spatial and temporal integration occur in the eye, (3) system response is linear with spatial invariance in the scan direction, (4) display noise is white, the MRTD can be expressed mathematically as

$$\mathrm{MRTD} = \frac{3\left[\dfrac{\mathrm{NE\Delta T}}{(\Delta F)^{0.5}}\right] f_T \left[\dfrac{\alpha}{\tau_d}\right]^{0.5} \beta^{0.5}}{\tilde{r}_s (T_e \dot{F})^{0.5}} \tag{7.24}$$

where

ΔF = electrical noise bandwidth in Hz;

f_T = target spatial frequency in cycles/mrad;

τ_d = detector dwell time in sec;

$\tilde{r}_s$ = system modulation transfer function (MTF) $\approx \exp(-2\pi^2\sigma_s^2 f_T^2)$ (7.25)

T_e = effective eye integration time $\approx$ 0.2 sec;

$\dot{F}$ = frame rate in sec^{-1}.

The factor $(T_e\dot{F})^{0.5}$ in the denominator of (7.24) represents the increase in S/N produced by the number of independent samples per visual display point due to temporal integration by the eye over 0.2 sec.

By substituting (7.17) for NEΔT, MRTD can be expressed in terms of fundamental sensor design parameters as

$$\mathrm{MRTD} = \frac{12}{D_o} \frac{F/\#f_T}{\tilde{r}_s(\tau_d)^{0.5}(T_e\dot{F})^{0.5}} \frac{1}{\displaystyle\int_{\lambda_L}^{\lambda_U} \tau_o \tau_a \frac{\partial W_b}{\partial T} D_d^* \, d\lambda} \tag{7.26}$$

In automatic target recognition, visual displays are not used and, hence, the $(T_e\dot{F})^{0.5}$ eye integration factor is set equal to unity. In this case, the MRTD becomes

$$\mathrm{MRTD} = \frac{12}{D_o} \frac{F/\#f_T}{\tilde{r}_s(\tau_d)^{0.5}} \frac{1}{\displaystyle\int_{\lambda_L}^{\lambda_U} \tau_o \tau_a \frac{\partial W_b}{\partial T} D_d^* \, d\lambda} \tag{7.27}$$

To minimize MRTD, the aperture, specific detector detectivity, and dwell time should be made as large as possible and $F/\#$ should be made small. This set of requirements typically increases the cost of the sensor, which must be balanced against the cost and threat posed by the target.

Modulation Transfer Function

The modulation transfer function provides a representation of the spatial dispersion produced by diffraction and aberrations in optical systems. Because of the dispersion, the image of a bright point source is not a point but rather is spread out over some area whose size is dependent on the diffraction and aberration effects through a point spread function. The Fourier transform of the point spread function is the optical transfer function (OTF) whose modulus or absolute value is the modulation transfer function and whose phase is the phase transfer function (PTF). In the absence of aberrations, the point spread function is the Airy diffraction pattern produced when the incident infrared energy is transmitted through a circular aperture. In most infrared sensors, however, both diffraction and aberration degrade the image of a point source. Aberrations are dominant in low $F/\#$ optics that are used in low-light-level imaging sensors where a high infrared energy collection efficiency and a large field of view are required [11]. The broadening due to aberrations is proportional to $1/(F/\#)^n$, where n is a function of the aberration type as shown in Table 7.6 [12].

Simply put, the modulation transfer function is the ratio of the actual output amplitude to the amplitude that would be obtained if the electro-optical system had perfect resolving capability [5]. Thus, the modulation transfer function can be expressed as the ratio of the modulation in image intensity to the modulation in object intensity at a particular spatial frequency. The modulation in image intensity is itself a ratio of the amplitude of a sinusoidal signal distribution to the mean background. The MTF provides a measure of how well the image is formed or the quality of the image as a function of spatial frequency.

While circular symmetrical test functions are appropriate for some optical systems, scanning infrared imaging systems normally do not have the same properties in all directions (i.e., they are anisotropic because of the scanning processes used to produce the real-time images). For this reason, Lloyd [10] defines the MTF

Table 7.6
Dependence of Aberrations on $1/(F/\#)^n$

Aberration Type	n
Astigmatism	1
Spherical	3–4
Coma	2
Field curvature (associated with astigmatism)	1
Chromatic	1

in the scanning direction as the amplitude response of a sensor system to a long infinitesimally narrow line source composed of a string of delta functions. The line source is oriented perpendicular to the scan direction and, by virtue of the delta functions, includes all spatial frequencies from zero to infinity. The spatial frequencies are oriented perpendicular to the long dimension of the line source (i.e., in the scan direction). As the analog signal corresponding to the line source passes through each component of the system, its frequency content is selectively attenuated. The image that appears at the output of the system is therefore not reproduced precisely as manifested by a spreading of the line image. This qualitative interpretation of the MTF can be expressed mathematically by redefining the MTF as the modulus of the Fourier transform of the line spread function. In practice, the MTF is measured by using 100% modulated sine wave test patterns representing several spatial frequencies. The modulation produced in the image plane is measured and plotted at a point that corresponds to the spatial frequency. The process is repeated for the other spatial frequencies in the series of test patterns.

The modulation transfer function of an infrared sensor system, such as that expressed by (7.25), includes the effects of the individual modulation transfer functions that represent the optics, spectral filters, detectors, amplifier electronics, display, and any other components in the photon detection and signal processing chain. Since the effects of the many individual modulation transfer functions combine to give the overall system modulation transfer function, the central limit theorem may be used to approximate the system modulation transfer function with a Gaussian model as was done in (7.25). If the modulation transfer function in Figure 7.14 is interpreted as applying to the entire electro-optical system including

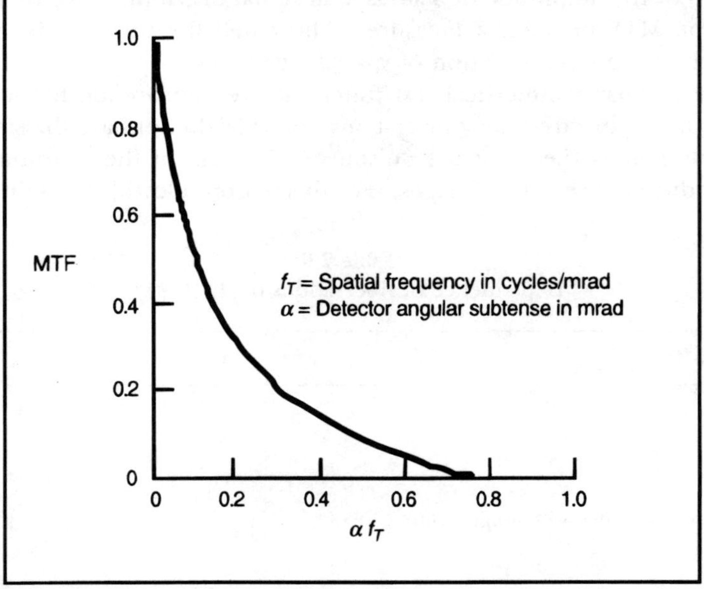

Figure 7.14 Modulation transfer function versus spatial frequency.

the electronic filters and amplifiers, then the vertical axis represents the decrease in amplitude observed in the electrical signal as the spatial frequency content of the target increases. By convention, the vertical axis is normalized to unity at or near-zero spatial frequency. In thermal imaging systems, the MTF is usually dominated by the detector angular subtense α, allowing the horizontal axis to be normalized with respect to α in dimensions of cycles.

The four conditions required for strict application of MTF theory are the following.

1. Detection is noncoherent;
2. Signal processing is linear;
3. Imaging is spatially invariant;
4. System mapping is single-valued (non-noisy).

Condition 1 does not usually present a problem because passive sensors use noncoherent detection. Condition 2 can be met with proper design of the signal-processing electronics. Condition 3 is technically violated if the background-clutter statistics change from one area to the next. To ensure conformance with this assumption, the regions of applicability can be made small enough to include only those where spatial invariance is satisfied. The fourth condition is only satisfied with ideal, noiseless electronics. However, its requirements can be met for practical purposes when the signal-to-noise ratio is large enough (i.e., where the signal is large). Even with the theoretical shortcomings of the MTF as applied to FLIR sensors, it retains its usefulness and has wide support in describing and evaluating FLIR imaging performance.

There are two types of modulation transfer functions, geometrical and diffraction limited [10,13] The geometrical MTF is calculated using ray tracing. It ignores diffraction effects and provides an accurate estimate of MTF if aberrations are large (i.e., the geometrical spot size is much greater than the diffraction-limited blur circle).

Diffraction-limited MTF is calculated using wave optics with a known aperture function (e.g., circular, square, and obscured circle). The diffraction-limited MTF is equal to the autocorrelation of the aperture transmittance function for noncoherent systems. With coherent systems, it is equal to the aperture function itself. The diffraction-limited MTF gives an upper limit to the MTF. Aberrations are assumed negligible in the diffraction-limited MTF calculation (i.e., the geometrical spot size is much less than the diffraction-limited blur circle). The diffraction-limited MTF is calculated only once for the entire system based on the limiting aperture (stop) for the overall optics train. It does not cascade on an element-by-element basis as does the geometrical MTF.

MRTD Calculation

The FLIR sensor design data in Table 7.7 are used to illustrate a sample calculation of the MRTD with (7.24). In this example, the FLIR sensor scans a 150-element

Table 7.7
FLIR Sensor Design Parameters Used for MRTD Calculation

Parameter	*Value*
Spectral band	8 to 12 μm
Peak detector detectivity D_d^*	2×10^{10} cm $\text{Hz}^{1/2}$ W^{-1}
Optical clear aperture D_o	2 cm
Number of detectors	150
Field of view	400 by 300 mrad
Square detector element dimensions	0.005 cm
Square detector angular subtense	1.0 mrad
Frame rate	30/sec

linear array through a 400-mrad by 300-mrad field of view using detectors whose angular subtense is 1 mrad by 1 mrad. The optical clear aperture is 2 cm and the other system design parameters are as shown in the table.

The focal length is calculated as 5 cm using (7.20) and the values for the detector element linear dimensions and their angular subtense. As described in Section 7.7.1, the system design parameters give a dwell time equal to 2.67×10^{-5} sec when the scan efficiency is 64%.

From (7.18),

$$\Delta F = \frac{\pi}{2}\frac{1}{2\tau_d} = 29.4 \text{ kHz} \tag{7.28}$$

and from (7.17)

$$\begin{aligned} \text{NE}\Delta\text{T} &= \frac{4f\sqrt{\Delta F}}{D_o^2\sqrt{\text{IFOV}}}\,\frac{1}{\displaystyle\int_{\lambda_L}^{\lambda_U}\tau_o(\lambda)\tau_\alpha(\lambda)\frac{\partial W_b(\lambda, T)}{\partial T}D_d^*(\lambda)\,d\lambda} \\ &= \frac{4 \times 5 \times \sqrt{29.4 \times 10^3}}{4 \times 1 \times 10^{-3}}\,\frac{1}{0.8 \times 1.97 \times 10^{-4} \times 2 \times 10^{10}} = 0.27^\circ\text{C} \end{aligned} \tag{7.29}$$

The product of the optical and atmospheric spectral transmission was set equal to 0.8 in calculating the result shown in (7.29). The derivative of the spectral emittance was found in Table 7.5 as 1.97×10^{-4} W cm^{-2} K^{-1} for the 8- to 12-μm wavelength band and a background temperature of 300K.

When the calculated values for ΔF and NEΔT, along with the frame rate $\dot{F}$ of 30 Hz and the eye integration time of 0.2 sec are inserted into (7.24), the MRTD becomes

$$\text{MRTD} = 0.38 \frac{f_T}{\tilde{r}_s} \text{°C} \tag{7.30}$$

when f_T is expressed in cycles/mrad. Inserting the system MTF data of Table 7.8 into (7.30) gives the MRTD plot shown in Figure 7.15.

Limitations of MRTD as a Performance Measure

MRTD measurement requires a human in the loop and hence is subjective. The measurement process includes phrases such as "adjust sensor under test until noise is *clearly visible* on the display" and "temperature difference is increased until it is *confidently resolved by an observer* viewing the display." The words that lead to a subjective evaluation have been italicized. These comments may not apply if automatic target recognition techniques are employed and the human is removed from the measurement process [14,15]. Magnification effects are not accounted for by the MRTD performance measure. This limitation does not apply to automatic target recognition where a human does not interpret the imagery. MRTD does not unequivocally correlate with field performance, but does supply a general indication of field performance. It is useful as a diagnostic tool when the same personnel make the measurements from sensor to sensor [10].

7.5.8 Minimum Detectable Temperature Difference

Minimum detectable temperature difference (MDTD) is generally defined as the minimum or threshold temperature difference between a square target and the background required for an observer to perceive the target through the electro-optical sensor under test. It is further assumed that the observer has unlimited

Table 7.8
Modulation Transfer Function Data [10]

f_T *(cycles)*	*System MTF* $\tilde{r}_s$
0	1
0.1	0.918
0.2	0.721
0.3	0.532
0.4	0.363
0.5	0.227
0.6	0.129
0.7	0.066
0.8	0.028
0.9	0.009
1.0	0

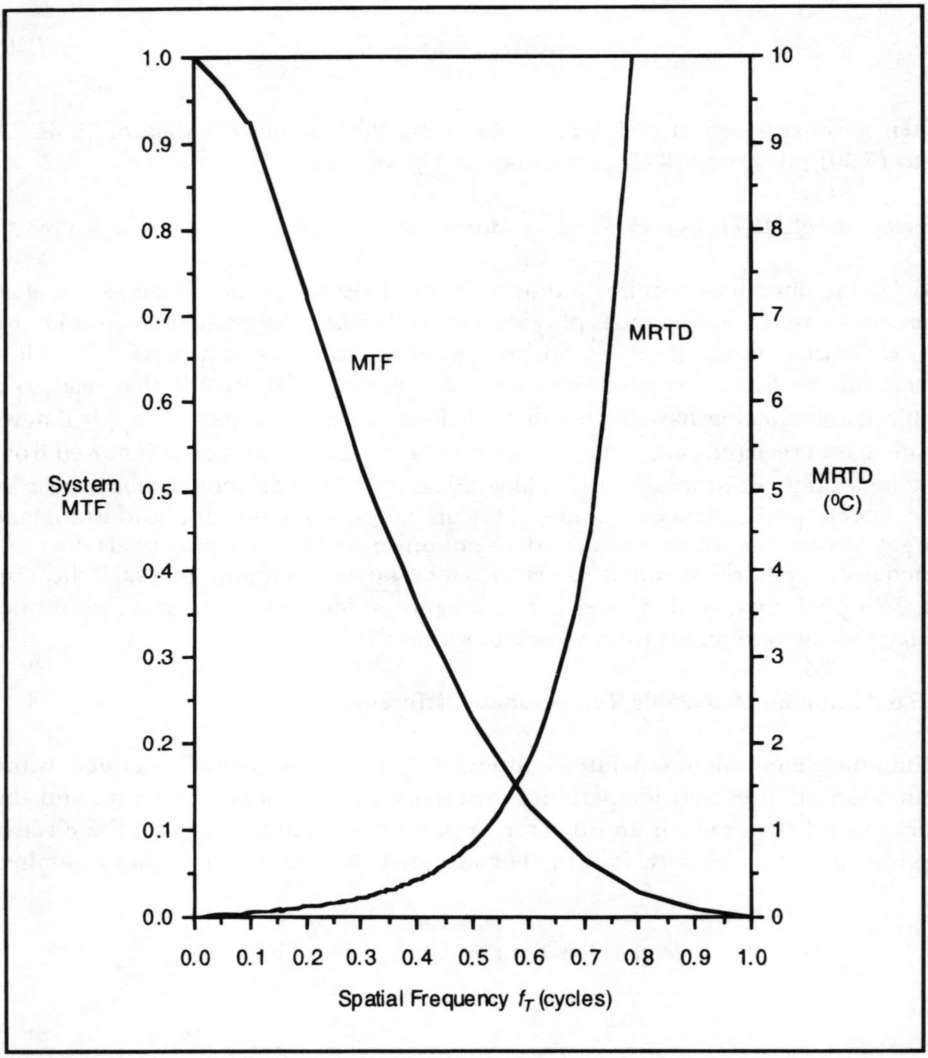

Figure 7.15 MRTD and MTF for FLIR design example.

time to make a decision and knows where to look for the target. MDTD thus expresses the magnitude of the signal that must be emitted by an arbitrarily sized target for it to excite the IFOV that corresponds to one pixel, as for example in star or hotspot detection. As a performance measure, MDTD is appropriate for IRST sensors. No common convention for its calculation presently exists, although Lloyd [10] and Shumaker and others [5] provide mathematical expressions that are suitable for its computation. MDTD is inversely proportional to the target size and directly proportional to NEΔT and the perceived *S/N*, a quantity that incorporates the perception of imagery by the eye-brain interface into the *S/N* required by an observer to detect the target. The relation of target size α, in mrad, to sensor MDTD, denoted by ΔT_D, is illustrated by Figure 7.16.

7.6 TARGET CLASSIFICATION MODELS

Models that support the calculation of the probability of correctly classifying targets have been developed by the Center for Night Vision and Electro-Optics (CNVEO). The models describe the relation of the target-to-background contrast temperature ΔT to the sensor resolution and to the range R_T between sensor and target at which the target can be detected, classified, or identified.

7.6.1 One-Dimensional Array FLIR Sensor Performance Model

Figure 7.17 shows the 1975 CNVEO target classification model that applies to first-generation FLIR sensors (i.e., those that scan one-dimensional detector arrays through a scene to create an image) [16]. Table 7.9 gives the average target-to-background contrast temperatures for a tank, truck, armored personnel carrier, and a man on a typical summer day under uniform background temperature conditions. Targets that suppress thermal signatures by controlling the heat distribution over the target area have lower contrast temperatures than those listed in Table 7.9, making them more difficult to detect.

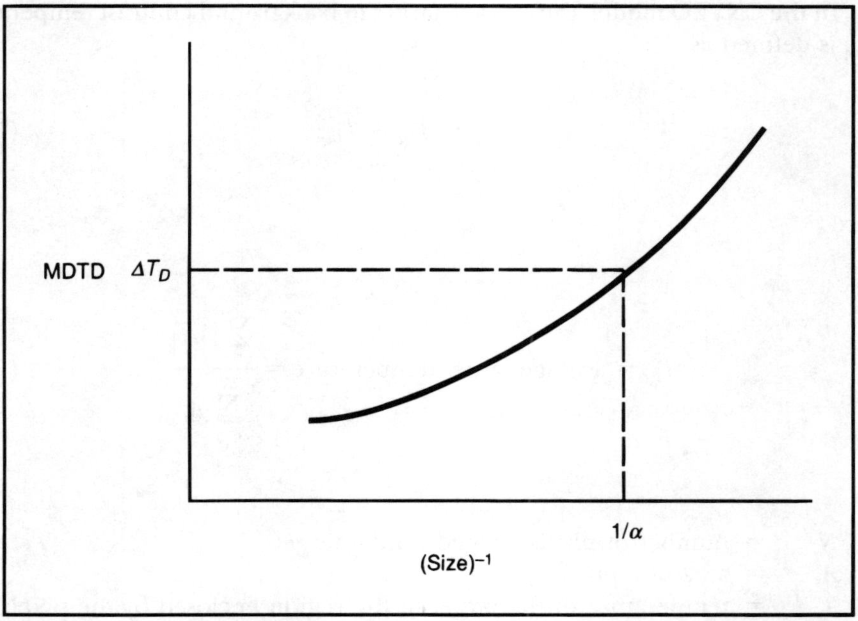

Figure 7.16 Minimum detectable temperature difference as a function of inverse target size.

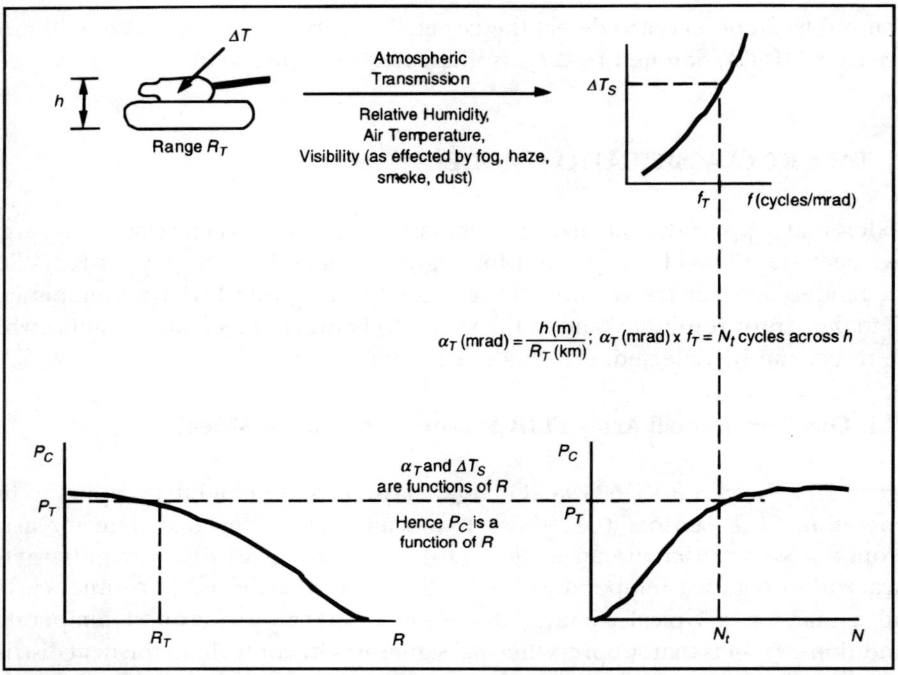

Figure 7.17 CNVEO model for calculating performance of first-generation thermal imaging systems.

In the CNVEO model, the average target-to-background contrast temperature ΔT_{avg} is defined as

$$\Delta T_{avg} = T_{tgt} - T_{bkg} \tag{7.31}$$

where

$$T_{tgt} = \text{average target temperature} = \frac{\sum_{i=1}^{N} A_i T_i}{\sum_{i=1}^{N} A_i} \tag{7.32}$$

N = number of pixels imaged on the target;
A_i = area of a pixel;
T_i = temperature of the target in the region enclosed by the pixel;
$T_{\rm bkg}$ = uniform background temperature.

The contrast ΔT is attenuated and its energy scattered by the intervening atmosphere between the sensor and target due to humidity and reduced visibility from, for example, fog, haze, smoke, and dust. As a result, the contrast temperature ΔT_S at the sensor is typically smaller than ΔT. Using the MRTD curve of the sensor, the resolvable spatial frequency f_T corresponding to ΔT_S may be found. Then, the number of line pairs across the minimum or critical dimension h of the target is calculated by multiplying the resolvable spatial frequency by the angle α_T subtended by the target. The probability of target classification P_C (or detection, etc.) is obtained from the empirical relationship between P_C and the number of line pairs as shown in Table 7.2 and Figure 7.5. Since the target angular subtense and ΔT_S are functions of range, the classification probability is also a function of range.

7.6.2 Two-Dimensional Array FLIR Sensor Performance Model

With the advent of two-dimensional detector arrays, the original target classification model was revised. The 1990 and 1992 CNVEO advanced FLIR sensor performance models apply to both first- and second-generation FLIRs [17]. They incorporate two-dimensional sampling and improved models for MTF, NEΔT, MRTD, MDTD, and noise. The eye-brain interface is treated as a matched filter having an integration time ≈0.1 sec (it was 0.2 sec) and a default S/N threshold = 2.5 (it was 2.25). It also allows for a user-specified eye-brain modulation transfer function. Stand alone menu operation and conversion routines for 1975 model files are provided in the newer models.

The second-generation FLIR model takes into account the number of pixels (equal to twice the number of line pairs) resolved within the boundaries of a target rather than the number of line pairs across the minimum dimension of the target [18]. An effective spatial frequency f_e is defined as

Table 7.9
Average Target-to-Background Contrast Temperatures [16]

Target	*Target-to-Background Contrast (°C)*	Target Dimensions ($m \times m$)
• Tank—side	5.25	2.7 × 5.25
• Tank—front	6.34	2.7 × 3.45
• 2-1/2 ton truck—side	10.40	0.83 × 4.22
• 2-1/2 ton truck—front	8.25	2.03 × 1.67
• APC—side	4.67	1.8 × 4.8
• APC—front	5.65	1.8 × 2.09
• Man	8.0	0.5 × 1.5

$$f_e = \sqrt{f_h f_v} \tag{7.33}$$

where

f_h = horizontal dimension spatial frequency;
f_v = vertical dimension spatial frequency.

The angular subtense α_e of the target is defined as

$$\alpha_e = \frac{\text{Target Area}}{(\text{Range})^2} \tag{7.34}$$

where the range is the distance between the target and the FLIR. (For the first-generation FLIR model, the angular subtense was defined as the minimum target dimension divided by the range.)

Number of pixels P is defined as

$$P = 4f_e^2 \frac{\text{Target Area}}{(\text{Range})^2} \tag{7.35}$$

Finally, the effective number of cycles N_e is expressed as

$$N_e = \frac{\sqrt{P}}{2} = f_e \frac{\sqrt{\text{Target Area}}}{\text{Range}} \tag{7.36}$$

The second-generation model requires that N_e be only approximately 75% of the number of cycles as given by the Johnson criteria to achieve a desired discrimination probability. Current copies of the model are available from the Director, U.S. Army CECOM CNVEO at Ft. Belvoir, VA 22060-5677.

7.6.3 Common Module FLIR Sensor

The common module FLIR design was initiated in the 1970s to reduce the cost of infrared imaging sensors. The system consists of the telescope, scanner, detector array, preamplifiers, postamplifiers, multiplexer, signal-conditioning electronics, and display shown in Figure 7.18.

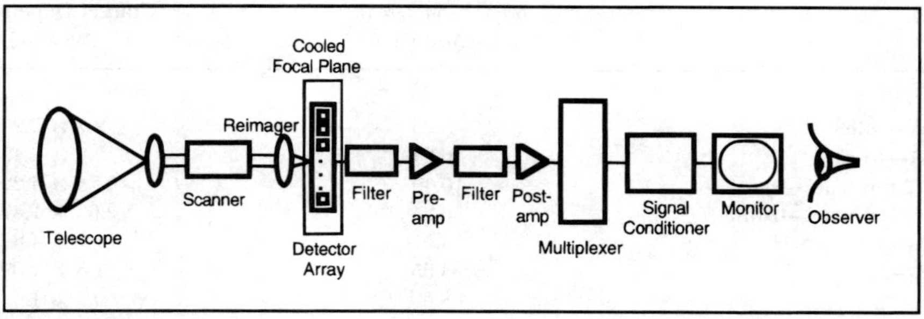

Figure 7.18 Common module FLIR.

The telescope optics collects infrared radiation and focuses it on the detector array. The scanner, using a mirror, repetitively moves the image of the scene over the detectors so that each part of the scene is registered by one or more detectors. The detectors convert the scene radiation into electrical signals that are filtered and amplified. Then the multiplexer combines the detector outputs into a single video signal that is amplified and filtered before being displayed on a monitor for human interpretation. The only design options were to modify the telescope optics to achieve the desired field of view, add electronic filtering to maximize the S/N, and (with optical multiplexers) alter the associated visual optics to present an appropriate image to the observer.

Two general types of multiplexing techniques were employed in common module FLIR design, optical and electronic. In optical multiplexing, the outputs of the individual detectors were amplified and used to modulate light-emitting diodes. The diodes were then scanned synchronously across the input of a conventional television with an IR scanner, thus converting the modulated light into standard television video. With electronic multiplexing, the output of the detectors were sequentially sampled in a direction normal to the scan direction. The time-based samples from the detectors were then concatenated (i.e., strung together with the beginning of the next sample attached to the end of the previous one). The process was repeated as the scene was scanned through the field of view of the detectors, with the concatenated samples from each pass through the detectors added to the end of the previous one to form a single video stream. The video stream could be further amplified and combined with synchronization signals to produce a composite video suitable for display [5].

The advent of charge-coupled devices (CCDs) introduced other methods of reading out the data produced by the scanning array. The outputs from the IR-sensitive detectors could be directly coupled into a CCD shift register mounted on the focal plane. The CCD operates by storing information (that is proportional to the output of the IR detector) in the form of packets of charge in an array of closely spaced capacitors. The information is electrically clocked through the array by pulsing the capacitors sequentially. This process transfers the charge packets from one capacitor to the next, eventually permitting them to be drawn off the array at an edge via the output node. Visible spectrum, and to some extent near-IR, imagery can be obtained directly using silicon CCD cameras that are sensitive to light in these wavelength bands.

Two forms of scanning and information readout are possible with one-dimensional IR arrays [19,20]. In the serially scanned FLIR system, shown in Figure 7.19(a), a horizontal linear array of IR detectors is raster scanned across the scene with the scan direction parallel to the array. The detector outputs are processed through a CCD serial shift register that functions as a time delay and integration (TDI) circuit, linearly summing the output from each detector as it is scanned through the scene. To avoid image smearing, the image is moved across the array in synchronization with the readout clock. Thus, as the charge shifts from capacitor to capacitor in the shift register, the same part of the image is always

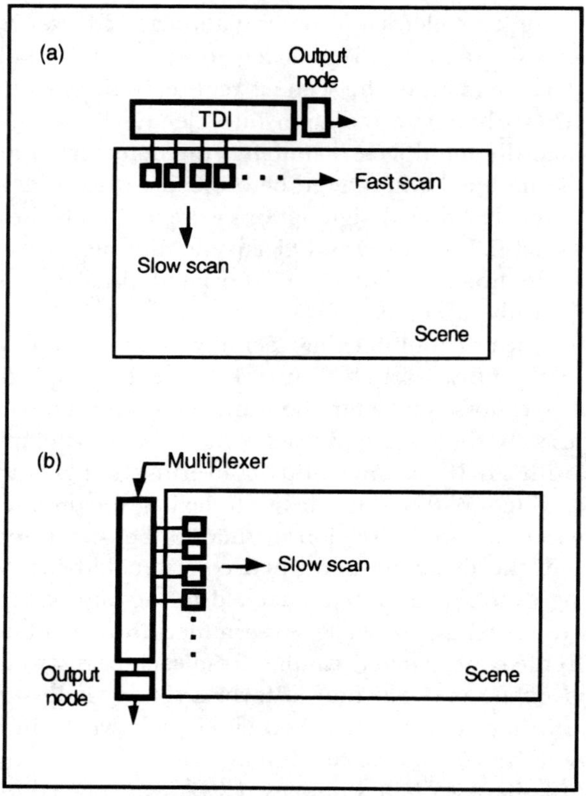

Figure 7.19 (a) Series and (b) parallel FLIR scanning designs that incorporate TDI and multiplexing.

being exposed to the next detector in the IR array. In the parallel-scanned FLIR, represented by Figure 7.19(b), a vertical linear array of IR detectors is slowly scanned across the scene. Here, the scan direction is perpendicular to the array. In this case, the CCD shift register functions as a multiplexer that reads out the information line by line.

The outputs from two-dimensional staring arrays of IR-sensitive detectors can also be directly coupled into a two-dimensional CCD array mounted on the focal plane. TDI is performed along each row of detectors, while multiplexing is used to read the outputs of the rows in the direction parallel to the columns.

7.6.4 Application of CNVEO Model to First-Generation FLIR Sensor Design

Suppose we wish to design the optics for a scanning FLIR sensor and calculate the corresponding range at which the sensor can detect and classify a tank target using the MRTD and MTF data illustrated in Figure 7.20 [21]. The other required design

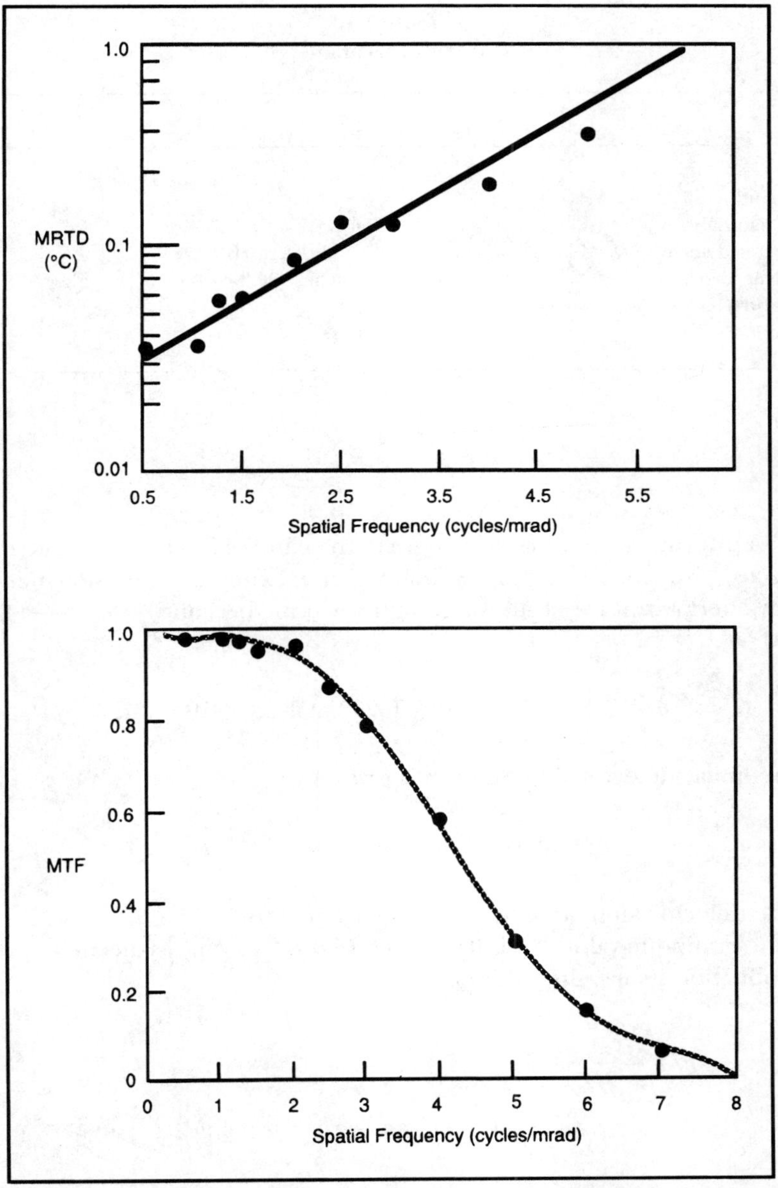

Figure 7.20 MRTD and MTF for FLIR design example. (*Source:* [21]. Reprinted with permission.)

parameters are given in Table 7.10. These data are representative of the common module FLIR.

The diffraction-limited telescope optics design is based on the diameter δ of the central Airy disk. When the detector angular subtense is set equal to the Airy disk diameter, the detector receives 84% of the available energy and its diffraction-limited angle resolution is

Table 7.10
FLIR Design Parameters

Parameter	*Value*
Spectral band	8 to 14 μm
Detector material	HgCdTe
Peak detector detectivity D_d^*	5×10^{10} cm Hz$^{1/2}$ W^{-1}
Optical clear aperture D_o	8 inches (20.3 cm)
Number of detectors	160
Frame rate	30/sec

$$\delta = \frac{2.44\,\lambda}{D_o}\ \text{rad} \tag{7.37}$$

For $\lambda = 14$ μm (the maximum wavelength in the band of interest is used to calculate the detector size since it gives the limiting or maximum dimension needed for optimum energy capture at all wavelengths within the band) and $D_o = 8$ inches (0.203m),

$$\delta = 2.44 \times 14 \times 10^{-6}/0.203 = 1.68 \times 10^{-4}\ \text{rad} \tag{7.38}$$

The linear detector dimension a is given by

$$a = (\delta)(\text{focal length}) = (\delta)(F/\#)(D_o) \tag{7.39}$$

where the detector angular subtense α is set equal to δ.

If the minimum value of a is 0.002 inch (5.08×10^{-5}m) as dictated by detector array production issues, then the $F/\#$ becomes

$$F/\# = \frac{a}{\delta D_o} \tag{7.40}$$

$$= (5.08 \times 10^{-5})/(1.68 \times 10^{-4} \times 0.203) = 1.49 \tag{7.41}$$

Practical lens manufacturing considerations such as speed, aberration control, and cost make an $F/\#$ of at least 2 desirable. Therefore, setting $F/\#$ equal to 2 and recalculating a from (7.39) gives

$$a = 1.68 \times 10^{-4} \times 2 \times 0.203 = 6.82 \times 10^{-5}\ \text{m} = 0.0027\ \text{inch} \tag{7.42}$$

for the linear detector dimension.

The detector angular subtense α is then

$$\alpha = \frac{a}{\text{focal length}} = \delta$$

$$= 6.82 \times 10^{-5}/0.406 = 0.17 \text{ mrad} \quad (7.43)$$

The maximum spatial frequency f_T is governed by the aliasing consideration such that

$$f_{Tmax} = \frac{1}{2\alpha}$$

$$= \frac{1}{2(0.17)} = 2.94 \text{ cycles/mrad} \quad (7.44)$$

Using a tank as the potential target, we set the minimum target dimension to 3.3 m and the target-to-background contrast temperature to 3°C. Selecting a 0.1°C contrast as the sensor MRTD design goal gives a resolvable target spatial frequency of approximately 2.50 cycles/mrad using the MRTD data of Figure 7.20. This frequency is realizable since it is less than the maximum allowed from the aliasing analysis. The MTF is adequate (0.89) at 2.50 cycles/mrad.

The range R_T, shown in Figure 7.21, at which the target can be detected, classified, or identified is found from the relation

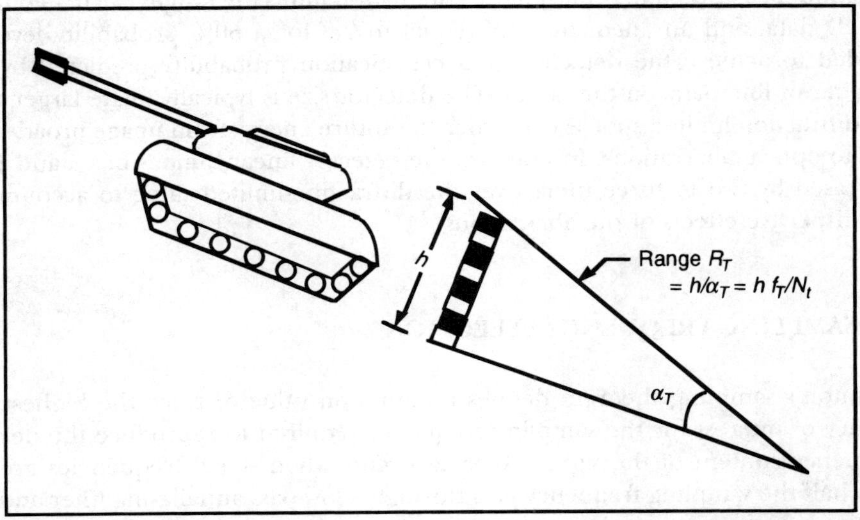

Figure 7.21 Angular subtense of target showing range, number of line pairs, and minimum target dimension.

$$R_T = h/\alpha_T = h\frac{f_T}{N_t} \tag{7.45}$$

where

h = minimum target dimension in meters;
α_T = angular subtense corresponding to the target critical dimension in mrad;
N_t = number of line pairs required to achieve a specific probability of discrimination = $\alpha_T f_T$ (7.46)
f_T = target spatial frequency in cycles/mrad.

From Table 7.2, an average of 1 line pair is needed for target detection and 4 line pairs for classification at a 50 percent probability level with a sensor that uses a one-dimensional scanning array. Thus, the maximum range R_T at which the tank target can be detected is

$$R_T = \frac{hf_T}{N_t} = (3.3\text{ m})(2.50\text{ cycles/mrad})/1\text{ line pair} = 8.2\text{ km} \tag{7.47}$$

and the maximum range at which it can be classified is

$$R_T = (3.3\text{ m})(2.50\text{ cycles/mrad})/4\text{ line pairs} = 2.1\text{ km} \tag{7.48}$$

Practical considerations further affect the performance of an imaging sensor and influence its design. Atmospheric attenuation limits the ranges calculated using MRTD data, and an adequate S/N (equal to 2.8 for a 50% probability level) is needed to achieve the detection and classification probability predicted by the number of line pairs on the target. The detector size is typically made larger than the diffraction-limited spot size in order to capture energy from image broadening due to optical aberrations. In practice, the detector linear dimensions a and b are increased by two to three times over the diffraction-limited value to account for the dispersive effects of the aberrations.

7.7 SAMPLING FREQUENCY SELECTION

Shannon's sampling theorem defines a minimum value of twice the highest frequency of interest for the sampling frequency required to reproduce the desired frequency content of the signal. Aliasing occurs when signal frequencies greater than half the sampling frequency pass through a low-pass antialiasing filter and are digitized. These frequencies create "image" frequencies centered around integer multiples of the sampling frequency that are folded back into the real signal frequency band. If the aliased frequencies are large enough in amplitude, they will distort the real signals and imagery produced by the sensor.

The frequency to which the image frequencies are aliased is given by

$$f_a = \left| f - \frac{(k+1)f_s}{2} \right| \tag{7.49}$$

when

$$\frac{kf_s}{2} \le f \le \frac{(k+2)f_s}{2}$$

where

f_a = aliased frequency;
f = real signal frequency;
f_s = sampling frequency;
k = odd integer.

Practically, the sampling frequency must be greater than the minimum value if the highest frequency signal components are to be reproduced without introducing aliasing. The reasons for this are that the three assumptions made in the derivation of the theorem are never fully satisfied by real signals or filters; namely the sampling theorem requires [22] (1) 100% band-limited signals to be sampled, (2) antialiasing filters to be ideal (i.e., have an infinitely steep rolloff with frequency), and (3) an interpolating function $\sin[2\pi f_m(t - kT)]/[2\pi f_m(t - kT)]$ to be summed over all k, $-\infty < k < \infty$, for perfect signal reconstruction where

f_m = frequency above which the spectrum of the signal vanishes;
T = time between samples = $1/(2f_m)$.

Obviously, the summation cannot extend to infinite limits because of time, memory, and computing power constraints. Practically it can cover a large enough number of samples so that the signal is reconstructed sufficiently well to meet the requirements of the application. However, the first two assumptions are not fulfilled in real systems.

7.7.1 Dwell Time

The dwell time is the time required for a detector to scan over a point. It can be set equal to the reciprocal of twice the signal bandwidth or $1/2f$. Thus, dwell time is also inversely proportional to the sampling frequency that determines the maximum frequency content of the signal that can be recovered through sampling. In an ideal system using a single detector scanning in the horizontal direction, the dwell time τ_d for a frame is expressed as [23,24]

$$\tau_d = t_f/N \tag{7.50}$$

where

t_f = time to scan a frame;
N = number of resolution elements or pixels in a frame.

The number of pixels in the frame can be found in several ways. One of the more convenient is to divide the solid angle of the frame by the solid angle of a pixel. This is the approach used in the MRTD example calculation described earlier. The number of pixels can also be computed as the number of lines in the image times the number of pixels per line.

Equation (7.50) is valid when the scanning velocity is constant over the total field of view and only the desired field of view is covered (i.e., the overscan ratio η_{os} is unity implying no overlap and no missed areas during scanning of the scene). If some areas in the scene are missed during the scan, then the overscan ratio is less than unity and the dwell time increases by the overscan ratio. If overlapping of scene areas by detector pixels occurs, then the overscan ratio is greater then unity and the dwell time decreases. Equation (7.50) also assumes that the scanning efficiency η_s is unity, implying a constant scan velocity and no dead time in retrace or at the edges such as from turnaround time. The scanning efficiency may be written as

$$\eta_s = \frac{\overline{v}}{v_{max}} \tag{7.51}$$

Here,

$\overline{v}$ = average scan velocity;
$v_{\max}$ = maximum scan velocity.

If there are m detectors operating in parallel (e.g., in the vertical direction), then the dwell time can be increased by a factor of m, leading to an increase in the S/N by a factor of $\sqrt{m}$. Incorporating the overscan ratio, scanning efficiency, and number of detectors into the definition for dwell time leads to

$$\tau_d = \frac{t_f \eta_s m}{N \eta_{os}} \tag{7.52}$$

In terms of dwell time, the maximum frequency of the signal information that can be recovered is

$$f_m = \frac{1}{2\tau_d} \tag{7.53}$$

The frequency content of the reconstructed signal in a sampling system can be increased by increasing the number of samples made in a dwell time, since decreasing the time between samples allows higher frequencies to be recovered. The impact of the number of samples per dwell time on image quality is discussed further in the next section.

7.7.2 Effect of Number of Samples per Dwell Time on Image Quality

The upper set of curves in Figure 7.22 illustrates the relation between image quality and normalized spatial frequency for one, two, and three samples per dwell and the limit of continuous sampling. With one sample per dwell, the highest frequency information is not reproduced as well as when the number of samples per dwell is increased to two. In fact, the image quality perceived with three samples per dwell is not appreciably different than when the sampling is continuous. These data were obtained using periodic images (sinusoids). Therefore, caution should be used when extending their application to real visual imagery as complex scenes may require more than two samples per dwell for "adequate" image quality. The lower curves relate the average probability of classification to the number of samples per dwell for signal-to-noise ratios of 4, 8, 16, and infinity (no noise). Larger numbers of samples per dwell are necessary to reach a given classification probability with the lower signal-to-noise ratios [25].

7.7.3 Effect of Array Configuration on Maximum Frequency Response of MTF

In theory, the MTF extends to frequencies of one-half the sample frequency. This limit can only be reached with the proper number of rows of detectors and interdetector spacing. The proper array design will allow recovery of the real signal if its amplitude is greater than the aliased signal amplitudes at frequencies equal to or less than $f_s/2$. This is achieved through overscanning by overlapping scene areas with adjacent detector pixels. As shown by (7.52), an overscan ratio greater than unity reduces the dwell time and hence increases the maximum frequency of the signal information that can be recovered.

Figure 7.23 illustrates the effect of detector configuration on the spatial frequency response of the MTF. A single row of detectors does not allow the upper frequency limit of the MTF to approach $f_s/2$ equal to $1/2[F/\# \times D_o/b]$, where b is the detector linear height dimension. When two rows of detectors are used such that detectors in alternate rows are tangent to each other (in the scan direction), the usable MTF frequency region extends further toward $f_s/2$ but still does not equal $f_s/2$. When the detectors in two rows overlap by half the length of a detector element, the useful MTF approaches $f_s/2$ but does not reach it. The factor of 1.4 to 1.7 shown in the third row, second column of Figure 7.23 is a function of detector height and interdetector spacing. It takes three rows of detectors (not shown in the figure) for the MTF to finally include $f_s/2$ without generating large amplitude aliased signals at frequencies $\leq f_s/2$ from the nonideal sampling process.

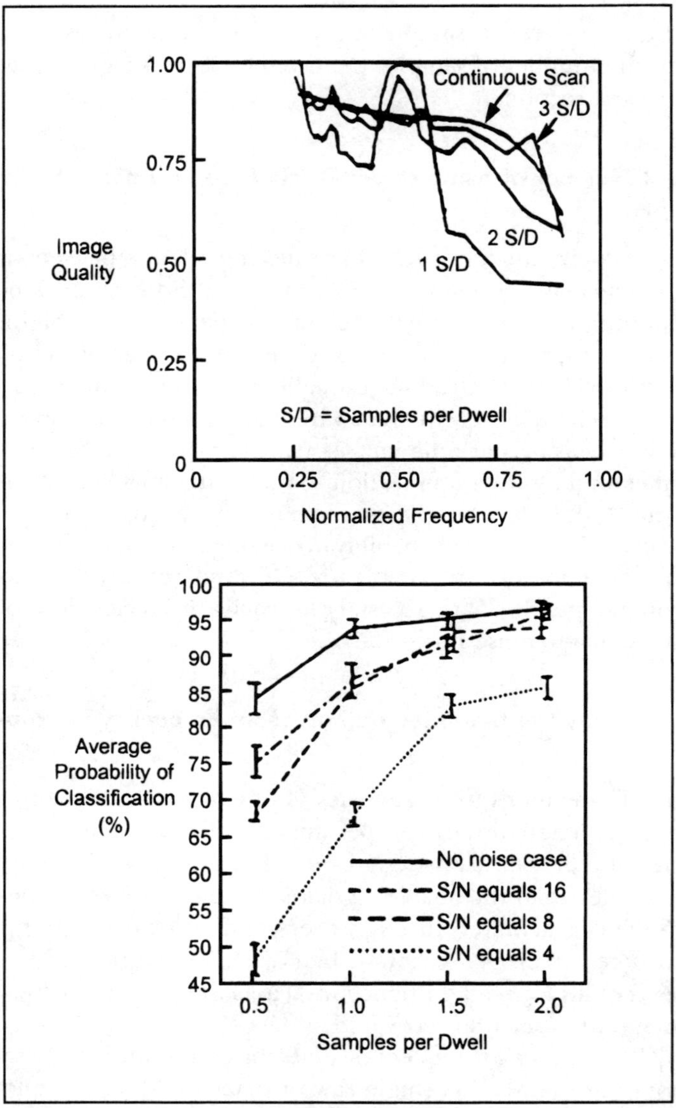

Figure 7.22 Image quality as a function of spatial frequency and samples per dwell, and average probability of classification as a function of number of samples per dwell and signal-to-noise ratio. (*Source:* [25]. Reprinted with permission.)

7.8 DIRECT DETECTION OF PHOTON ENERGY

In this section we analyze the detection of photon energy that is incident on a detector whose output is filtered with a post-detection filter as represented in

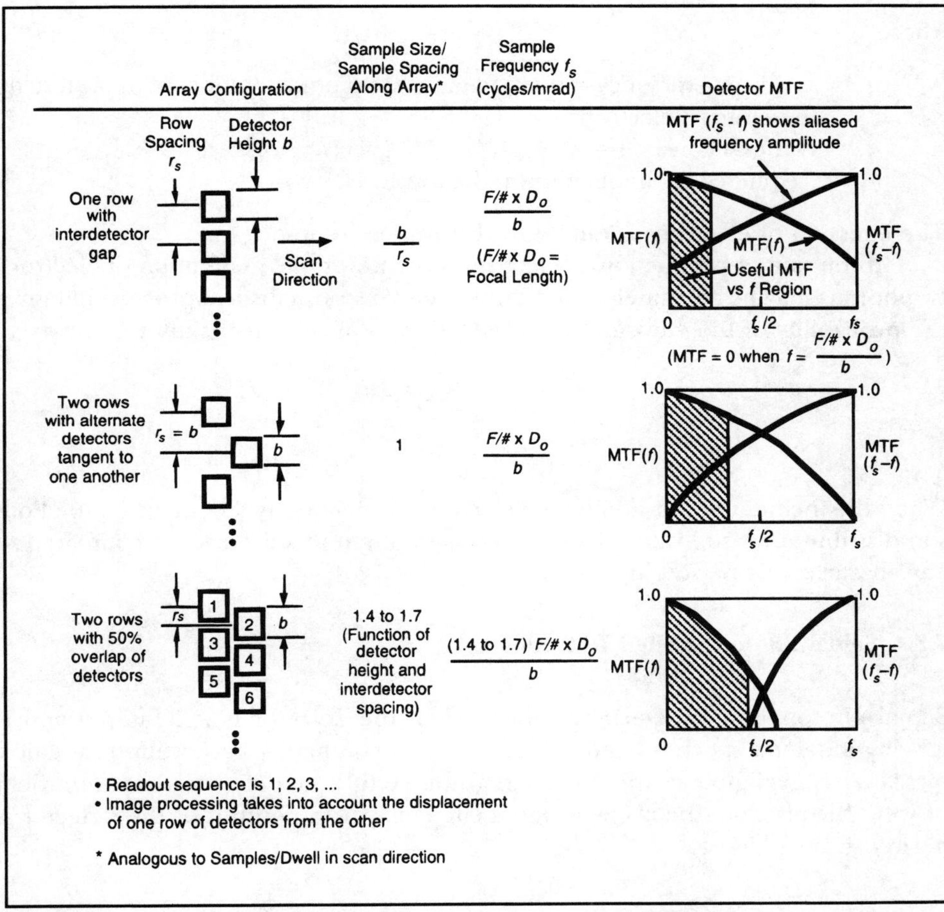

Figure 7.23 Modulation transfer function for one- and two-row detector arrays.

Figure 7.24. If radiation of constant power P is incident on an ideal photon detector, then electrons will be produced at an average rate $\bar{r}$ given by

$$\bar{r} = \eta P / h\nu \tag{7.54}$$

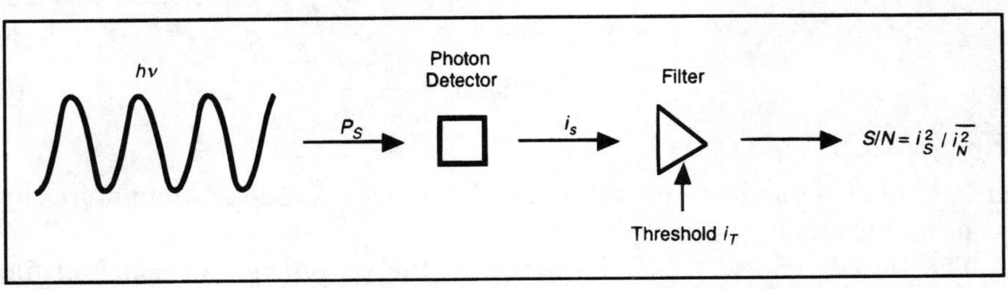

Figure 7.24 Direct detection of photon energy by a detector.

where

η = quantum efficiency = fraction of incident power P effective in producing the emitted electrons;
h = Planck's constant = 6.6256×10^{-34} J sec;
ν = frequency of the incident radiation in Hz;

The emission of electrons is randomly distributed in time.

If the rate of production of electrons is constant, the generation of electrons by photons may be accurately represented as a Poisson statistical process that gives the probability of the emission of k electrons in a measurement interval τ as

$$p(k, \tau) = \frac{(\bar{r}\tau)^k e^{-\bar{r}\tau}}{k!} \tag{7.55}$$

When the incident power and hence electron production is time-varying, the Poisson distribution is still valid provided the sampling interval τ is short compared to any characteristic period in the power variation [26].

7.8.1 Signal Photon-Limited Detection

Signal photon-limited detection applies when the detector is well-filtered and is viewing a narrow spectral band source, or when competing background radiation or scatter is negligible as when the signal source fills the detector's apparent field of view. In this case, the average signal current i_s at the output of the detector is derived from (7.54) as

$$i_s = \frac{q\eta P_s}{h\nu} \tag{7.56}$$

where q = electron charge = 1.6021×10^{-19} coulomb and P_s = incident signal power.

If the filter bandwidth B is equal to the bandwidth of the input signal power envelope, the mean square noise current $\sqrt{\overline{i_N^2}}$ at the output of the filter is

$$\sqrt{\overline{i_N^2}} = \frac{qi_s}{\tau} = 2qi_sB = \frac{2q^2\eta P_sB}{h\nu} \tag{7.57}$$

where $\tau = 1/2B$ is the time interval over which the many independent noise current measurements are averaged.

The signal-to-noise power ratio S/N at the output of the matched filter is given by the ratio of the squared currents as

$$S/N = \frac{i_s^2}{\sqrt{\overline{i_N^2}}} = \frac{(q\eta P_s)^2}{(h\upsilon)^2}\frac{h\upsilon}{2q^2\eta P_s B} = \frac{\eta P_s}{2h\upsilon B} \tag{7.58}$$

The noise-equivalent power *NEP* is defined as the input signal power P_s required to produce a signal-to-noise ratio equal to unity. When (7.58) is solved for P_s with $S/N = 1$, the *NEP* for signal photon-limited (PL) detection is obtained as

$$(NEP)_{PL} = \frac{2h\upsilon B}{\eta} \tag{7.59}$$

Substituting $B = 1/2\tau$ gives

$$(NEP)_{PL} = h\upsilon/\eta\tau \tag{7.60}$$

This implies that, on the average, one signal photon may be detected per unit measurement time when $\eta = 1$. However, the Poisson statistics still give the actual probability of detection during a larger but finite measurement interval.

7.8.2 Background-Limited Detection

When thermal radiation from the background is greater than from the signal source or from the detector itself, the detector input is dominated by the background radiation and the detector is said to be operating as a background-limited infrared photodetector (BLIP). In BLIP performance, the background photon source is the dominant source of noise [26].

If the background energy is constant (i.e., the *fluctuation* of the background radiation does not fall within the bandwidth of the post detection filter), the mean square fluctuation of the output current from the filter, including both signal and background components, is

$$\sqrt{\overline{i_N^2}} = 2qiB = \frac{2\eta q^2(P_s + P_B)B}{h\upsilon} \tag{7.61}$$

where P_B is the power generated by the background radiation. The average current due to the signal is still given by (7.56).

The signal-to-noise power ratio formed by i_s^2 over $\sqrt{\overline{i_N^2}}$ is now

$$S/N = \frac{(\eta q P_s)^2}{(h\upsilon)}\frac{h\upsilon}{2\eta q^2(P_s + P_B)B} = \frac{\eta P_s^2}{2h\upsilon B(P_s + P_B)} \tag{7.62}$$

BLIP performance is achieved when the power emitted by the background is much greater than the signal power or power from internal noise sources (i.e., the background photon source is the dominant noise source). Therefore, under BLIP conditions the noise equivalent power is

$$(NEP)_{BLIP} = \sqrt{\frac{2h\upsilon BP_B}{\eta}} \tag{7.63}$$

In deriving (7.63), the factor of $P_s + P_B$ in the denominator of (7.62) was collapsed to P_B since $P_B >> P_s$ under BLIP conditions.

With background-limited noncoherent infrared detectors, the *NEP* is proportional to the square root of the bandwidth rather than to the bandwidth as with signal photon-limited systems. The direct proportionality to bandwidth occurs in a photon-limited system because the output power (proportional to i_s^2) in a photon detector is proportional to the square of the input power as shown by (7.56). If the bandwidth is doubled, the output noise power is doubled as given by (7.57). But only a square root of 2 increase in signal is required to obtain the same output signal-to-noise ratio since S/N is proportional to $P_s^2/\overline{P_N}$, where $\overline{P_N}$ has been used to denote the output noise power. There is a corresponding proportionality between *NEP* and the square root of bandwidth for direct detection laser radars and between NEP and bandwidth for heterodyne detection laser radars as described in Section 8.6.

Since the detectivity D_d of a photon detector is equal to the inverse of the *NEP*, the detectivity for background-limited performance becomes

$$D_{BLIP} = \sqrt{\frac{\eta}{2h\upsilon BP_B}}\ \mathrm{W}^{-1} \tag{7.64}$$

By substituting the irradiance E_B for P_B/A_D, where A_D is the area of the detector in cm^2,

$$D_{BLIP} = \sqrt{\frac{\eta}{2h\upsilon E_B}}\frac{1}{\sqrt{A_D B}} \tag{7.65}$$

Defining the specific detectivity D_d^* as the value of D_d normalized to a detector area of 1 cm^2 and a bandwidth of 1 Hz gives

$$D_{BLIP}^* = \sqrt{A_D B}\, D_{BLIP} = \sqrt{\frac{\eta}{2h\upsilon E_B}}\ \mathrm{cm\ (Hz)^{1/2}\ W^{-1}} \tag{7.66}$$

Thus, when background radiation is the limiting noise source, (7.66) gives the absolute upper limit for the sensitivity of the detector as its performance is limited by noise associated with the incident photons. Even if the background is reduced to an insignificant level, the signal photon stream still presents a fundamental limit to detector D_d^*. This value is found by substituting the signal irradiance E_s for E_B.

7.8.3 Detection and False Alarm Probability

The signal current produced by a target in the field of view of the infrared sensor must compete with internal noise generated by the sensor and with extraneous signals produced by photons emitted by the background. The ability to detect the target signal is represented by a stochastic process whose statistics are dependent on the probability density functions of the desired signal and the competing noise.

The current at the output of the detector caused by white noise has a fluctuating or ac component that is Gaussian distributed with zero mean and a probability density given by [27]

$$p(i_N) = \frac{1}{\sqrt{2\pi \overline{i_N^2}}} \exp[-i^2/(2\overline{i_N^2})] \tag{7.67}$$

Examples of white noise (i.e., noise of constant spectral density) include noise produced by electronic thermal energy (Johnson noise) and shot noise containing a large number of photoelectron events per measurement interval. In the presence of a noncoherent signal current i_s, the probability density is still Gaussian-distributed but now has a mean value of $\overline{i_s}$ as shown in Figure 7.25. The area under $p(i_N)$ to the right of i_T represents the false alarm probability, while the area under $p(i_N + i_s)$ to the right of i_T that does not also lie under $p(i_N)$ represents the detection probability.

If the signal current exceeds a threshold current i_T, a detection is said to have occurred. The bandwidth B of the postdetection filter is usually matched to the

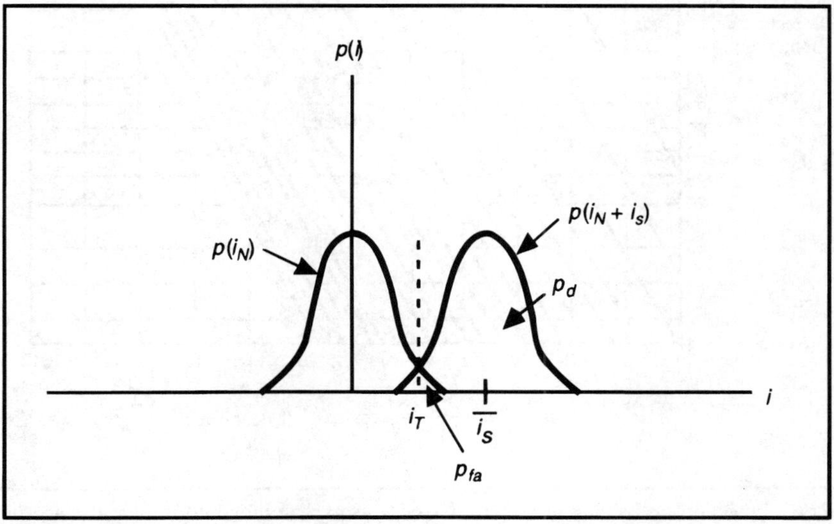

Figure 7.25 Probability density for Gaussian noise alone and signal in Gaussian noise.

spectral bandwidth of the signal and determines the value of the mean square noise $\overline{i_N^2}$. Under these conditions the false alarm probability p_{fa} is expressed as

$$p_{fa} = p(i_N > i_T) = \frac{1}{\sqrt{2\pi \overline{i_N^2}}} \int_{i_T}^{\infty} \exp(-i^2/2\overline{i_N^2})\, di \tag{7.68}$$

and the detection probability p_d as

$$p_d = p(i_s > i_T) = 1 - \frac{1}{\sqrt{2\pi \overline{i_N^2}}} \int_{i_T - i_s}^{\infty} \exp(-i^2/2\overline{i_N^2})\, di \tag{7.69}$$

These probabilities are plotted in Figure 7.26 as a function of signal-to-noise ratio.

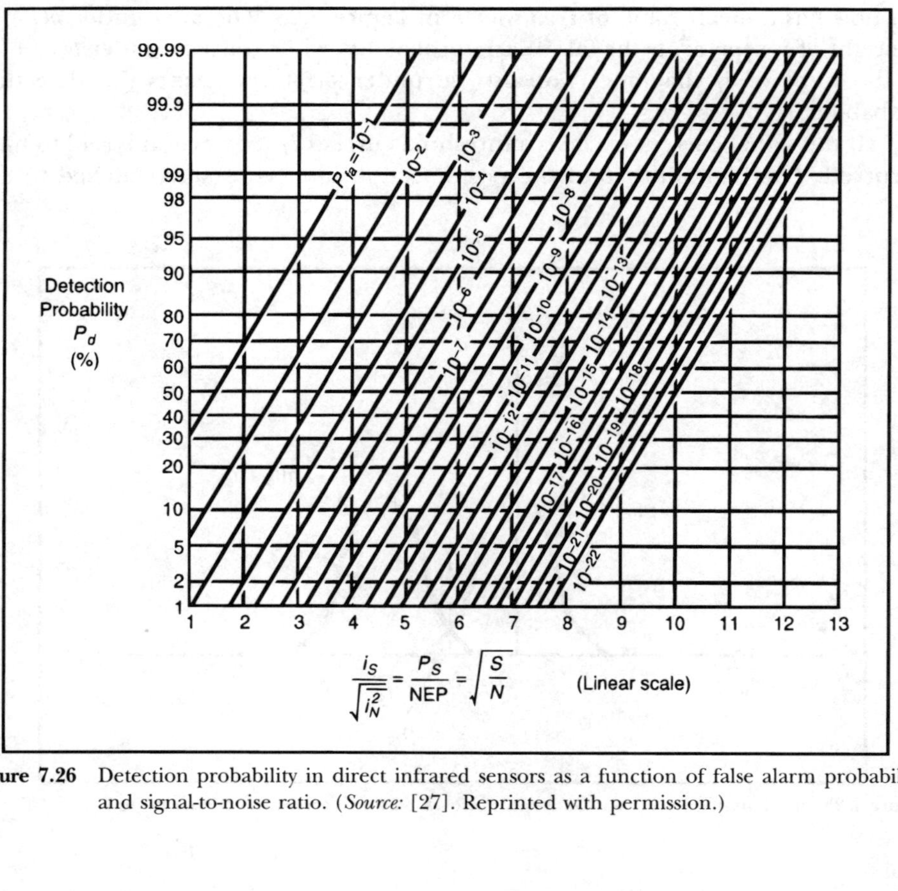

Figure 7.26 Detection probability in direct infrared sensors as a function of false alarm probability and signal-to-noise ratio. (*Source:* [27]. Reprinted with permission.)

In the limiting case of small p_{fa} and large p_d,

$$p_{fa} = \frac{1}{\sqrt{2\pi \overline{i_N^2}}} \int_{i_T}^{\infty} \exp(-i^2/2\overline{i_N^2})\ di \tag{7.70}$$

$$= \frac{1}{\sqrt{2\pi \overline{i_N^2}}} \exp(-i_T^2/2\overline{i_N^2}) \int_0^{\infty} \exp[-(x^2 + 2i_T x)/2\overline{i_N^2}]\ dx$$

For $i_T^2 >> \overline{i_N^2}$, (7.70) becomes

$$p_{fa} \approx \frac{1}{\sqrt{2\pi \overline{i_N^2}}} \exp(-i_T^2/2\overline{i_N^2}) \int_0^{\infty} \exp[-(i_T/\overline{i_N^2})x]\ dx \approx \frac{1}{\sqrt{2\pi \overline{i_N^2}}} \exp(-i_T^2/2\overline{i_N^2}) \frac{\overline{i_N^2}}{i_T} \tag{7.71}$$

Thus, for small false alarm probability,

$$p_{fa} = \frac{\sqrt{\overline{i_N^2}}}{\sqrt{2\pi}\, i_T} \exp(-i_T^2/2\overline{i_N^2}) \tag{7.72}$$

For example, when $i_T/\sqrt{\overline{i_N^2}} = 6$, the false alarm probability is

$$p_{fa} \approx 10^{-9} \tag{7.73}$$

Inserting similar approximations into the detection probability equation of (7.69) gives

$$p_d \approx 1 - \frac{\sqrt{\overline{i_N^2}}}{\sqrt{2\pi}(i_s - i_T)} \exp[(i_s - i_T)^2/2\overline{i_N^2}] \tag{7.74}$$

From (7.57) and (7.58),

$$\frac{i_s}{\sqrt{\overline{i_N^2}}} = \frac{P_s}{NEP} \tag{7.75}$$

When $(i_s - i_T)/\sqrt{\overline{i_N^2}} = 2$ and $i_T/\sqrt{\overline{i_N^2}} = 6$, (7.75) becomes

$$\frac{i_s}{\sqrt{\overline{i_N^2}}} = \frac{P_s}{NEP} = 8 \tag{7.76}$$

from the addition of the two current ratios. Then, from (7.74),

$$p_d \approx 98 \text{ percent.} \tag{7.77}$$

When the signal is generated by a Rayleigh-fluctuating target (i.e., one whose signal fluctuates with an exponential probability density from integration cycle to integration cycle), the general solution for the detection probability is given by [27]

$$p_d = p_d(i_s) = \int_0^\infty p(i_s)\, di_s \tag{7.78}$$

where

$$p(i_s) = \frac{1}{\bar{i}_s} \exp(-i_s/\bar{i}_s) \tag{7.79}$$

The integral equation of (7.78) represents the detection probability averaged over the signal power present during the integration time. When the average signal current $\bar{i}_s$ is much greater than the root mean square (rms) value of the noise current such that

$$\bar{i}_s >> \sqrt{\overline{i_N^2}}$$

noise current contributions can be ignored. When i_T is much greater than the noise current, the fluctuations of i_s due to the noise are small compared to i_T. In this case, the detection probability p_d is

$$p_d = p(i_s > i_T) = \int_{i_T}^\infty \frac{1}{\bar{i}_s} \exp(-i_s/\bar{i}_s)\, di_s = \exp(-i_T/\bar{i}_s) \tag{7.80}$$

In terms of a threshold input optical power P_T, (7.80) becomes

$$p_d = \exp(-P_T/\overline{P}_s) \approx \left(1 - \frac{P_T}{\overline{P}_s}\right) \quad \text{when} \quad \overline{P}_s >> P_T \tag{7.81}$$

Using the same threshold-to-noise current ratio as before, namely $i_T/\sqrt{\overline{i_N^2}} = 6$, gives the same false alarm probability as before ($p_{fa} \approx 10^{-9}$).

However, with $\bar{i}_s/\sqrt{\overline{i_N^2}} = 8$, the detection probability now decreases to

$$p_d = \exp(-6/8) = 47\% \tag{7.82}$$

The degradation of detection probability experienced with slow-fluctuating targets can be reduced if rapid fluctuations occur during some of the integration cycles, or the bandwidth is increased (but the noise current is still kept small relative to the threshold and signal currents), or spatial decorrelation of the signal occurs.

7.9 HETERODYNE DETECTION OF PHOTON ENERGY

Heterodyne detection of coherent photon energy in noise is described by the same detection probability curves as used with MMW radars. Although heterodyne detection typically occurs in active infrared sensors such as some laser radars, their detection probability is described here in order to consolidate the discussion. When thermal or shot noise competes with the desired signal, the curves in Figure 7.27(a) are used to compute the detection probability of nonfluctuating targets and those in Figure 7.27(b) to compute the detection probability of Swerling 2 targets.

In heterodyne systems, the detector output power rather than the output current is proportional to input power. The square of the signal-induced current i_s is therefore proportional to P_s. Thus, a signal-to-noise current ratio of 8 corresponds to an input signal-to-noise power ratio of 64.

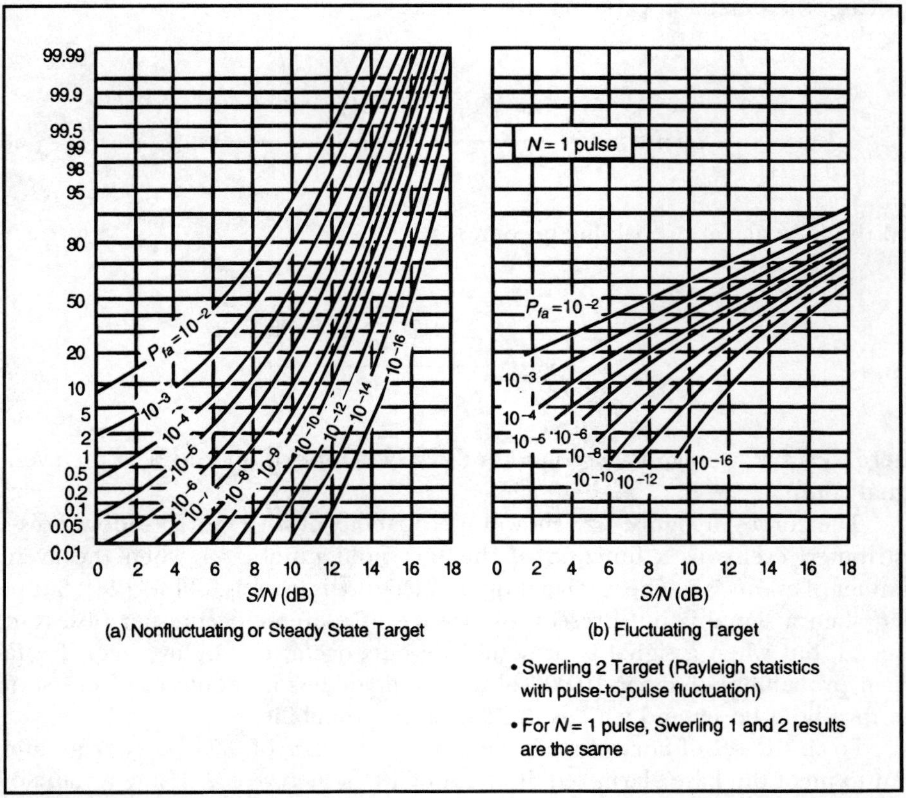

Figure 7.27 Detection probability in heterodyne infrared sensors as a function of false alarm probability and signal-to-noise ratio. (*Source:* [27]. Reprinted with permission.)

7.10 DETECTION OF PHOTON ENERGY IN QUANTUM NOISE

When the signal and noise are small, a small number of photons (of the order of 1 to 100) arrives at the detector in any given short time interval. The number of photons in any one interval is independent of the number of photons in another interval. Under these circumstances, the Poisson distribution describes the probability that in a given time interval the average number of photons arriving or events occurring will be exactly equal to *k*, where

$$p(\overline{n} = k) = \frac{(\overline{n})^k \exp(-\overline{n})}{k!} \tag{7.83}$$

where $\overline{n}$ is the average number of photons that arrive during the time interval, equal to $\overline{r}\tau$ as used in (7.55).

By counting the number of photons n during the selected interval, a threshold number n_t can be established for deciding whether or not a signal is present. In this case, the detection probability becomes

$$p_d(n \geq n_t) = \frac{\sum_{k=n_t}^{\infty} (\overline{n_b} + \overline{n_s})^k \exp[-(\overline{n_b} + \overline{n_s})]}{k!} \tag{7.84}$$

and the false alarm probability becomes

$$p_{fa}(n \geq n_t) = \frac{\sum_{k=n_t}^{\infty} (\overline{n_b})^k \exp[-(\overline{n_b})]}{k!} \tag{7.85}$$

where $\overline{n_b}$ is the average background and dark-current count and $\overline{n_s}$ is the average signal count.

The curves in Figure 7.28 show the probability of n_t or more random Poisson-distributed events as a function of the threshold number n_t when the average number of events is $\overline{n}$. The information in the figure is used as follows [28]. Suppose that when a signal is not present an average of 2 events occurs per observation interval, but when a signal is present 16 events occur, on the average. If a false alarm probability of 1 per 10,000 observations or less is required, where should the threshold be set and what is the detection probability?

From the set of curves on the left, a threshold n_t of 9.5 events or greater is seen to meet the false alarm requirement of 10^{-4} when $\overline{n} = 2$. Using a value of $n_t = 10$ (the next higher integer) in the curves on the right shows that a detection probability of 0.95 can be achieved when $\overline{n} = 16$ in the presence of a signal.

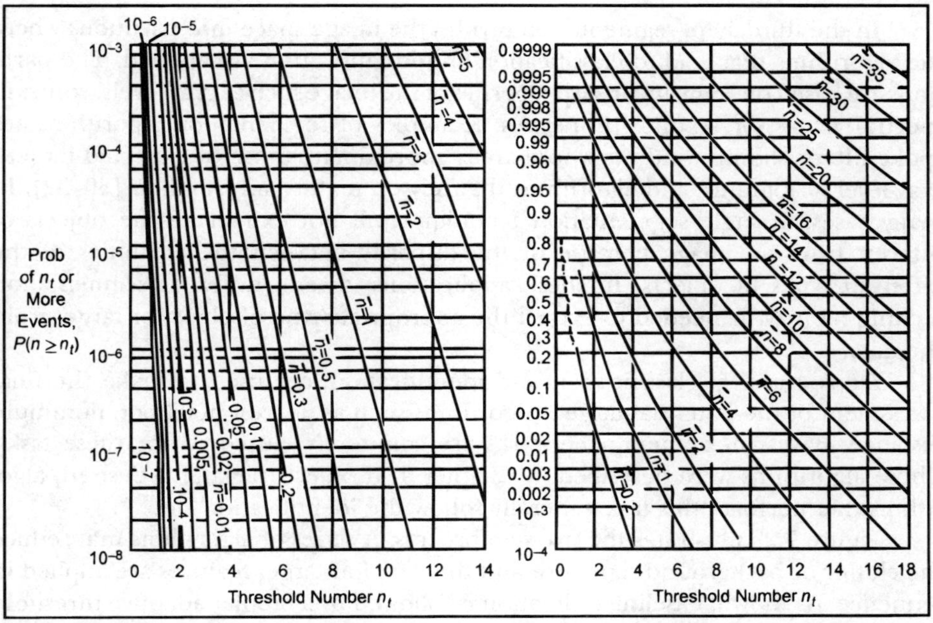

Figure 7.28 Detection probability of signals in quantum noise as a function of threshold number and false alarm probability. (*Source:* [28]. Reprinted with permission of BURLE INDUSTRIES.)

7.11 SIGNAL PROCESSING OF MULTIPIXEL IMAGERY FOR AUTOMATIC TARGET RECOGNITION

The extraction of target classification and identification information from multipixel imagery can be described as consisting of five general steps: (1) image enhancement, (2) detection, (3) segmentation, (4) classification, and (5) identification. In the first step, enhancement techniques are applied to all input pixel data to improve signal quality. Included in this step are hardware-based discriminants such as spectral filtering, noise suppression, and resolution control, and firmware-based features such as baseline estimation, feature space transformation, and contrast enhancement. The latter techniques include linear remapping of pixel values to "stretch" the intensity range, adjustment of pixel intensity values according to the local (neighborhood) statistics of pixel intensities, and temporal averaging of multiple images [29].

The detection process of step 2 acts on all the data as well. Here, one or more thresholds limit and segregate data that are passed on to the rest of the algorithms. It is undesirable to severely limit the number of potential targets during detection, for once data are removed they cannot be recovered. Therefore, false detections are permitted at this stage since false targets themselves are not declared at the conclusion of detection processing. Rather, algorithms contained in subsequent steps in the processing flow are relied on to eliminate false targets. Therefore, detection processing is not used to commit a weapon, for example, to a target.

In the third step, segmentation divides the image space into partitions where the algorithms that lead to classification and identification are applied. The partitions are based on some intrinsic properties of the image such as gray levels, contrast, spectral values, or textural properties. Examples of segmentation algorithms are spoke filters, sliding windows whose areas approximate those of expected targets, gray-level histograms, and algorithms that rely on texture information [30–32]. In many cases, a single segmentation technique will not extract all the objects of interest from an image because of the disparity between the attributes of the different types of objects. In these applications, several different segmentation techniques may be needed to extract the appropriate type of object or target from the data.

Steps 4 and 5, classification and identification, are used to make the final assessment of the potential target. Algorithms such as nearest neighbor, minimum distance, Bayesian, or neural networks are commonly employed for these tasks. These algorithms were described in Chapter 4. If object tracking is desired, algorithms that perform this task generally follow the identification step.

Figure 7.29 illustrates the five-step process. A series of algorithms that reduce the clutter or background signature and then exploit target features are applied in sequence. As shown, baseline estimation is followed by a bipolar adaptive threshold algorithm. The bipolar threshold is useful when a target can be cooler or warmer than the background, depending on time of day, weather, and operational state of the target. This type of threshold allows pixels with varying target-to-background temperature differences to be passed on to a series of algorithms, denoted by algorithm 1 to algorithm *n* in the figure, that define the shape and size of the object, and perhaps some of the features enclosed by its boundary.

Further examples of algorithms that may be represented in the sequence 1 to *n* include the following. A median filter may be used to eliminate large, spike-like clutter objects from further processing. Edge-based segmentation may be applied to find the contrast that often marks region boundaries. Gradient algorithms are an example of this type of algorithm. Once the edges are identified, ratios of relative

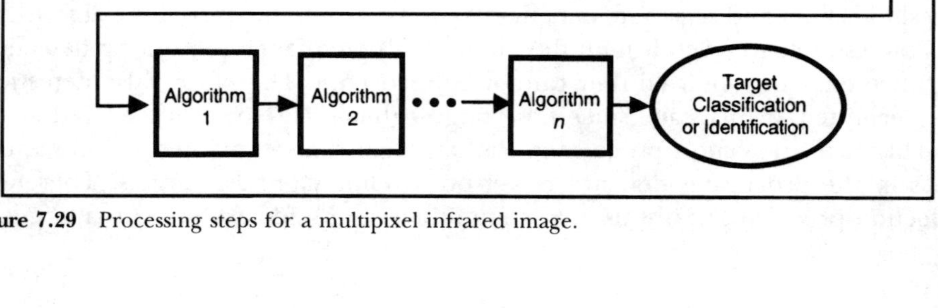

Figure 7.29 Processing steps for a multipixel infrared image.

length to width can be formed to determine the shape of the objects. Pixel-clustering algorithms are used to pass only groups of pixels that approximate the size of expected target features or the size of the expected target itself. They allow isolated hotspots or areas that are smaller than the dimension of the targets of interest to be removed from the scene. This may be helpful in eliminating countermeasures such as flares. Still other types of data may require that edge location algorithms used in the data segmentation process have adjustable sensitivities. Edge location algorithms may lead to oversegmentation of the data, causing the object to break up into many parts. Conversely, lower sensitivities may cause data to become undersegmented, causing multiple objects to be connected as a single object. Tuning the parameters of the segmentation algorithm may not produce the ideal or required segmentation in all the data sets of interest. In these situations, region merging and splitting techniques are frequently used to combine or differentiate appropriate areas. Region merging, used to combine over segmented data, relies on recognizing region characteristics that are homogeneous for a given set of properties such as average intensity and texture. Region merging may also be performed by measuring the ratio of object size to shared boundary. Region splitting is used to break apart areas that should be multiple objects. Convex segmentation that breaks regions into convex blobs is one method used for region splitting [33].

Since targets are often moving and are scanned at arbitrary oblique angles, moments that are invariant with respect to rotation, translation, reflection, and changes in object size are powerful types of algorithms that can be applied to classify and identify targets [34,35]. If range information can be obtained from another sensor, then true object size can be found, further enhancing target identification. Once an object is classified or identified as a target, other algorithms can be used to identify the ideal point of impact for a munition fired at the target. These algorithms identify warm areas that lie within the boundaries of the target. It is important that the impact point be selected within this region to avoid aiming the munition at a hotspot that may lie outside the physical target boundary. Such outlying areas can be produced by hot exhaust gases.

Figure 7.30 illustrates the application of automatic target recognition image processing to simulated and real data. The simulated data allow development of prototype algorithms even before real data are available. In this manner, an algorithm set is available for evaluation against real data when the latter are gathered. There are two approaches for simulating target and background signatures [36,37]. The first is suitable for predicting the probability of target detection in a clutter background. Here, target signatures are modeled in a deterministic manner and background signatures treated as stochastic processes. The deterministic model for targets is used because under a given set of conditions (that include viewing angle, time of day, weather, and thermal distribution on the surfaces of the target), the signature is rather well-defined and leads to a unique characterization. The same is not true for backgrounds, which are many and varied. Thus, the general approach for modeling backgrounds is to select a data set from a homogeneous scene, extract pertinent statistics such as the mean temperature, standard deviation, characteristic

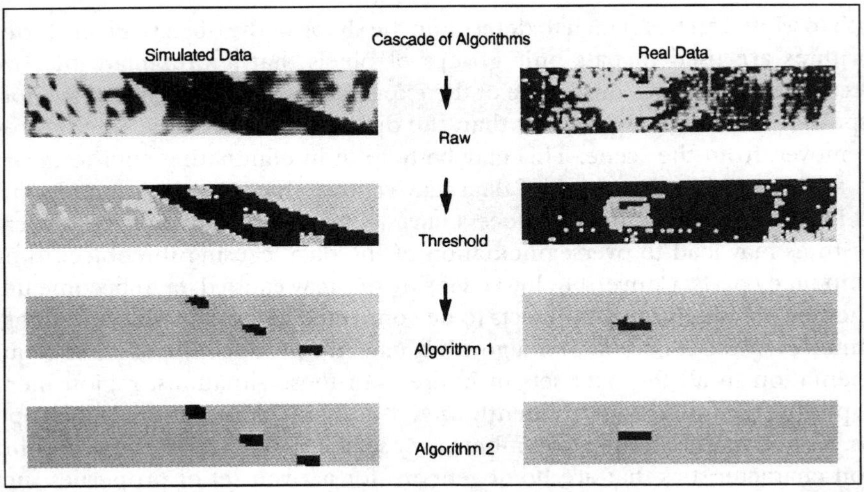

Figure 7.30 Application of multipixel infrared signal processing to simulated and real data. (*Source:* Aerojet, Azusa, CA. Reprinted with permission.)

length such as the spatial extent of the correlation function, and power spectral density, and then to develop a model that can simulate the background class (from which the data were drawn) with these same statistics. The size of each homogeneous background class produced from the model can be larger than the pixel resolution of the sensor and still satisfy the purpose of this type of simulation.

The second type of simulation is more suitable for assessing the sensor's capability to reject false targets. In this case, the background IR signature model includes specific individual scene elements that may confuse the target detection logic, such as the transition region between different textured background classes. Examples of backgrounds that have different textures are forested areas, plowed fields, and paved roads. The measured background signatures extracted from the data set and used in the model have to be compatible with the mission profile and sensor design in terms of viewing direction, sensor wavelength band, and scene pixel size. One approach to creating scene data that replicate the statistics of measured data is to use a time series model. This model improves upon models that calculate pixel-to-pixel temperatures drawn from normal distributions because the time series approach naturally forecasts outlier temperatures that represent the false alarms found in the data. The time series models apply autocorrelation functions to forecast the temperature of succeeding rows and columns of pixels based on the temperature of the first pixel. Some of the time series models do not adequately model spatial variation in the measured data, but do provide simulated data that are consistent with the statistics of the measured data such as mean temperature, standard deviation, and energy spectrum [36]. The application of Weibull and log-normal distributions, as discussed in Chapter 4, is another approach that can be used to model pixel-to-pixel temperature distributions when larger numbers of false alarms are desired.

The top photo on the left side of Figure 7.30 shows unprocessed or raw simulated data of three tanks on a road in an otherwise high-IR clutter background. The scene was simulated using a combination of deterministic models for manmade objects and stochastic processes for pixel-to-pixel background variations, transitions from one clutter type to another, and texture [37]. A radiance map was used to model the emission from the elements in the scene and a topographic map to model global features, such as major hills and valleys, and smaller scale features corresponding to each type of background class. A projection model simulated obscuration of one object by another, perspective effects, and range-dependent phenomena such as haze and fog. Background trees and shrubs were modeled using both deterministic and stochastic processes based on "growth rules" that emulated the way these objects grow in nature. Texture was simulated using a two-dimensional autocorrelation function.

In most fire-and-forget weapons applications, the sensor scans across many square meters of background before a target is encountered. During this time, the sensor output signal is processed by algorithms that generally include an adaptive threshold that is proportional to the average of the sensor output signal. In the second row of Figure 7.30, such a threshold is used to remove the clutter that is not in the vicinity of the targets. Then, a series of additional algorithms, as discussed above, are used to further isolate the targets from the remaining clutter background. When the real data were obtained and processed as shown on the right, the same set of algorithms performed as well. Although a set of algorithms produced from simulated data may not always achieve the same performance against a real data set, the simulated data do allow initial algorithms to be evaluated and can speed up the development of the final algorithm set.

7.12 SUMMARY

A process for designing infrared sensors has been presented. It begins by specifying the sensor design goals and the scenario in which the sensor will function, and continues through the selection of the spectral band, detector material, resolution, and signal-processing concept. Often, one or more iterations occur between specifying the goals and finalizing the sensor design. The sensor resolution required for target detection, orientation, classification, and identification has been defined for first- and second-generation FLIRs in terms of the number of pixels enclosed by an object. The probability of correct target detection, classification, and identification increases as the number of pixels and the signal-to-noise ratio increase. Measures of performance for nonimaging and imaging infrared sensors were defined and discussed. For nonimaging devices, these are signal-to-noise ratio, noise-equivalent temperature difference, and minimum detectable temperature difference. For imaging sensors, the minimum resolvable temperature difference and modulation transfer function are more descriptive of their imaging capability. Some performance measures, such as the noise-equivalent temperature difference, are not constant for a given detector material, IFOV, or focal plane temperature, but rather are functions of the wavelength band of operation and the background temperature. The effects of the detector array configuration on aliased frequencies and the effect of sampling frequency on the image quality and probability of classification were illustrated. As expected, the image quality and classification probability improve as the sampling frequency increases. Methods of calculating the detection probability and false alarm probability of infrared sensors operating in direct detection, heterodyne detection, and quantum noise-limited modes were described. Finally, a general five-step method for processing multipixel imagery for automatic object recognition applications was discussed and an example of the process was given. The steps are image enhancement, detection, segmentation, classification, and identification.

References

[1] Pinson, L. J., *Electro-Optics*, New York, NY: John Wiley & Sons, 1985.

[2] Johnson, J., "Analysis of Image Forming Systems," *Image Intensifier Symposium*, Fort Belvoir, VA, Report AD-220160, Oct. 1958, pp. 249–273.

[3] Johnson, J. and W. Lawson, "Performance Modeling Methods and Problems," *Proc. IRIS Imaging Systems Group*, Jan. 1974.

[4] Shumaker, D., "Target Detection and Recognition Revisited," Technote 95-03, *Spectral Reflections*, Infrared Information Analysis Center, ERIM, Ann Arbor, MI: Infrared Information Analysis Center, 1995.

[5] Shumaker, D. L., J. T. Wood, and C. R. Thacker, *Infrared Imaging Systems Analysis*, ERIM, Ann Arbor, MI: Infrared Information Analysis Center, 1993.

[6] Rosell, F. A., and R. H. Willson, "Recent Psychophysical Experiments and the Display Signal-to-Noise Ratio Concept," Chapter 5 in *Perception of Displayed Information*, L. M. Biberman, (Ed.), New York, NY: Plenum Press, 1973.

[7] Schmieder, D. E. and M. R. Weathersby, "Detection Performance in Clutter With Variable Resolution," *IEEE Vol. AES-19*, No. 4, July 1983, pp. 622–630.

[8] Gordon, G. A., R. L. Hartman, and P. W. Kruse, "Imaging-Mode Operation of Active NMMW Systems," Chapter 7 in *Infrared and Millimeter Waves, Vol. 4: Millimeter Systems*, K. J. Button and J. C. Wiltse, (Eds.), New York, NY: Academic Press, 1981.

[9] Seyrafi, K., *Electro-Optical Systems Analysis*, Los Angeles, CA: Electro-Optical Research Company, 1973.

[10] Lloyd, J. M., *Thermal Imaging Systems*, New York, NY: Plenum Press, 1982.

[11] Schnitzler, A. D., "Analysis of Noise-Required Contrast and Modulation in Image-Detecting and Display Systems," Chapter 4 in *Perception of Displayed Information*, L. M. Biberman, (Ed.), New York, NY: Plenum Press, 1973.

[12] Smith, W. J., "Optical Elements-Lenses and Mirrors," Chapter 9 in *The Infrared Handbook, Revised Edition*, W. L. Wolfe, and G. J. Zissis, (Eds.), IRIA Center ERIM, 1989.

[13] Boreman, G. D., "Modulation Transfer Function in Optical and Electro-Optic Systems," SPIE Short Course Notes, Univ. of Central Florida, Orlando, Jan. 1991.

[14] McHugh, S. W., "High-Performance FLIR Testing Using Reflective-Target Technology," *Photonics Spectra*, July 1991, pp. 112–114.

[15] Sapir, E. and J. Eliason, "Automated FLIR Testing for Efficiency, Reliability," *Photonics Spectra*, June 1992, pp. 119–122.

[16] *Night Vision Laboratory Static Performance Model for Thermal Viewing Systems*, ECOM Report 7043, AD A011212, April 1975.

[17] Scott, L. and L. Conduff, "CNVEO Advanced FLIR Systems Performance Model," *SPIE Vol. 1309,* Paper 17, April 1990.

[18] Kennedy, H. V., "Modeling Second Generation Thermal Imaging Sensors," *SPIE Vol. 1309,* Paper 1, April 1990.

[19] Steckl, A. J., "Chapter 12: Charge-Coupled Devices," in *The Infrared Handbook, Revised Edition,* W. L. Wolfe and G. J. Zissis, (Eds.), IRIA Center, ERIM, 1989.

[20] Prettyjohns, K., "Improved CCDs Meet Special Demands of Spectroscopy," *Laser Focus World,* Oct. 1992, pp. 127–136.

[21] Buck, A., "Wide Dynamic Range Forward Looking Infrared Set (WDRF)," *SPIE Vol. 344,* 1982, pp. 89–96.

[22] Freeman, H., *Discrete-Time Systems—An Introduction to the Theory,* New York, NY: John Wiley & Sons, 1965.

[23] Wolfe, W. L., "Chapter 10: Optical-Mechanical Scanning Techniques and Devices," in *The Infrared Handbook, Revised Edition,* W. L. Wolfe, and G. J. Zissis, (Eds.), IRIA Center, ERIM, 1989.

[24] Wolfe, W. L., "Chapter 19: Imaging Systems," in *The Infrared Handbook, Revised Edition,* W. L. Wolfe, and G. J. Zissis, (Eds.), IRIA Center, ERIM, 1989.

[25] Chow, S., and T. Jones, "Army's FLIR/ATR Evolution Path," *Infrared Systems and Components III,* SPIE, Vol. 1050, 1989, pp. 42–49.

[26] Dereniak, E. L., and D. G. Crowe, *Optical Radiation Detectors,* John Wiley & Sons, NY, 1984.

[27] Kingston, R. H., *Detection of Optical and Infrared Radiation,* Springer-Verlag, NY, 1978.

[28] *Electro-Optics Handbook TP-135,* BURLE INDUSTRIES, INC., Tube Products Div., Lancaster, PA, 1974.

[29] Dawson, B. M., "Processing Extends Scientific Vision," *Laser Focus World,* Feb. 1991, 121–131.

[30] Weszka, J. S., C. R. Dyer, and A. Rosenfeld, "A Comparative Study of Texture Measures for Terrain Classification," *IEEE Vol. SMC-6,* No. 4, Apr. 1976, pp. 269–285.

[31] Trivedi, M. M., et al., "Object Detection Based on Gray Level Coocurrence," *Computer Vision, Graphics, and Image Processing,* Vol. 28, 1984, pp. 199–219.

[32] Moigne, J. L., and J. C. Tilton, "Refining Image Segmentation by Integration of Edge and Region Data," *IEEE Trans. on Geos. and Rem. Sens.,* Vol. 33, No. 3, May 1995, pp. 605–615.

[33] Eggleston, P., "Scientific Image Analysis Development Tools," *Advanced Imaging,* Oct. 1995, pp. 48–53.

[34] Dudani, S. A., K. J. Breeding, and R. B. McGhee, "Aircraft Identification by Moment Invariants," *IEEE Vol. C-26,* No. 1, Jan. 1977, pp. 39–46.

[35] Reddi, S. S., "Radial and Angular Moment Invariants for Image Identification," *IEEE Vol. PAM 1-3* No. 2, March 1981, pp. 240–242.

[36] Strenzwilk, D. F., M. P. Meredith, and W. T. Federer, "A Two Dimensional ARMA Model for the Simulation of IR Backgrounds," *Commun. Statist.—Part B: Simula.,* Vol. 18, No. 4, 1989, pp. 1539–1555.

[37] Schmieder, D. E., et al., "3-D IR Clutter Model," *Proc. Sixth Annual KRC Symposium on Ground Vehicle Signatures,* Houghton, MI, Aug. 21-22, 1984.

CHAPTER 8

Laser Radar

The advent of small, rugged, and uncooled semiconductor laser diodes has contributed to the growth of solid state laser radars in military and commercial applications. Military applications include laser designators that illuminate targets with a beam of light to guide munitions to them, laser range finders that measure the distance from an observer to an object, and laser radar sensors that are integrated into munitions to identify and autonomously guide the munition to the desired target. Commercial uses vary from compact disk players, laser printers, short-distance fiber-optic communications, and optical data storage to traffic sensors that measure vehicle speed and length and have the potential to classify vehicles from a high-resolution image of the vehicle's profile.

Although not as compact as the solid state laser, carbon dioxide gas lasers suit applications where heterodyne operation can be applied to measure the speed of objects through the Doppler frequency shift they produce. These lasers can also be used as a differential absorption laser (DIAL) radar to monitor atmospheric pollutants appearing as trace gases. This application requires the trace gas to have an absorption line in the emission wavelength band of the laser. If the laser wavelength is tuned to the absorption line, the received signal is proportional to the total atmospheric attenuation at that wavelength, including attenuation by the trace gas. Tuning the laser to an emission wavelength other than the absorption line of the trace gas allows measurement of the atmospheric characteristics in the absence of the trace gas. With knowledge of the atmospheric attenuation at the two wavelengths, the strength of the trace gas attenuation and hence its concentration can be calculated. Carbon dioxide lasers can be designed with output wavelengths that range from 9.2 to 10.8 μm [1]. This chapter describes the military applications of laser radars and the methods used to calculate the signal-to-noise ratio and detection probability performance measures that characterize them. Signal detection theory for infrared and millimeter-wave sensors are compared as part of this discussion.

8.1 TARGET DESIGNATION AND RANGING

Figure 8.1 contains examples of laser radar target designators as used in a ground battlefield scenario. In the upper portion of the figure, a military aircraft carrying Low-Altitude Navigation and Targeting Infrared System for Night (LANTIRN) is shown illuminating a target. Information that includes the target bearing angle and range is received by the aircraft and is programmed into the munitions that are launched against the targets. Ground-based command and control systems can supply similar data to the aircraft as represented by the communications truck in the lower left area of the figure. Other munitions such as laser-guided bombs can be directed to the target by using a ground-based laser designator to paint the target. The same laser designator can also guide ground-launched weapons such as Copperhead to the targets. Other munitions, such as some models of the air-launched Maverick and Hellfire, contain built-in lasers that locate and help guide the weapons to the target. Several types of target designators are in the military inventory, four of which are listed in Table 8.1 with their operating parameters [2]. Most of these systems require two people to set up and operate, as shown in Figure 8.2.

Laser rangefinders are smaller and lighter in weight than laser target designators. However, their effective range is a factor of ten less. They resemble a pair of binoculars, as shown in Figure 8.3. Laser rangefinders are made by several countries for use by their military, as indicated in Table 8.2 [2].

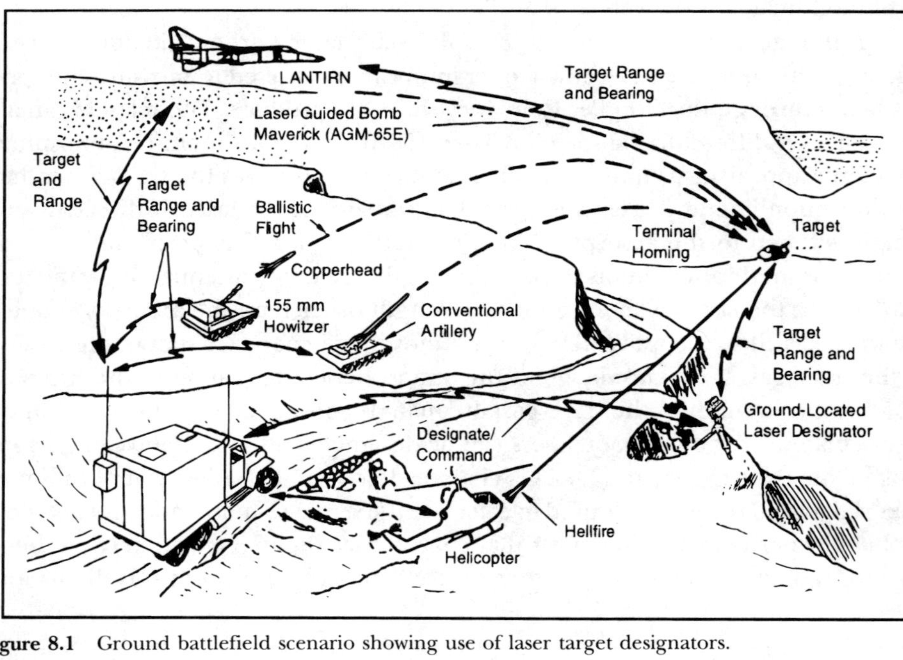

Figure 8.1 Ground battlefield scenario showing use of laser target designators.

Table 8.1
Laser Target Designators

	Laser Designator Model			
	LWTM[1]	*MULE*[2]	*G/VLLD*[3]	*Type 306*
Manufacturer	Litton (U.S.)	Hughes (U.S.)	Hughes (US)	Ferranti (UK)
Wavelength (μm)	1.064[4]	1.064	1.064	1.064
Sight magnification	7	10	13	8
Beam divergence (mrad)	0.5	0.25	0.13	0.25
Weight (lb/kg)	12.3/5.6	37.9/17.2	52.0/23.6	21.1/9.6
Range finder range (km)	10	10	10	10
Duty cycle	1 min on/ 2 min off	2 min on/ 2 min off	Continuous	2 min on/ 3 min off

[1] Lightweight target marker.
[2] AN/PAQ-3 Modular Universal Laser Equipment.
[3] AN/TVQ-2 Ground/Vehicular Laser Locator-Designator.
[4] 1.540 eyesafe wavelength option is available.

Figure 8.2 Laser radar target designators. (*Source:* Hughes Aircraft Company, Los Angeles, CA. Reprinted with permission.)

8.2 SOLID STATE LASERS

Near-infrared band (0.87 to 0.92 μm) laser diodes made of gallium arsenside (GaAs) and gallium aluminum arsenside (GaAlAs) were among the first semiconductor materials used in military applications [3,4]. Their emitted laser wavelength

AN/GVS-5 (RCA) LH-83

Figure 8.3 Laser rangefinders. (*Source:* [2] *Jane's Infantry Weapons 1987–1988.* Reprinted with permission.)

depends on the bandgap of the active layer, as was illustrated in Figure 2.21 where the bandgap is the energy needed for an electron in the valence band to jump to the conduction band and be free to carry current. Current is carried by both the free conduction band electrons and holes in the valence band created by the absent electrons. Pure GaAs has a bandgap energy of 1.35 eV (electron volts) at room temperature, corresponding to a wavelength of 905 nm. Adding aluminum increases the bandgap energy, moving the laser emission to shorter wavelengths since energy is inversely proportional to wavelength as described by Planck's law. For aluminum concentrations x from 0 to 0.45, $Ga_{1-x}Al_x As$ is a direct bandgap material supporting emission at wavelengths as short as 620 nm. Changes in the energy-level structure, however, reduce laser efficiency at shorter wavelengths, limiting lifetime of the laser. The shortest wavelength offered commercially is 750 nm. Structures needed to achieve continuous wave operation require some aluminum in the active layer, limiting the longest wavelength to about 870 nm. Pulsed lasers with pure GaAs active layers extend to 905 nm. Materials such as zinc selenide (ZnSe) (480 to 525 nm), AlGaInP (630 to 680 nm), GaInP (670 nm), GaAlAs (620 to 895 nm), and frequency-doubled titanium (Ti): sapphire (700 to 1050 nm) are used to manufacture visible wavelength (400 to 700 nm) semiconductor laser diodes.

In addition to the development of GaAlAs as a near-infrared laser diode material, recent research and development activities have also focused on mid-infrared band devices. The mid-infrared band has more wavelengths available with which to match a particular absorption, emission, or transmission feature and, at wavelengths greater than $\approx 1.4\ \mu m$, make the laser light safe to the human eye. As shown in Figure 8.4, the transmission of the vitreous humor of the eye is very low at wavelengths greater than $1.4\ \mu m$ [5]. Hence these wavelengths cannot be focused on the retina and are thus noninjurious to the human eye. Table 8.3 lists the operating wavelengths and other characteristics of several mid-infrared solid state laser materials [6,7].

Table 8.2
Laser Rangefinder Characteristics

	Laser Rangefinder Model					
	LRR-104 Mark V	*AN/GVS-5*	*LH-83*	*TM-18*	*LP 7*	*ELAM*
Manufacturer	Litton (U.S.)	RCA and OEC (U.S.)	OEC (US), United Sci. Instr. (UK)	CILAS (FR)	Simrad (NOR)	Armscor (SO. AF.)
Wavelength (μm)	1.064	1.064	1.540 (eyesafe)	1.064	1.064	1.064
Range (m)	50 to 9995	200 to 9995 (100 m option)	100 to 9995	150 to 9990	150 to 9995	200 to 8000
Ranging accuracy (m)	±3	±10 (±5 m option)	±5	±10	±5	±5
Measurements per battery charge at 24°C	2000 with lithium	700	700 with lithium	600	≃600	600
Ranging probability	Not available	99% to 5-km target with 5-km visibility	Not available	Not available	Not available	Not available
Energy output (mJ)	4	10	20	30	10	8
Sighting optics:						
• Magnification	5	7	7	6	7	6
• FOV (degrees)	2	7	7	7	7	6
• Display (in eyepiece)	• Range (4 digits) • Battery charge • Multiple targets	• Range (4 digits) • Battery charge • Multiple targets	• Range (4 digits) • Ready to range • Multiple targets	• Range	• Range (4 digits) • Battery charge	• Range
Size (mm)	69 × 150 × 182	89 × 152 × 203	65 × 165 × 198	75 × 150 × 180	93 × 202 × 215	70 × 125 × 160
Weight (lb/kg)	2.2/1.0	4.9/2.2	4.2/1.9	4.2/1.9	4.9/2.2	3.3/1.5
Operating temperature (°C)	−30 to +55	−46 to +71	−25 to +55	−30 to +55	−30 to +55	Not available

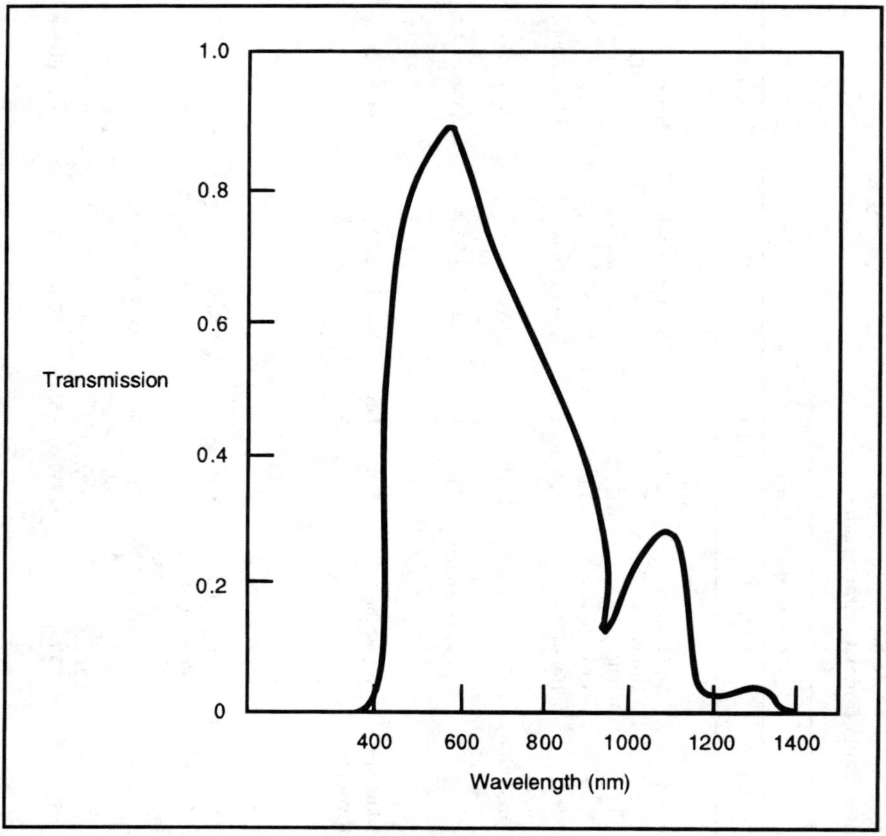

Figure 8.4 Infrared transmission through the human eye.

8.3 LASER RADAR IMAGERY

In military applications, compact near-infrared laser radars autonomously locate and classify targets and guide munitions to them. The three-dimensional imagery produced from the range information can be analyzed in real time using signal-processing algorithms that recognize features that are characteristic of the desired targets. By improving target classification, the three-dimensional images reduce susceptibility to naturally occurring false alarms and manmade countermeasures. Laser radars that use direct detection can produce three general types of imagery: (1) range images, (2) elevation images, and (3) intensity images. Range images are developed from the high-resolution range data obtained by processing the backscattered signals from the scatterers on the targets. Elevation images add an artificial groundplane reference and remove angular distortions that occur in processing scanned data. Elevation images permit the height profile along each scan line to be analyzed for features that lead to recognition of objects in the radar's field of view. Intensity images are produced by the differences in the reflectivity

Table 8.3
Mid-Infrared Solid State Laser Material Characteristics

*Material**	*Wavelength (μm)*	*Transition*	*Lifetime (msec)*	*Lower Level (cm^{-1})*	*Possible Sensitizers*
Er:YLF	1.73	$^4S_{3/2} \rightarrow {}^4I_{9/2}$	0.08	≈12,500	
Tm:YAG	2.32	$^3H_4 \rightarrow {}^3H_5$	0.8	≈8,200	Cr
Er:YAG	2.94	$^4I_{11/2} \rightarrow {}^4I_{13/2}$	0.08	≈6,700	
Er:YSGG	2.80	$^4I_{11/2} \rightarrow {}^4I_{13/2}$	1.4	≈6,700	Cr
Dy:YLF	4.34	$^6H_{11/2} \rightarrow {}^6H_{13/2}$	0.00083	≈3,500	
Er:YAG	1.66	$^4I_{13/2} \rightarrow {}^4I_{15/2}$	≈6	525	Yb
Tm:YAG	1.88, 2.01	$^3F_4 \rightarrow {}^3H_6$	15	588	Cr, Er
Ho:YLF	2.04 to 2.06	$^5I_7 \rightarrow {}^5I_8$	10	≈300	Er, Am, Cr
Ho:YAG	2.09 to 2.10	$^5I_7 \rightarrow {}^5I_8$	5.5	462	Er, Tm, Cr
Ho:YSGG	2.09	$^5I_7 \rightarrow {}^5I_8$	1.2 to 8.0	≈500	Er, Tm, Cr

*Er = erbium; Tm = thulium; Dy = dysprosium; Ho = holmium; YAG = yttrium-aluminum-garnet; YLF = yttrium-lithium-fluoride; YSGG = yttrium-scandium-gallium-garnet ($Y_3Sc_2Ga_3O_{12}$).

(i.e., power reflection coefficient) of the objects in the field of view. The ground typically has a relatively high reflectivity, while manmade targets typically have lower reflectivity, as shown in Table 8.4 [8,9].

8.3.1 Solid State Laser Radar Imagery

A near-infrared laser radar, similar to the one in Figure 8.5 built by Schwartz Electro-Optics (SEO), was used to produce the imagery shown in Figures 8.6 to 8.9. This single-channel laser employs a time-of-flight range finder that measures the range to noncooperative diffuse targets. The radar's pertinent characteristics are given in Table 8.5 [10]. The transmitter consists of a GaAlAs laser diode, optics, and pulse-forming electronics. Newer laser diodes can be produced in arrays to increase the scan area covered by the field of view and hence decrease the time required to scan a scene. They operate with greater pulsed output power of up to 120W and

Table 8.4
Near-Infrared Reflectivity Values [8,9]

Object	*Reflectivity*
Dry ground	0.39 ($1\sigma \approx 0.04$)
Wet ground	0.29 ($1\sigma \approx 0.03$)
Dry target	0.06
Wet target	0.025

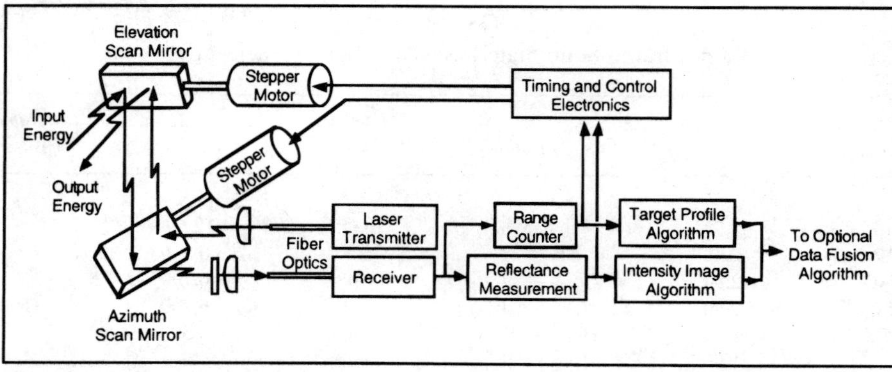

Figure 8.5 Single channel direct detection laser radar.

higher repetition frequencies of up to 25 kHz. These characteristics allow smaller features with smaller cross-sections to be observed and operation at greater ranges to be achieved. Laser arrays with the highest output powers require water cooling to keep the array from self-destruction. Since semiconductor diode laser radars emit energy into a large solid angle, the optical system designed by SEO collimates the received energy into a narrow beam. Furthermore, in order to gather as much of the backscattered energy as possible, the receiver optics create a large collection cone. This requires fast optics or equivalently a low F/number. These requirements are satisfied by a 2-inch clear aperture F/1.4 lens that collects approximately 50% of the transmitted energy. The scanner consists of two front-surface silver flat mirrors driven by microcontroller-actuated stepper motors. The motion of the mirrors creates a raster formatted image of a 4 by 8-degree field of regard.

The upper part of Figure 8.6 shows a photograph of a tank to the right of some shrubbery. By scanning the laser beam across the scene, a real-time (30-Hz frame rate) raster scanned image is produced from the scatterer range data as shown in the lower part of the figure. The resolution of the laser image with 2/3 pixel overlap is sufficient for the human eye to discern the tank. If the pixels are imaged in a contiguous manner without overlap, some resolution is lost in the visual imagery, but the autonomous signal-processing algorithms that are discussed below are still capable of recognizing the targets.

In Figure 8.7, an artificial ground reference plane has been used to convert the range image into an elevation image. Elevation images add the capability to place profile slices through the image along various scan lines. At each profile slice, such as the one shown by the black line through the turret of the tank, the range data are plotted as a function of their location along the profile, as illustrated in the lower part of the figure. A second profile through the gun is shown in Figure 8.8. From these and other profile slices, it becomes apparent that the profiles of clutter, such as the shrubbery, exhibit more change from profile to profile than do the profiles from man made objects such as targets. Thus target features such as shape, height, and rates of change of height can be constructed from which to classify targets. With the tank target, an algorithm can be developed to seek out

Figure 8.6 Photograph and range image of tank parked next to shrubbery. (*Source:* Schwartz Electro-Optics, Orlando, Florida. Reprinted with permission.)

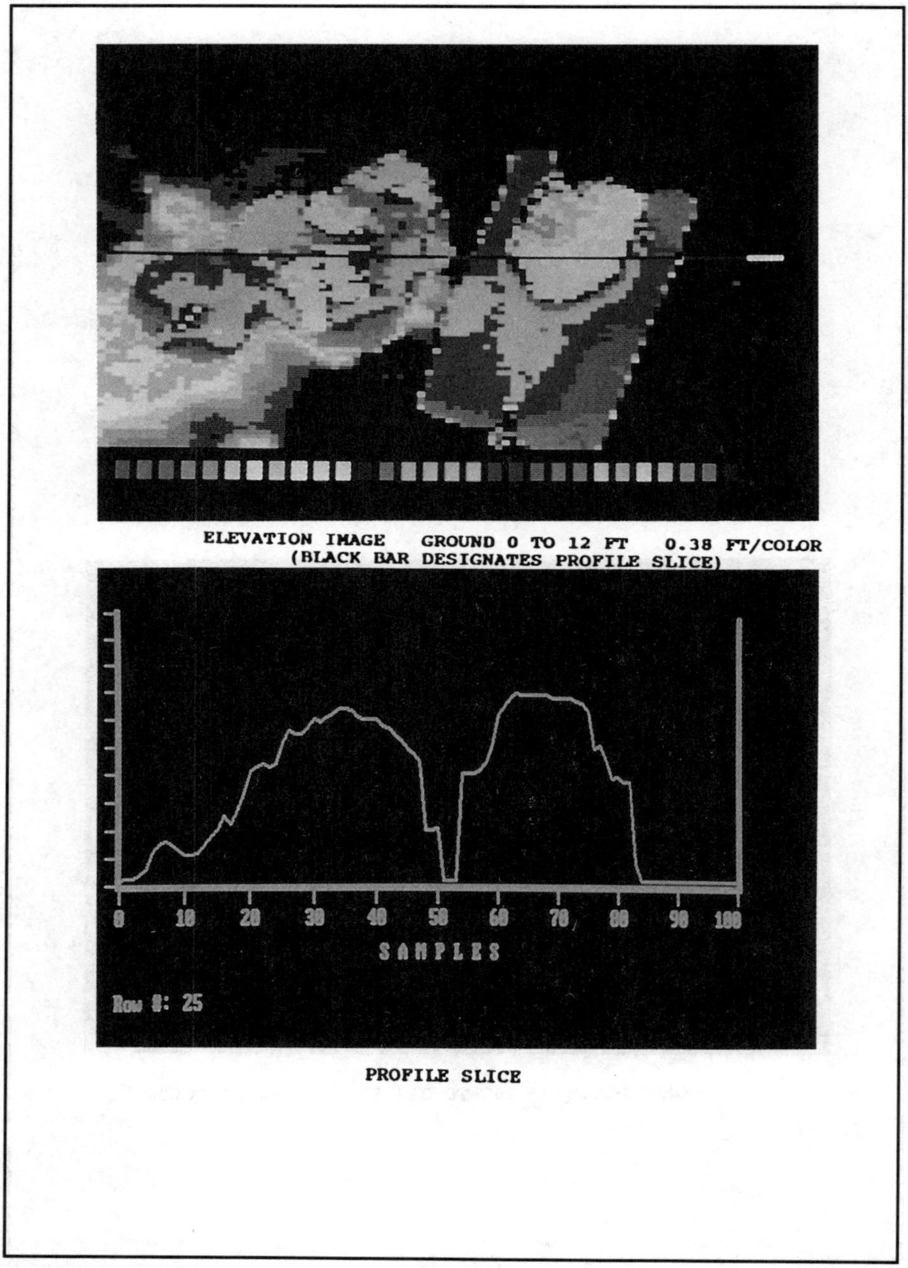

Figure 8.7 Elevation image of tank and shrubbery showing profile slice signature through hatch area of tank. (*Source:* Schwartz Electro-Optics, Orlando, Florida. Reprinted with permission.)

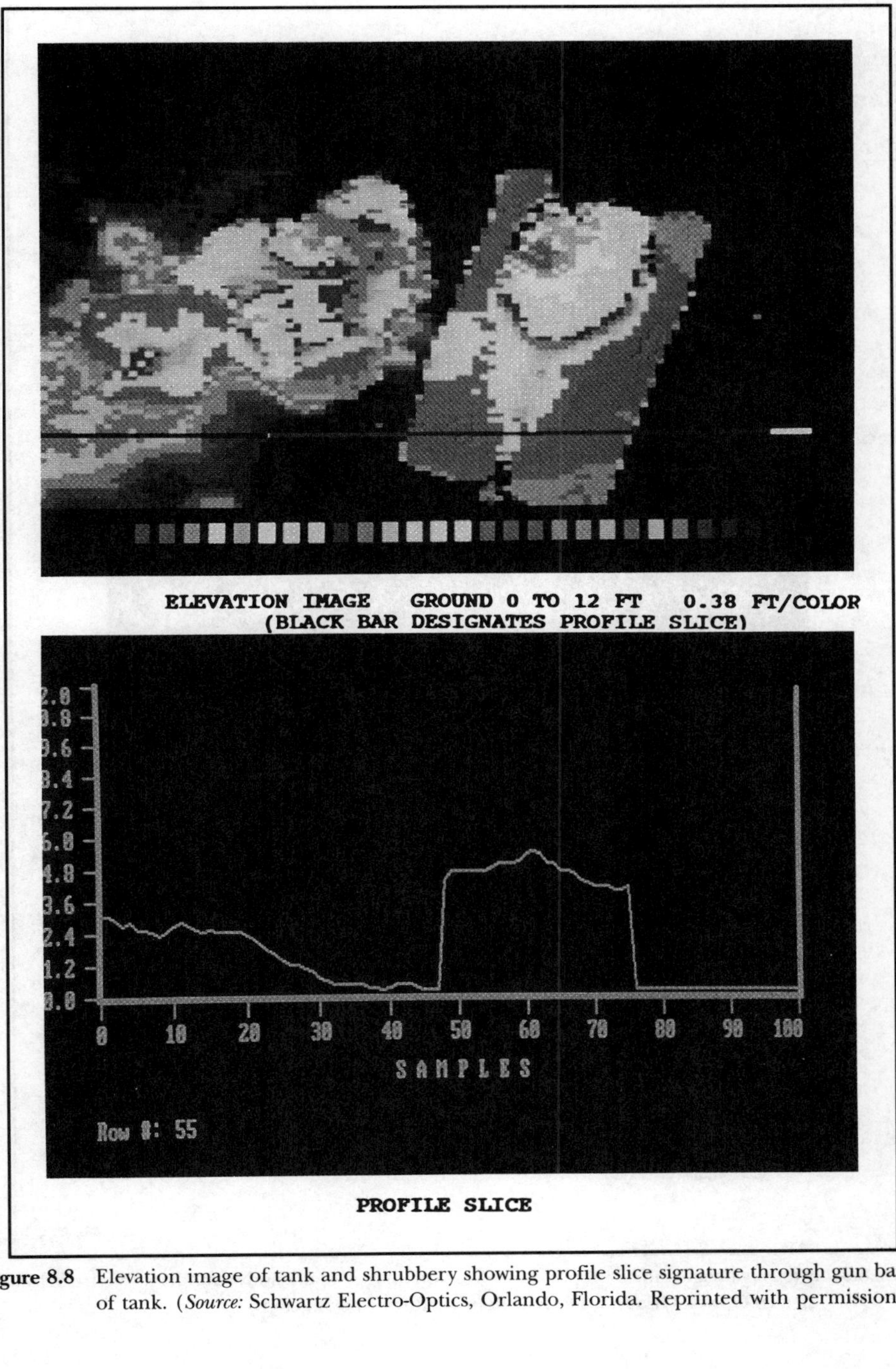

Figure 8.8 Elevation image of tank and shrubbery showing profile slice signature through gun barrel of tank. (*Source:* Schwartz Electro-Optics, Orlando, Florida. Reprinted with permission.)

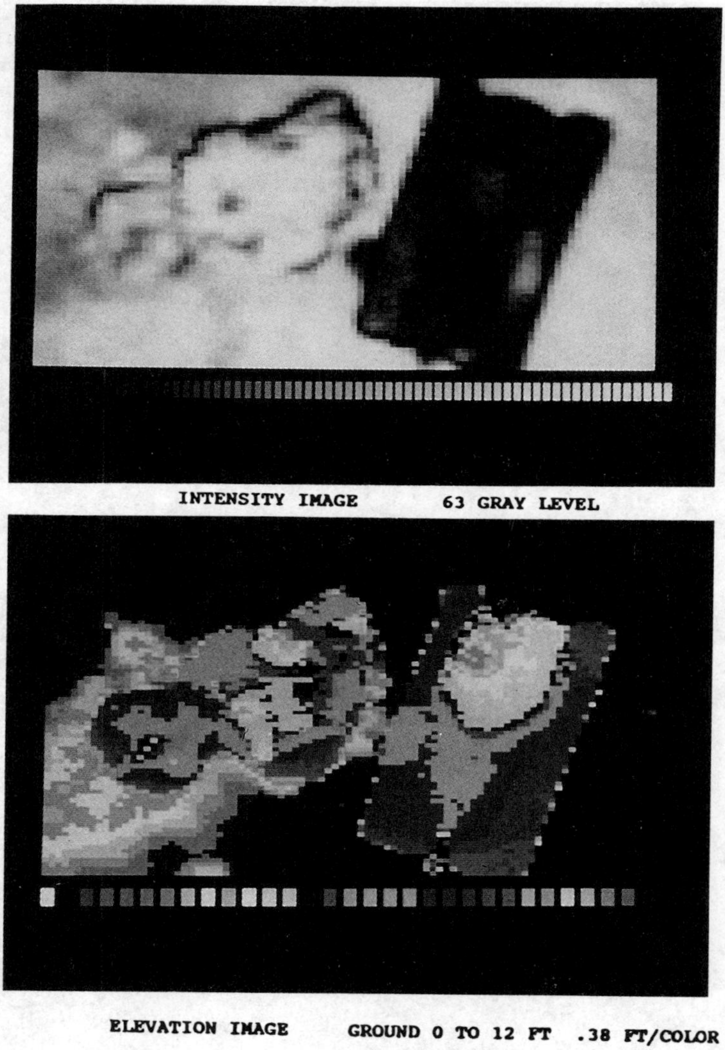

Figure 8.9 Intensity image and corresponding elevation image of tank and shrubbery. (*Source:* Schwartz Electro-Optics, Orlando, Florida. Reprinted with permission.)

Table 8.5
Schwartz Electro-Optics Near-Infrared Laser Radar Characteristics

Parameter	*Value*
Laser diode material	GaAlAs
Detector	Si avalanche photo-diode
Wavelength	850 nm
Output power	≈50 W from 6-nsec pulse
Pulse repetition frequency	1.5 kHz
Pixel size	5 mrad × ≈ 5 mrad
Raster scanning option	2/3 pixel overlap
Observation platform	300-ft (91-m) high tower
Slant range	275 to 325 ft (84 to 99-m)
Range resolution	10 cm
Field of regard	4 degrees by 8 degrees

features produced by profiles that, for example, pass through the engine deck, turret, and gun barrel.

The target classification algorithm initially developed by Schwartz Electro-Optics contains three major parts: a preprocessor, profile encoder, and a classifier [10]. The preprocessor employs a median filter, described in Chapter 4, to help eliminate large spike-like clutter from the data. The reference groundplane that converts the range data into an elevation image and compensates for scanning-induced angular distortion is also added at this time. A CFAR (constant false alarm rate) algorithm containing four tests is then used to divide the data into potential targets or background terrain. In general, CFAR allows the number of false alarms to remain constant even as the background clutter increases. The first test in this CFAR algorithm examines the average height or elevation of the digitized data in a target gate whose size can be adjusted to contain all of the samples belonging to a desired target. To pass the test, the average height must be less than a threshold that is adjustable to account for different target heights and hilly terrain. The second test compares the average sample height in the target gate with a minimum threshold to confirm that it represents a valid target. If the first two tests that act on the average sample height are passed, then a third test is applied to each sample in the gate to see if they fall within the maximum and minimum height thresholds. A sufficient number of samples, dependent on target size, in the target gate must satisfy this requirement for a valid target to be present. The final test checks the variance of the samples in the target gate against two constants related to the type of terrain that is expected.

If all CFAR tests are passed, the profile encoder portion of the classification algorithm operates on the data in the target gate. The encoder creates a table containing an array of relative elevation changes between two adjacent samples. Five features are measured to characterize the elevation change: a plateau, a rising or falling edge, and a positive or negative peak. The table also contains the number of samples and the elevation of each plateau.

The classifier makes the final decision as to whether the object is a valid target by using rules that operate on the feature set composed of the total number of plateaus, maximum elevation of a plateau, relative sizes of the plateaus, and the orientation of the plateaus. If an autonomous munition is equipped with such a laser radar and signal processing, the algorithm can also locate an aimpoint on the target that will maximize the munition's effectiveness. For example, by calculating the midpoint of the highest and longest plateau, aimpoints are found on the turret of a tank and the canopy area of a truck.

In addition to range data, the laser radar produces intensity images from the differences in the near-infrared reflectivity of the objects in the field of view. The upper part of Figure 8.9 shows the intensity image of a metal tank and background. The lower reflectivity tank appears darker than the background. The intensity data can be combined with the elevation data in a fusion algorithm to increase the probability of correct target classification by using pixel-level or feature-level fusion as defined in Chapter 2. The elevation image corresponding to the intensity image is shown in the lower part of the figure.

Even more robust target classification algorithms can be applied to three-dimensional imagery formed by a multielement laser diode array and a matching number of avalanche photodiode detectors [10]. In addition to the preprocessor described above, the three-dimensional algorithm uses a recognizer to scan the elevation image and identify all objects and subregions. The subregions give information about the internal structure of the object. For example, a tank object can have subregions consisting of the turret and engine deck, and a truck object can have subregions consisting of the canopy and cab areas. A classifier uses the information output by the recognizer to classify each object by evaluating features such as surface area, ratio of surface area of the subregions to the total surface area of the object, ratio of perimeter of the object to the total surface area (this feature eliminates objects with irregular shapes), and three rotation-based features found by rotating the object parallel to a major axis. The rotation-based features are the length-to-width ratio, average object width, and ratio of the number of samples in the object to the number of samples that fill a rectangular region that totally encompasses the object. The last feature helps eliminate irregular-shaped objects as these produce a lower valued ratio because a larger region is needed to enclose them.

Another set of GaAs laser radar imagery was produced by Perkin-Elmer using both continuous wave (CW) and amplitude-modulated (AM) transmitted energy. The laser was mounted in an aircraft such that it scanned the ground in a forward-looking mode as shown in Figure 8.10. The laser radar's characteristics are given in Table 8.6. The Perkin-Elmer radar package was 18 inches by 16 inches by 12 inches (46 cm by 41 cm by 30 cm) and weighed 80 lb (36 kg). It was designed to be mounted on the door of the RF4 aircraft on which it flew.

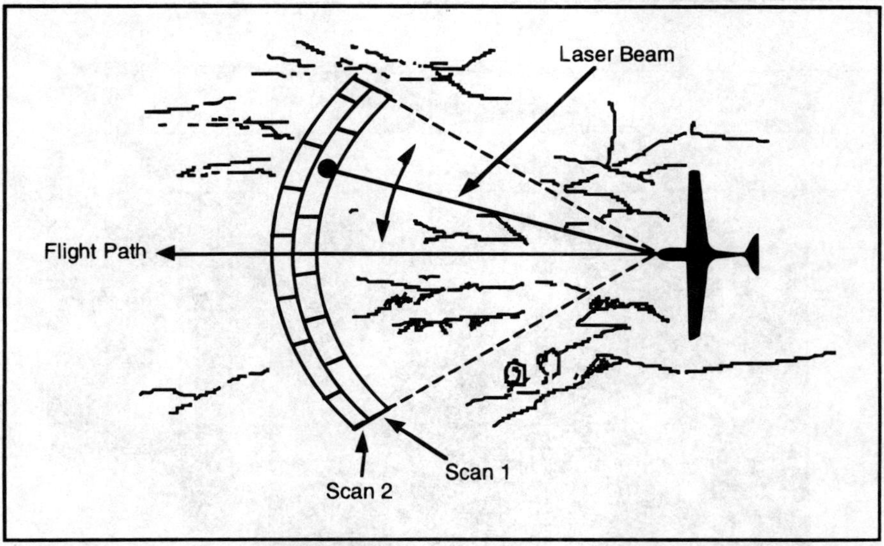

Figure 8.10 Scanning profile of Perkin-Elmer laser radar [11].

Table 8.6
Perkin-Elmer Near-Infrared Laser Radar Characteristics

Parameter	*Value*
Laser diode material	GaAs
Detector	Si avalanche photo-diode
Wavelength	850 nm
Output power	1 W
Receiver area	116 cm^2
Scan time	1/500 sec
Resolution	0.75 mrad
Azimuth scan angle	120 degrees
Altitude of aircraft	500 ft (152 m)
Aircraft velocity	100 kn (51 m/sec)

With CW transmission and direct detection, intensity images such as that shown in Figure 8.11 are obtained by synchronizing the raster scan of the received laser signal with the scanning of an electron beam in a cathode ray tube [11]. The magnitude of the received laser signal from each pixel in the scene is superimposed on the cathode ray tube's brightness axis. This particular image shows several aircraft parked on blacktop at a private airport. The aircraft and hanger on the left appear lighter in color, and the blacktop appears dark due to differences in the reflectivity of the objects. Since the optic's diffraction-limit angular resolution is much smaller than the receiver's resolution of 0.75 mrad, the image is nearly speckle-free because

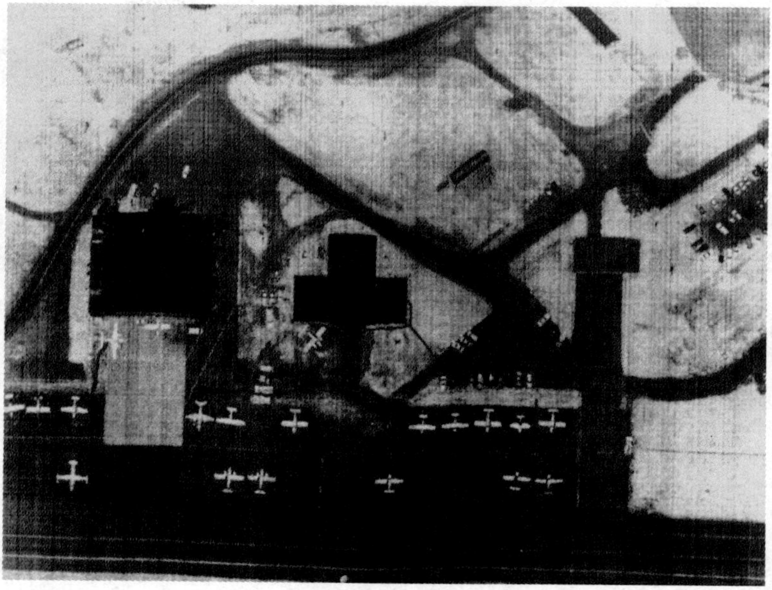

Figure 8.11 Direct detection intensity image produced with a near infrared wavelength laser. (*Source:* [11]. Copyright © 1986 American Institute of Physics. Reprinted with permission.)

of spatial averaging across the aperture. The larger receiver resolution was achieved by using an oversize detector. The result is a reduction in intensity variations due to speckle by about two orders of magnitude.

With AM transmission and direct detection, range images such as that shown in Figure 8.12 are obtained by superimposing the phase shift between the modulated transmitted beam and the received signal on the brightness axis of the cathode ray tube. The phase shift is proportional to the round trip transit time of the laser signal to each pixel in the scene. In the relative range topology image that is produced, pixels that have a zero phase shift, relative to a known reference, are assigned maximum brightness. Those pixels that have a 180-degree phase shift are assigned zero brightness. Pixels with a given brightness in the same phase shift cycle are at the same range from the laser sensor. The minimum range variation that can be measured with this technique is inversely proportional to the product of the modulation frequency and the signal-to-noise power ratio of the sensor. The laser diode and modulation frequency used in the radar flown over the airport shown in Figure 8.12 supported a range resolution of one foot (30 cm). Newer diodes and modulation techniques are expected to yield six inch (15 cm) resolution. The intensity striations parallel to flight path were caused by the change in sensor-to-ground range produced by the scan in azimuth angle. Improvements in the modulation technique that use two sine wave frequencies to achieve 100 percent

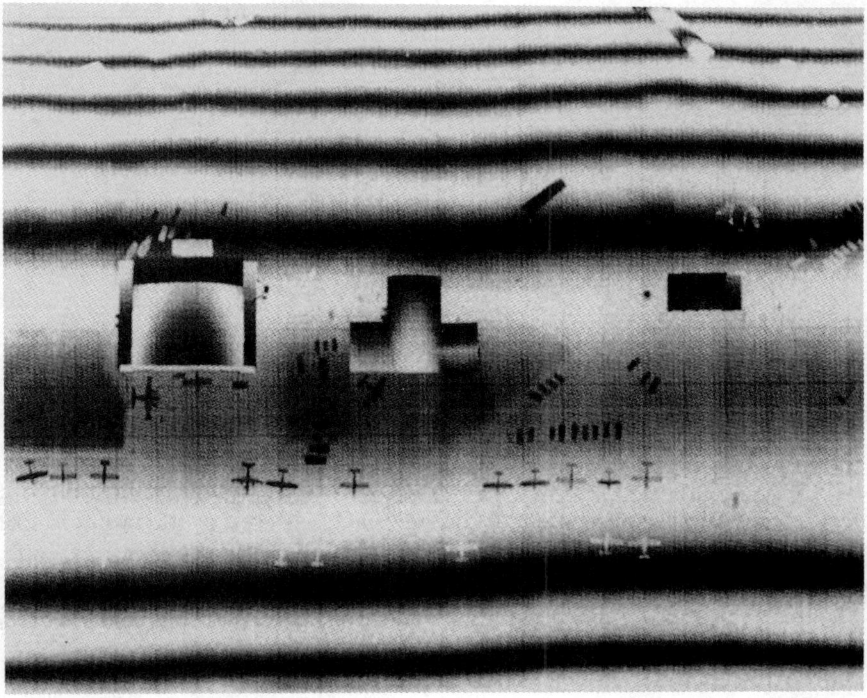

Figure 8.12 Direct detection range image produced with a modulated beam from a near infrared wavelength laser. (*Source:* [11]. Copyright © 1986 American Institute of Physics. Reprinted with permission.)

peak-to-peak modulation in the received signal can extend the unambiguous range and remove the striations from the imagery. An advantage of the range image is that it depicts objects that have insufficient reflectivity contrast, with respect to the background, to appear in the intensity image. An example of such an object is the antenna on the right side of the hanger toward its rear.

8.3.2 Carbon Dioxide Laser Radar Imagery

Heterodyne detection allows coherent mixing of laser beams produced by different frequency sources as illustrated by the simplified heterodyne laser configuration in Figure 8.13. With coherent homodyne detection, a portion of the laser's transmitted beam is diverted into the frequency shifter to function as the local oscillator frequency and mix with the received laser signal. The local oscillator frequency is

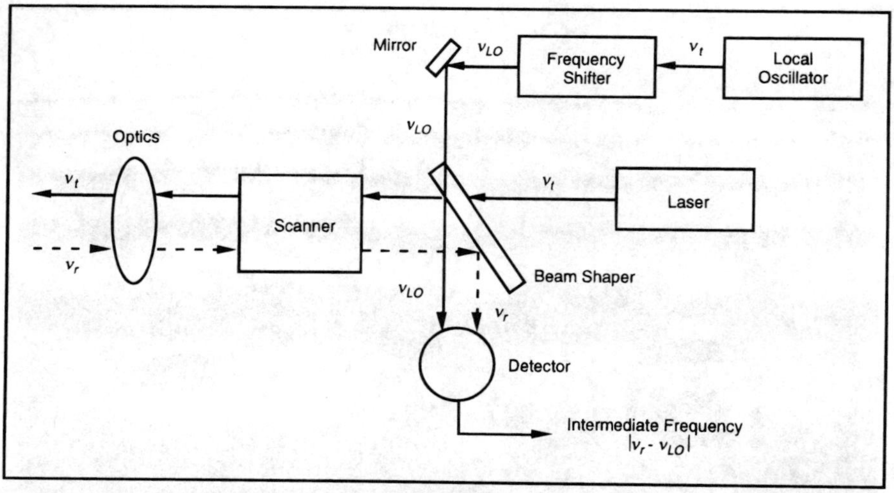

Figure 8.13 Heterodyne laser (simplified configuration).

denoted by ν_t, frequency-shifted local oscillator frequency by ν_{LO}, transmitted laser frequency by ν_t, and received signal frequency backscattered by the target ν_r. Coherent laser operation not only provides intensity and range information, but also radial target speed based on the Doppler effect.

The long-term frequency stability obtained with a carbon dioxide laser at 10.6 μm makes it a popular choice for use in a coherent laser system [11]. An example of the imagery and data available from a carbon dioxide laser is shown in Figure 8.14. A telephoto picture, taken from a range of 2.7 km, of aircraft parked outside a hanger is shown in Figure 8.14(a). The intensity image from the same range produced by reflectivity differences in the objects is shown in Figure 8.14(b). The hanger door is closed (it was open in the photograph), which accounts for the light color in the region occupied by the door. Small-diameter objects such as power lines that are not very apparent in the visible photograph appear clearly in the laser intensity image because of the difference in reflectivity between the wires and the background. Relatively smooth surfaces, such as aircraft skin and asphalt runways, show specular reflector behavior at shallow angles and hence appear black in the image. The corresponding range image in Figure 8.14(c) provides a three-dimensional aspect to the scene and enhances the ability to detect small objects. Although shown here in black and white, each range increment in the scene can be assigned a false color and the intensity of the range signal can be indicated with shades of the color.

Moving vehicles in Figure 8.14(d) illustrate the Doppler data available from the radar. A small vehicle is shown on the left and a truck on the right. Although this photo was reproduced in black and white, objects that lie within the same radial velocity increment can be color coded. In this manner, the higher radial speed of the truck tires can be made to appear as a different color than the truck body. In color imagery, those returns whose Doppler shift is less than some threshold

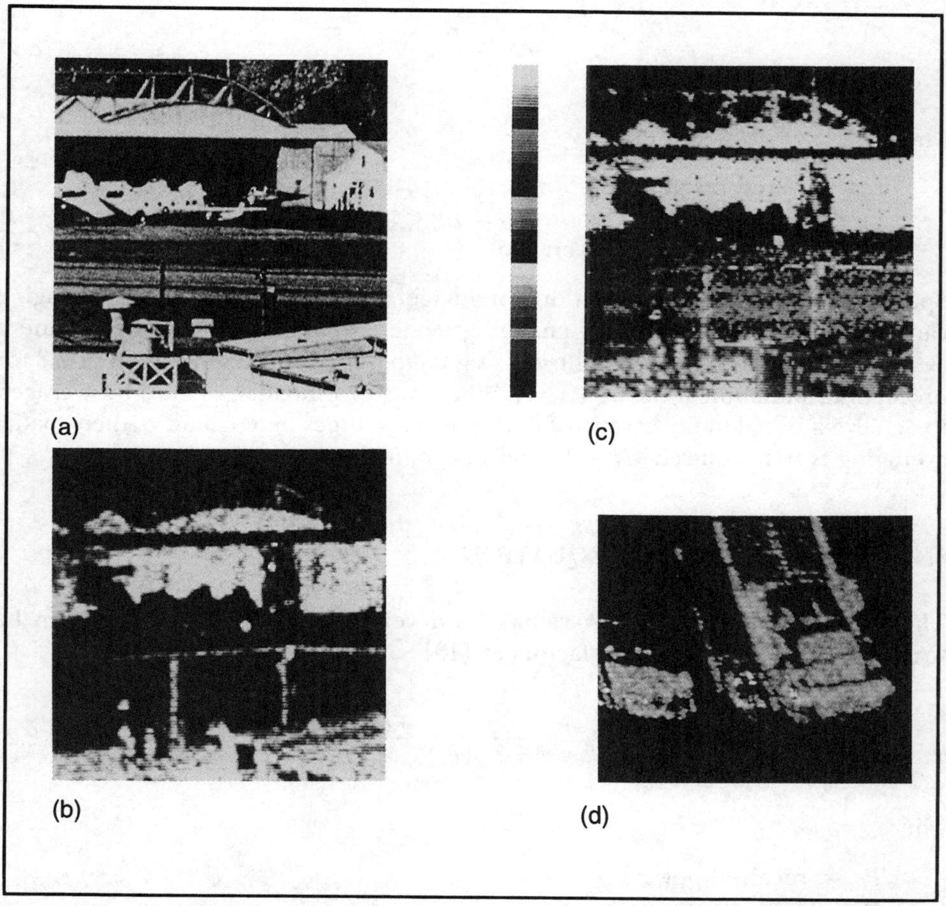

Figure 8.14 Coherent laser radar images obtained with 10.6-μm laser: (a) telephoto picture, (b) intensity, (c) range, and (d) Doppler. (*Source:* [11]. Copyright © 1986 American Institute of Physics. Reprinted with permission.)

can be made to appear black to eliminate clutter introduced by receiver noise. Speckle appears in these laser images because neither spatial nor temporal pixel averaging was applied.

8.3.3 Speckle

Speckle is produced by the random phase interference of returns from multiple scatterers in the beam. Since higher resolution sensors produce less speckle, a laser radar image of an object such as a tank has less speckle than a MMW radar image of the same vehicle [12]. Speckle is reduced by spatial and temporal averaging that increases the signal-to-noise ratio as shown by

$$S/N = \frac{r_I}{r_O}\sqrt{N} \tag{8.1}$$

where

r_I = desired image resolution;
r_O = sensor optics resolution ($r_O \geq r_I$);
N = number of images averaged.

Spatial averaging trades resolution for image quality, while temporal averaging can sacrifice details in a rapidly changing scene. Equation (8.1) can determine if averaging is needed to satisfy Johnson's signal-to-noise ratio requirement of 2.8 for object discrimination at the 50% level. When $r_O \approx r_I$, nine images must be averaged to achieve a signal-to-noise ratio of 3. If r_O is three times better than r_I, then spatial averaging is not required ($N = 1$) and image quality is improved.

8.4 LASER RADAR RANGE EQUATION

The received power for MMW radars, as developed in earlier chapters, can be written as the product of four factors as [13]

$$P_R = \frac{P_T G_T}{4\pi R^2}\frac{\sigma}{4\pi R^2}\frac{\pi D_o^2}{4}\frac{1}{L_a} \tag{8.2}$$

where

P_R = received power;
P_T = peak transmitted power;
G_T = gain of transmitting antenna;
R = radar-to-target range;
λ = wavelength of transmitted energy;
σ = effective radar cross-section of target;
D_o = effective diameter of receiving aperture
= $(4A_R/\pi)^{1/2}$;
A_R = effective area of receiving aperture;
L_a = atmospheric and other system losses ($L_a \geq 1$).

The first factor describes the local power density at a spherical target produced by a transmitter of power P_T having an aperture of gain G_T. The second factor represents the portion of the incident power intercepted by a spherical target and its isotropic reradiation by the target. Factor three describes the capture of the reradiated power by the area of the receiving aperture. Factor four represents losses in the atmosphere and the radar system. This form of the radar range equation applies in the far field of the aperture defined where the range

$$R > 2D_o^2/\lambda \tag{8.3}$$

A comparison of the far-field range for 35- and 94-GHz MMW radars and a 1-μm laser radar is shown in Table 8.7 for comparable apertures of 5.5 inches (14 cm) such as might be found in a 6-inch (15 cm) diameter missile. Since the far field for a laser radar begins some tens of thousands of meters from the 5.5-inch aperture, laser radars often do not operate in the far field, but instead may function in the near field or in the intermediate Fresnel region. Even with a 2-inch (5 cm) aperture as in the SEO laser radar, the far field does not begin until some 5,000 meters from the aperture. The near-field region is defined as the immediate neighborhood of the aperture where approximations to the Fresnel-Kirchhoff integral equation that describes the diffraction-limited field do not apply [14]. The Fresnel region extends from the near-field region to the start of the far field.

8.4.1 Near-Field Radar Range Equation

For near-field applications, the radar cross-section of the target appearing in (8.2) must be modified. Radar cross-section is defined by

$$\sigma = \frac{4\pi}{\Omega} \rho \, dA \tag{8.4}$$

where

Ω = scattering solid angle of target in sr;
ρ = target reflectivity;
dA = illuminated area of target.

The scattering solid angle and radar cross-section for isotropic, Lambertian, and hemispherical scattered energy distributions are given in Table 8.8. An isotropic surface scatters energy equally in all directions over 4π steradians. A Lambertian surface is a perfect rough surface that scatters energy in a half-plane extending from 0 to π steradians. When the backscattered energy distribution follows Lambert's law,

Table 8.7
Far-Field Range for MMW and Laser Radars

Wavelength	D_o	$2D_o^2/\lambda$
0.86 cm (35 GHz)	5.5 in (14 cm)	4.56 m (0.0025 nmi)
0.32 cm (94 GHz)	5.5 in (14 cm)	12.25 m (0.0066 nmi)
1 μm (300 THz)	5.5 in (14 cm)	39,200 m (21.2 nmi)
1 μm (300 THz)	2.0 in (5 cm)	5,161 m (2.8 nmi)

Table 8.8
Scattering Solid Angle and Radar Cross-Section for Isotropic, Lambertian, and Hemispherical Scatterer Distributions

Target Scatterer Distribution	Ω	*Application*	*Radar Cross-Section* (m^2)
Isotropic	4π	MMW radar diffuse targets	$\rho\ dA$
Lambertian	π	Laser radar diffuse targets	$4\rho\ dA$
Hemispherical	2π	Other laser radar diffuse targets	$2\rho\ dA$

the angular variation is only dependent on the product of $\cos\theta_i \cos\theta_s$, where θ_i and θ_s are the angles the incident and scattered power vectors P_i and P_s, respectively, form with the normal to the surface as shown in Figure 8.15. Thus, a Lambertian surface represents the opposite extreme from a perfect flat surface that scatters energy only in the specular direction. A hemispherical surface scatters energy throughout a 2π steradian region.

Target reflectivity appearing in (8.4) is a function of wavelength, generally decreasing as the wavelength increases [8,15,16]. It is defined as the ratio of reflected power to incident power. In terms of emissivity ϵ, the reflectivity is

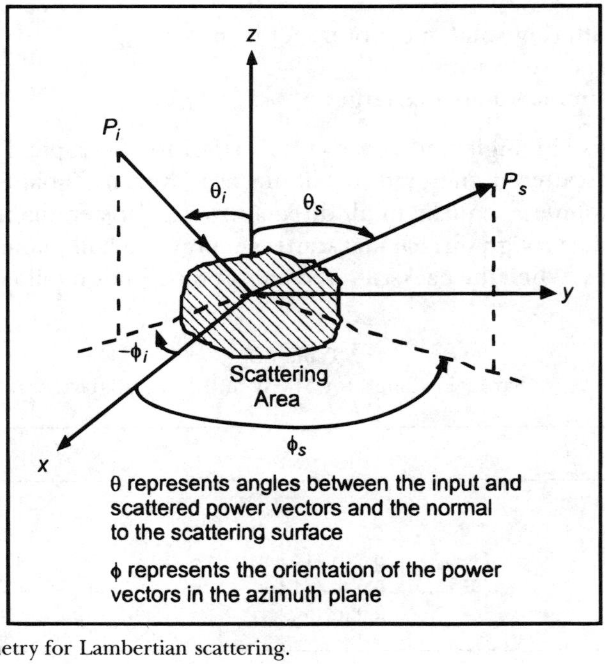

Figure 8.15 Geometry for Lambertian scattering.

$$\rho = 1 - \epsilon \tag{8.5}$$

Emissivity is equal to the absorptivity of a surface, where absorptivity is the ratio of absorbed power to incident power. Unpainted metal is quite reflective in the near- through far-infrared wavelength bands, as well as the MMW region. Therefore, the infrared signature of a highly reflective unpainted metal target will be predominantly that of the reflected background, which is often the sky. Water at a shallow grazing angle exhibits similar reflective properties. The emissivity of water in the 2- to 15-μm region is greater than 0.90 for incidence angles up to 60 degrees as measured from the normal to the surface. It then falls sharply to approximately 0.4 at an incidence angle of 85 degrees [17].

The laser radar range equation assumes different forms for resolved, unresolved, and ground-clutter targets. These are developed below.

8.4.2 Laser Radar Range Equation for a Resolved Target

A resolved target is one that is larger than the laser beam footprint at a given range as illustrated in Figure 8.16. Laser radars with this resolution configuration are used for imaging applications. The laser radar range equation for resolved targets can be derived directly from the MMW radar range equation by incorporating the effects of near-field operation on radar cross-section and aperture illumination on aperture gain.

For a MMW radar, the gain G_T of a lossless antenna with one major lobe may be approximated by

$$G_T = \frac{4\pi}{\theta_e \theta_a} \tag{8.6}$$

where θ_e, θ_a = two-way 3-dB elevation, azimuth beamwidths of the antenna expressed in radians.

If the elevation and azimuth beamwidths are equal to each other and are denoted by θ_b, the MMW antenna gain becomes

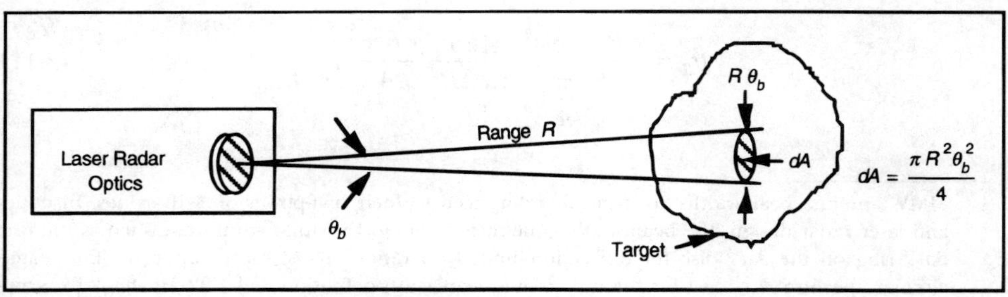

Figure 8.16 Laser beam configuration for resolved target.

$$G_T = \frac{4\pi}{\theta_b^2} \tag{8.7}$$

For the laser radar, the optical gain is expressed as

$$G_T = \frac{4\pi}{\left(\frac{\pi}{4}\right)\theta_b^2} \tag{8.8}$$

The different expressions for the gain are apparently due to the use of different aperture illumination functions (i.e., distribution of energy across the aperture) for the MMW antenna and laser optics. If θ_b is written as

$$\theta_b = k\lambda/D_o \text{ rad} \tag{8.9}$$

where k is a function of the aperture illumination, the laser radar gain equation implies $k = \sqrt{\pi/4} = 0.88$. This corresponds to the 3-dB beamwidth for a uniformly illuminated rectangular aperture. If a uniformly illuminated circular aperture is assumed, then $k = 1.02$ [14].[1] MMW radar gain generally includes the specific value for the aperture illumination designed into the antenna. In (8.6), $k = 1$ is assumed. When the illumination is tapered to reduce sidelobes, higher gain results corresponding to values of $k < 1$.

The laser radar range equation for resolved targets is found from (8.2) by substituting for transmitter gain G_T using (8.8), for radar cross-section σ using (8.4), and for incremental target area dA using the definition in Figure 8.16 of

$$dA = \frac{\pi R^2 \theta_b^2}{4} \tag{8.10}$$

With these substitutions, the laser radar range equation for the received power becomes

$$P_R = \frac{P_T}{4\pi R^2} \frac{4\pi}{\frac{\pi}{4}\theta_b^2} \frac{4\pi\rho}{4\pi\Omega R^2} \frac{\pi R^2 \theta_b^2}{4} A_R \frac{1}{L_a} \tag{8.11}$$

1. MMW antenna beamwidths are typically referenced to their half-power or 3-dB values. In optics and laser radar design, the beamwidth sometimes refers to the null-to-null value such as the first dark ring on the Airy disk for diffraction-limited operation. Thus, for a uniformly illuminated circular aperture, $k = 2.44$ for the null-to-null beamwidth definition and 1.02 for the 3-dB beamwidth definition.

or

$$P_R = \frac{P_T \rho A_R}{\Omega R^2 L_a} \tag{8.12}$$

If a Lambertian power scattering distribution is present, $\Omega = \pi$. If the power-scattering distribution is hemispherical, $\Omega = 2\pi$.

The received power from a resolved target is inversely proportional to the square of radar-to-target range, unlike the fourth power dependence observed with MMW radars operating against far-field targets.

8.4.3 Laser Radar Range Equation for an Unresolved Target

An unresolved target is one that is smaller than the laser beam footprint at a given range as illustrated in Figure 8.17. The laser radar range equation for the unresolved target is derived from (8.11) by replacing the fourth factor, representing the area intercepted by the laser beam, by the laser radar cross-section σ_T of the unresolved target.

Thus for Lambertian power scattering, the received power is

$$P_R = \frac{4 P_T A_R \rho \sigma_T}{\pi^2 R^4 \theta_b^2 L_a} \tag{8.13}$$

When θ_b is written as λ / D_o and D_o is expressed in terms of A_R as $(4A_R/\pi)^{1/2}$, the expression for received power becomes

$$P_R = \frac{16 P_T A_R^2 \rho \sigma_T}{\pi^3 R^4 \lambda^2 L_a} \tag{8.14}$$

For hemispherical power scattering from a target,

$$P_R = \frac{2 P_T A_R \rho \sigma_T}{\pi^2 R^4 \theta_b^2 L_a} \tag{8.15}$$

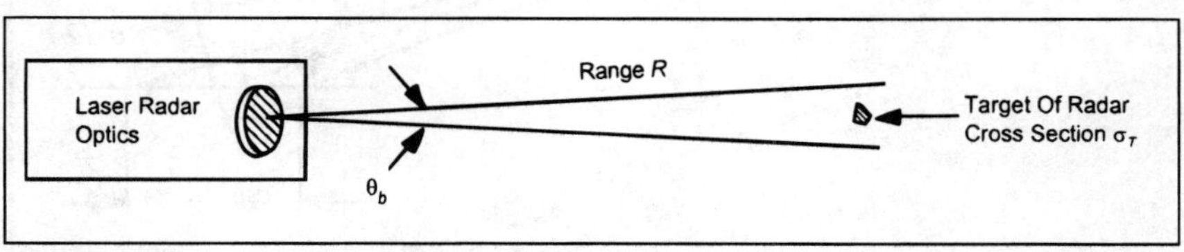

Figure 8.17 Laser beam configuration for unresolved target.

or

$$P_R = \frac{8P_T A_R^2 \rho \sigma_T}{\pi^3 R^4 \lambda^2 L_a} \tag{8.16}$$

Here, the received power is inversely proportional to the fourth power of the radar-to-target range, similar to that of MMW radars operating with far-field targets.

8.4.4 Laser Radar Range Equation for Power Backscattered From the Ground

The power backscattered to a laser radar from ground scatterers is found from the radar range equation for an unresolved target by replacing the target radar cross-section σ_T by the intercepted ground area A_c where

$$A_c = R\theta_b \frac{c\tau}{2} \tag{8.17}$$

and τ is the width of the transmitted laser pulse, as illustrated in Figure 8.18.

If the pulse width produces a larger ground intercept than the beamwidth θ_b, then the $c\tau/2$ factor is replaced by the footprint-limiting $R\theta_b/\sin\beta$ factor, resulting in

$$A_c = R\theta_b \frac{R\theta_b}{\sin\beta} = \frac{(R\theta_b)^2}{\sin\beta} \tag{8.18}$$

where β is the depression angle.

Ground-clutter reflectivity ρ as used in (8.13) through (8.16) is a function of terrain type, roughness, wetness, and laser wavelength as discussed earlier.

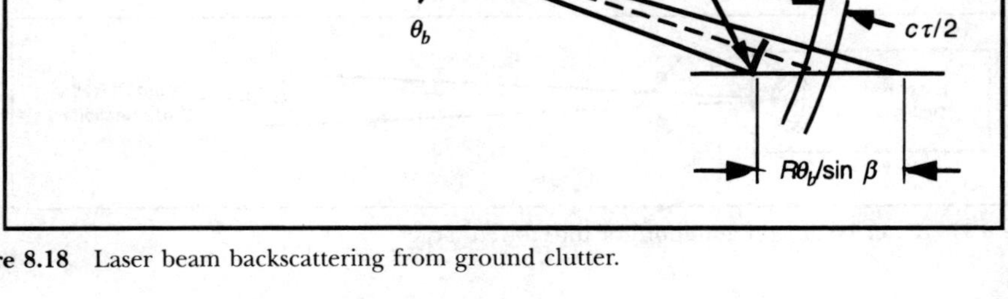

Figure 8.18 Laser beam backscattering from ground clutter.

8.5 EFFECTIVE RADAR CROSS-SECTION OF LASER TARGETS

The radar cross-section of spherical targets is a function of the ratio of the target radius r_{tgt} to the wavelength λ of the laser energy [18]. When the radius of the target is smaller than the wavelength, Rayleigh scattering occurs and the radar cross section σ becomes proportional to

$$\sigma \; \alpha \; \lambda^{-4} \quad \text{for} \quad 2\pi r_{tgt}/\lambda << 1 \tag{8.19}$$

where α indicates a proportional relation. This type of scattering generally occurs with precipitation particles illuminated by energy at microwave wavelengths. When the radius of the target is greater that the laser wavelength, geometric ray trace techniques are used to find the radar cross-section. In this case, the radar cross-section dependence on wavelength is a function of the radii of curvature of the target as shown in Table 8.9. Under these conditions, referred to as the optical region, a spherical target has a radar cross-section given by

$$\sigma \approx \pi r_{tgt}^2 \quad \text{for} \quad 2\pi r_{tgt}/\lambda >> 1 \tag{8.20}$$

Retroreflectors have the ability to backscatter relatively large amounts of energy compared to their size due to internal capture and reflection of the energy by the reflector [19]. However, the angle over which the energy can be received or captured is relatively narrow. Dihedral retroreflectors allow double reflection of the incident energy and trihedral reflectors support triple reflection. The number of internal reflections has implications for target classification by radars that have separate receivers for the odd and even bounce returns as discussed in Chapter 4. The angle θ_M between the beam direction corresponding to maximum energy reflection, and the direction in which the radar cross-section decreases by a factor of two is approximately

Table 8.9

Wavelength Dependence of Radar Cross-Section on Radii of Curvature of a Target for Radii Greater Than Laser Wavelength

Radar Cross-Section Wavelength Dependence	*Radii of Curvature of a Target*
λ^{-2}	Two infinite radii of curvature (flat plate)
λ^{-1}	One infinite, one nonzero (cylinder)
λ^{0}	One infinite, one zero (wedge)
λ^{0}	Two nonzero finite (spheroid)
λ^{1}	One nonzero finite, one zero (curved edge)
λ^{2}	Two zero (apex of a cone)

$$\theta_M \approx 7.5\,\lambda/r_g \text{ degrees} \tag{8.21}$$

where r_g is the radius of gyration of the aperture about an axis through the aperture's center of gravity and perpendicular to the plane in which the deviation from the center of the beam is taken [20].

Dihedral retroreflectors present a large radar cross-section to radars lying in the plane perpendicular to the dihedral axis (the line of intersection between the two plates that comprise the dihedral corner reflector). The radar cross-section of dihedral retroreflectors is given by

$$\sigma = \frac{8\pi l^2 m^2}{\lambda^2} \tag{8.22}$$

where l and m are the lengths of the edges of the reflector as defined in Figure 8.19.

Trihedral retroreflectors provide a strong specular reflection in both elevation and azimuth relative to the trihedral symmetry axis and over a wider range of aspect angles than dihedral reflectors. The radar cross-section of trihedral retroreflectors is given by

$$\sigma = \frac{4\pi l^4}{3\lambda^2} \tag{8.23}$$

where l is the length of the edge of the reflector as defined in Figure 8.20.

Thus, the radar cross-section of both dihedral and trihedral corner reflectors increases inversely with the square of the wavelength, allowing one to scale the corner reflector cross section for radar calibration applications at different wavelengths (or frequencies).

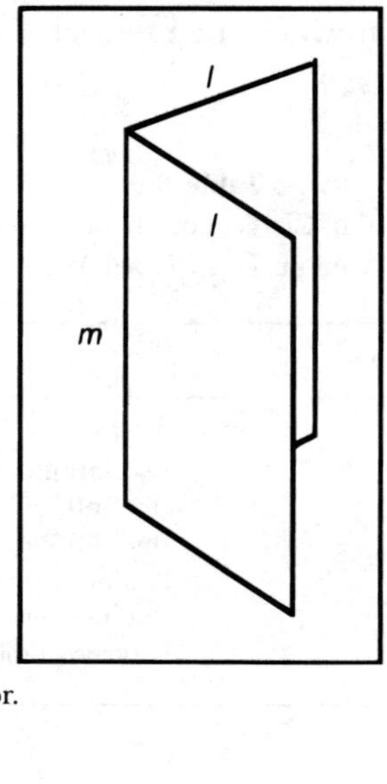

Figure 8.19 Dihedral retroreflector.

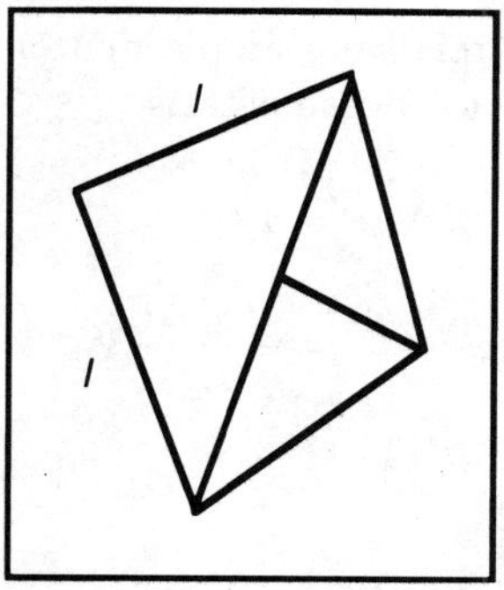

Figure 8.20 Trihedral retroreflector.

8.6 NOISE SOURCES IN LASER RADARS

In order to calculate the signal-to-noise ratio, and hence the detection probability, the sources of noise in a laser radar must be evaluated. The dominant noise sources are different for direct detection and heterodyne lasers. In direct detection laser sensors, the noise sources shown in Table 8.10 are usually present [21,22]. In heterodyne detection, the sensor is designed so that the local oscillator photon shot noise dominates all other noise sources in the detector and preamplifier [23]. Therefore, the noise-equivalent power is simply equated to the local oscillator photon noise contribution when computing the signal-to-noise ratio in heterodyne lasers.

8.6.1 Signal-To-Noise Ratio for Direct Detection Laser Radar Sensors

The noise-equivalent power *NEP* in a direct detection laser includes the noise contributions from all the noise sources listed in Table 8.10. Since *NEP* can be

Table 8.10
Noise Sources in Direct Detection Laser Systems

Noise Source	*Noise Current Mechanism*
Signal photons	Shot noise
Background photons	Shot noise
Dark current	Shot noise
Johnson	Thermal noise
Amplifier	Thermal noise
Generation-recombination	Thermal noise
$1/f$	Surface effects that vary with detector
Quantization	Only discrete signal levels allowed

written in terms of D_d^*, it is a relatively simple matter to compute the noise power from a measurement of detector detectivity as

$$NEP = 1/D_d \tag{8.24}$$

$$= \frac{\sqrt{A_D \Delta F}}{D_d^*} \text{ W} \tag{8.25}$$

where

D_d = detector detectivity in W^{-1};
D_d^* = specific detector detectivity in cm $Hz^{1/2}$ W^{-1};
A_D = detector area in cm^2;
ΔF = receiver noise bandwidth in Hz.

Therefore, the signal-to-noise ratio for an imaging laser radar illuminating a target with Lambertian scattering can be written as the ratio of (8.12) to (8.25). Thus

$$S/N = \frac{P_T \rho A_R D_d^*}{\pi R^2 \sqrt{A_D \Delta F} L_a} \tag{8.26}$$

8.6.2 Signal-to-Noise Ratio for Heterodyne Detection Laser Radar Sensors

In heterodyne laser radar sensors, the *NEP* is dominated by photon shot noise and is expressed as

$$NEP = \frac{\sqrt{\overline{i_N^2}}}{R_c(\lambda)} \tag{8.27}$$

where

$\sqrt{\overline{i_N^2}}$ = rms value of photon shot noise current contributed by the local oscillator
= $\sqrt{2qR_c(\lambda)P_{LO}\Delta F}$;
q = electron charge = 1.6021×10^{-19} coulomb;
$R_c(\lambda)$ = current responsivity of the detector = output current/optical input power
= $(\eta q)/(hv)$ for a photon detector;
η = quantum efficiency (η at 850 nm $\approx$ 0.5 to 0.8, η at 10.6 μm $\approx$ 0.1 to 0.2);
h = Planck's constant = 6.6256×10^{-34} J sec;
v = transmitted frequency in Hz;
P_{LO} = local oscillator output power (set equal to *NEP*) in W.

Therefore, for a heterodyne receiver,

$$NEP = \frac{2h\upsilon\Delta F}{\eta}\,\text{W} \tag{8.28}$$

and the signal-to-noise ratio for an imaging laser radar illuminating a target with Lambertian scattering is

$$S/N = \frac{P_T \rho A_R \eta}{2\pi R^2 h\upsilon\Delta F L_a} \tag{8.29}$$

where (8.12) has been used for the received power and (8.28) for the noise power.

8.7 LASER RADAR TARGET FLUCTUATION CHARACTERISTICS

Laser radar targets are divided into two general categories, glint and speckled. Glint targets contain a single dominant scatterer. Returns from these targets do not fluctuate and are equivalent to the nonfluctuating targets defined for MMW radar. Speckled targets contain multiple independent scatterers with no single scatterer dominant. The speckled target is equivalent to either the Swerling 1 or 2 targets defined for MMW radars in Chapter 4. Detection probabilities of Swerling 1 and 2 targets are the same when one pulse is noncoherently integrated. The method used to calculate the detection probability for laser radars with noncoherent integration of multiple laser pulses is the same as that used with multiple pulse integration of echoes from MMW radars.

8.8 DETECTION PROBABILITY CALCULATION FOR LASER RADAR SENSORS

The detection probability calculation procedure for laser radars is summarized in Figure 8.21. The first step is to recognize whether the target is resolved or unresolved and select the appropriate equation,(8.12) or (8.14) or (8.16), to calculate the received power backscattered from the target. If the detection process is noise-limited, the noise-equivalent power is calculated from the dominant noise source depending on whether direct or heterodyne detection is used. If detection is clutter-limited, then the power backscattered from the background is used as the limiting interference in the signal-to-interference ratio calculation. Since a laser radar receives large numbers of photons, it generates large numbers of photoelectrons in the detector material. By the central limit theorem, the noise probability density function of large numbers of photoelectrons can be represented by a Gaussian probability density function. Thus, the detection of signals in noise in an optical

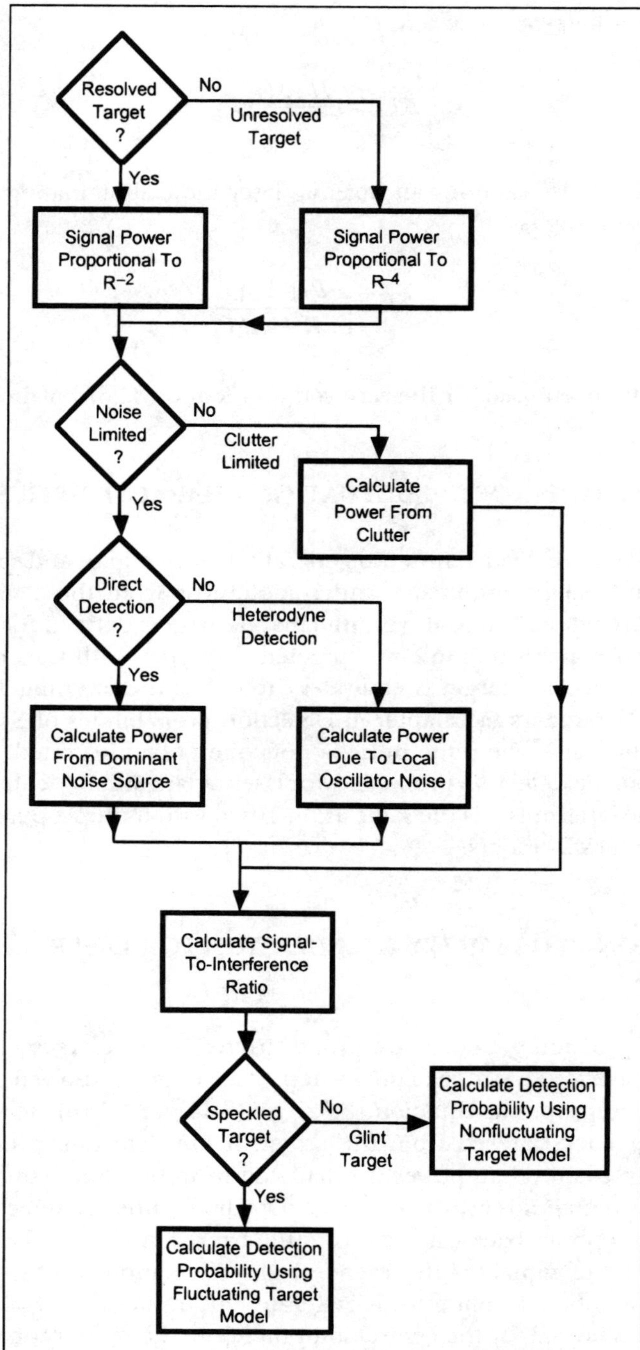

Figure 8.21 Detection probability calculation process for laser radars.

receiver reduces to conventional signal detection in noise as in a MMW radar receiver. Depending on whether the target is glint or speckled, the detection probability is found using the appropriate graphs or models developed for MMW radars [22].

8.9 COMPARISON OF MILLIMETER-WAVE AND INFRARED SIGNAL DETECTION THEORY

At this point, it is worthwhile to summarize the target detection phenomena for active and passive millimeter-wave and infrared sensors as shown in Table 8.11. MMW sensors usually incorporate a mixer to increase their sensitivity. In noise-limited operation, such as searching for airborne targets in clear weather, signal detection is governed by statistics derived from white Gaussian noise interference.

Table 8.11
Millimeter-Wave and Infrared Signal Detection Phenomena

Millimeter-Wave Phenomena	*Infrared Phenomena*
Sensors are either heterodyne or homodyne, i.e., all use a mixer	Sensors are either • Direct detection as with passive sensors and GaAs laser radar, or • Heterodyne or homodyne detection as with CO_2 laser radar
Targets behave as fluctuating or nonfluctuating	Targets behave as either • Fluctuating to laser radars and passive sensors when speckle targets are observed, or • Nonfluctuating to laser radars when glint targets are observed and to passive sensors
Noise statistics are usually Gaussian • Electronic noise is dominant interference for air-to-air combat applications	Dominant noise source is a function of whether detection process is direct, heterodyne, or photon limited
Ground clutter statistics vary from relatively benign Rayleigh to severe log-normal • Ground clutter is dominant interference when searching for ground-based targets	Ground clutter statistics must be evaluated for the operating scenario taking into account factors such as wavelength and angle of incidence
Detection probability for passive and active sensors can be calculated from the same theory	• Detection probability for passive sensors is a function of noise statistics (white or quantum) • Detection probability for active sensors corresponds to MMW radar theory
Processing is either • Noncoherent as with radiometers and some radars, or coherent as with other radars	Processing is either • Noncoherent as with passive sensors and GaAs laser radar, or • Coherent as with CO_2 laser radar

Active IR sensors, such as direct detection and coherent heterodyne laser radars, share the same signal detection phenomenology as MMW sensors because large numbers of photons are received. Hence, the same type of detection probability calculation is applicable for both the active IR and MMW sensors. Noise is also commonly the limiting factor in passive IR sensor operation. However, the statistics that describe the detection probability for passive IR sensors are not the same as for MMW sensors because of the smaller number of photons that are detected. When MMW and IR sensors operate against ground targets, ground clutter may limit the performance. In this case, the interference may not be described by Gaussian or Poisson statistics but by Rayleigh, Weibull, or log normal scatterer amplitude distributions. Both MMW and IR sensors encounter fluctuating and nonfluctuating targets and both can utilize noncoherent and coherent processing techniques.

8.10 SUMMARY

Target designator, range finder, and target detection and classification military applications of laser radars have been discussed. Solid state gallium arsenide lasers are well-suited for these applications because of their small size, ruggedness, and output power characteristics. By extending the transmitted wavelength to greater than 1.4 μm through appropriate designs, eye-safe operation can be achieved. This has motivated the development of mid-infrared band solid state lasers. Direct detection laser radars produce imagery that appears well-suited to the application of automatic target recognition algorithms such as those that analyze the elevation of features along scan lines through the image. When the transmitted signal is modulated, smaller features appear that may not be apparent in sensors that transmit unmodulated signals. In addition to range data, laser radars produce another independent set of information that is a function of the reflectivity of the objects in the scene illuminated by the laser. These two sets of data can be input to a data fusion algorithm to enhance the target detection and classification performance of the sensor. Carbon dioxide coherent lasers add yet another dimension to the target signature by providing target speed through the measurement of the Doppler shift. Speckle, sometimes seen in laser radar imagery, can be reduced by spatial and temporal averaging that increase the signal-to-noise ratio but sacrifice spatial and temporal resolution.

The laser radar range equation was developed for resolved targets, unresolved targets, and ground backscatter by extending the definition of radar cross-section to near-field operation and by incorporating the appropriate transmission gain provided by the laser optics. Radar cross-sections of spherical objects were shown to be a function of the ratio of the target radius to the wavelength. The radar cross-sections of dihedral and trihedral retroreflectors scale inversely with the square of the wavelength. In preparation for the calculation of signal-to-noise ratio and detection probability, the process of noise generation in direct and heterodyne lasers was discussed. Direct detection lasers are subject to a variety of shot, thermal, and other noise sources, while heterodyne laser noise is dominated by the local oscillator photon shot noise. Direct detection and coherent heterodyne laser radars share the detection probability theory developed for nonfluctuating and Swerling fluctuating targets introduced for MMW radars. Finally, a comparison was provided of signal detection theory as applied to passive and active MMW and IR sensors.

References

[1] Nordstrom, R. J., "The CO_2 Laser in Remote Sensing," *Photonics Spectra*, Feb. 1992, pp. 89–92.

[2] *Jane's Infantry Weapons 1987–1988*, Thirteenth Edition, London, 1987. Manufacturer's data sheets were also used to obtain the information shown.

[3] Hecht, J., "Diode-laser performance rises as structures shrink," *Laser Focus World*, May 1992, pp. 127–143.

[4] Hecht, J., "Gallium arsenide lasers offer an array of options," *Laser Focus World*, July 1993, pp. 83–92.

[5] Petricevic, V., A. Seas, and R. R. Alfano, "Eye safety of forsterite lasers disputed," *Laser Focus World*, Feb. 1991, p. 25.

[6] Barnes, N. P., and K. I. Schepler, "Solid-State Lasers for the Mid-Infrared Region," *SPIE Vol. 1062*, 1989, pp. 2–8.

[7] Hecht, J, "Rare earths create useful long-wavelength lasers," *Laser Focus World*, Nov. 1993, pp. 135–142.

[8] Suits, G. H., "Natural Sources," Chapter 3 in *The Infrared Handbook*, W. L. Wolfe, and G. J. Zissis, (Eds.), Infrared Information Analysis Center, Environmental Research Institute of Michigan, 1989.

[9] Humphrey, R. G., W. C. Ruff, and D. E. Robinson, *Armored Target Reflectance Near 1 μm: Survey of Existing Data and Report of New Measurements*, Report HDL-TM-88-8, Harry Diamond Laboratories, Att'n: SLCHD-ST-SP, Adelphi, MD, Sept. 1988.

[10] Gustavson, R. L., and T. E. Davis, "Diode-laser radar for low-cost weapon guidance," *SPIE Vol. 1633 Laser Radar VII*, 1992, pp. 21–32.

[11] Keyes, R. J., "Heterodyne and nonheterodyne laser transceivers," *Rev. Sci. Instrum.*, April 1986, pp. 519–528.

[12] Gordon, G. A., R. L. Hartman, and P. W. Kruse, "Imaging-Mode Operation of Active NMMW Systems," in *Infrared and Millimeter Waves*, Chap. 7 in Vol. 4: *Millimeter Systems*, K. J. Button and J. C. Wiltse, (Eds.), New York, NY: Academic Press, 1981.

[13] Hovanessian, S. A., *Introduction to Sensor Systems*, Norwood, MA: Artech House, 1988.

[14] Sherman III, J. W., "Aperture-Antenna Analysis," Chapter 9 in *Radar Handbook*, M. Skolnik, (Ed.), New York, NY: McGraw-Hill, 1970.

[15] Wolfe, W. F. *Handbook of Military IR Technology*, Washington, DC: U.S. Government Printing Office, 1965.

[16] LaRocca, A. J., "Artificial Sources," Chap. 2 in *The Infrared Handbook*, W. L. Wolfe and G. J. Zissis, (Eds.), Infrared Information Analysis Center, Environmental Research Institute of Michigan, 1989.

[17] Shumaker, D. L., J. T. Wood, and C. R. Thacker, *Infrared Imaging Systems Analysis*, Environmental Research Institute of Michigan, Infrared Information Analysis Center, Ann Arbor, MI, 1993.

[18] Crispin, J. W., and K. M. Siegel, *Methods of Radar Cross-Section Analysis*, New York, NY: Academic Press, 1968.

[19] Kell, R. E., and R. A. Ross, "Radar Cross Section of Targets," Chap. 27 in *Radar Handbook*, M. Skolnik, (Ed.), New York, NY: McGraw-Hill, 1970.

[20] Brown, D. R., R. J. Newman, and J. W. Crispin, Jr., "RCS Enhancement Devices," Chapter 8 in *Methods of Radar Cross-Section Analysis*, J. W. Crispin, Jr. and K. M. Siegel, (Eds.), New York, NY: Academic Press, 1968.

[21] Jelalian, A. V., "Laser Radar Theory and Technology," Chapter 24 in *Radar Technology*, E. Brookner, (Ed.), Norwood, MA: Artech House, 1985.

[22] Johnson, C. M., "Laser Radars," Chap. 37 in *Radar Handbook*, M. Skolnik, (Ed.), New York, NY: McGraw-Hill, 1970.

[23] Koehler, T, "How detectors can meet the demands of lidar systems," *Laser Focus World*, March 1992, pp. 133–137.

CHAPTER 9

Windows and Domes for Dual-Wavelength MMW and IR Sensors

Windows and domes protect the enclosed focal plane structure of IR sensors or the antenna on MMW sensors from the effects of the environment. Earth-based window and dome materials must endure temperature extremes, humidity, rain, ice, snow, hail, lightning, wind, blowing dust and sand, salt spray, soot and smog, ultraviolet radiation, and seismic vibration. Airborne window materials must also withstand aerodynamic heating and thermal shock, aerodynamic drag forces, acceleration and vibration, and high impact with precipitation and airborne particles. Environmental conditions, such as wind, temperature, and weight of rain, snow, and ice, primarily influence the required strength of the window or radome wall. Other environmental conditions primarily affect the outer surface of the window. These include humidity, rain, lightning, salt spray, and ultraviolet radiation. If it is not possible to find a single material to withstand all of the environmental factors that are likely to be encountered in a particular application, then it is necessary to apply outer layers of special materials to a window or radome. Such layers can be used to block ultraviolet radiation, seal against moisture, divert lightning, ablate heat, and prevent rain erosion.

A common aperture, dual millimeter-wave (MMW) and infrared (IR) sensor design is used to illustrate the need for dual-wavelength window and dome materials. The index of refraction, loss tangent, absorption coefficient, and other dielectric properties of materials are defined in terms of the relative complex dielectric constant of the material. Window and radome wall configurations are described to show how these affect their strength, erosion resistance, and electromagnetic energy transmission. The physical properties of candidate materials and their impact on performance and applications are discussed. The physical properties include flexure strength, specific heat, index of refraction, melting temperature, hardness, variation of dielectric constant with temperature, and resistance to thermal shock and moisture.

9.1 GENERAL REQUIREMENTS FOR A DUAL-WAVELENGTH SENSOR WINDOW

Figure 9.1 shows an example of a common aperture, dual-mode MMW and IR seeker requiring a window or a dome that transmits dual-wavelength electromagnetic radiation. Another common aperture dual-wavelength sensor configuration is described in Chapter 10. The dual sensor in Figure 9.1 provides IR imagery through Cassegrain optics augmented with refractive elements. Special coatings are used to increase the energy reaching the separate MMW and IR focal planes. The common aperture approach reduces scan hardware, optimizing the volume available for the MMW antenna and IR optical apertures, and maintains boresight alignment. In the sensor model illustrated in the figure, the primary IR optical element is a polished aluminum parabolic mirror surface that also serves as the primary MMW antenna element. The aperture at the center of the aluminum reflector consists of a germanium compensating lens with a coating that reflects MMW energy (to enhance antenna gain) and transmits IR energy. The secondary optical element is made of quartz that is transparent to electromagnetic energy at millimeter wavelengths. It is coated to reflect IR energy through the germanium lens and onto the IR focal plane. The coating also enhances transmission of MMW energy by the MMW feedhorn onto the primary reflector. The MMW feedhorn is placed at the focus of the primary aluminum surface [1].

Common aperture window requirements for this configuration are summarized in Table 9.1. The numerical values are from [2]. In general, the low reflection requirement is typically satisfied in the IR spectrum but not the MMW, while the low absorption requirement is satisfied in the MMW spectrum but not as well in the IR.

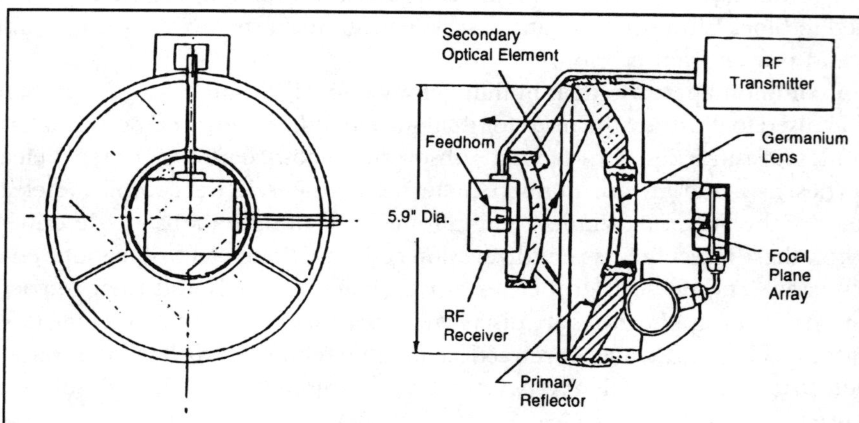

Figure 9.1 Common aperture dual-mode MMW/IR seeker. (*Source:* [1]. Reprinted with permission.)

Table 9.1
Window Requirements for Common Aperture MMW and IR sensors

Requirements
Transparent to both MMW and IR. Implies low reflection and low absorption
Optical transmission of 90% desirable
Flat within 1/10 wavelength
High strength
High hardness
Low coefficient of thermal expansion. Operating temperatures of 500°C or higher
Resistance to thermal shock
Resistance to rain erosion at aircraft speeds of Mach 2 at impact angles of 15 to 90 degrees

9.2 DIELECTRIC THEORY

IR window materials are frequently selected for dual-wavelength use at MMW frequencies, where their dielectric properties determine their suitability for the MMW application. Normal incidence is an important condition for window and radome applications. At this angle, maximum energy is transmitted at minimum window thickness when the thickness is equal to any integer multiple of one-half wavelength in the material. Although a one-half wavelength-thick window may be ideal for energy transmission and minimizing weight, it may not be strong enough to withstand forces such as stresses induced by high speeds, thermal shock from a launch, or rain erosion. When more strength is required, either stronger materials are selected (perhaps with degradation in transmission properties), or integer multiples of half-wavelength thickness are used.

Based on the relative complex dielectric constant e_r of the window material, the square of the Fresnel reflection coefficient, or reflectivity, at normal incidence at the air-material boundary is written as

$$|R|^2 = \left|\frac{1 - \sqrt{e_r}}{1 + \sqrt{e_r}}\right|^2 \tag{9.1}$$

when the window is nonmagnetic; that is, the relative magnetic permeability

$$\mu_r = \frac{\mu}{\mu_0} = 1 \tag{9.2}$$

Here, μ is the magnetic permeability of the window material and μ_0 is the permeability of free space equal to $4\pi \times 10^{-7}$ henry/m. Thus, the reflection coefficient decreases as the relative complex dielectric constant decreases toward unity or, equivalently, that of free space.

The relative complex dielectric constant e_r is defined as [3,4]

$$e_r = \frac{e_c}{e_0} = e_r' - je_r'' \tag{9.3}$$

where

e_c = complex dielectric constant or complex permittivity

$$= e - j\frac{\sigma_c}{\omega} \tag{9.4}$$

$$\equiv e' - je'' \tag{9.5}$$

σ_c = conductivity of the material in mhos/m;

$\omega = 2\pi f$ = radian frequency;

$e' = e$ = real part of the complex dielectric constant;

e'' = imaginary part of the complex dielectric constant;

e_r' = real part of the relative complex dielectric constant;

e_r'' = imaginary part of the relative complex dielectric constant;

$e_0 = 8.854 \times 10^{-12}$ farad/m = permittivity of free space.

The real part of the relative complex dielectric constant may also be written as

$$e_r' = \left(\frac{c_0}{c}\right)^2 \tag{9.6}$$

where

c_0 = speed of light in free space = 3×10^8 m/sec;

$c = \dfrac{1}{\sqrt{\mu e'}}$ = speed of light in the dielectric window material.

The loss tangent is defined as

$$\tan\delta = \frac{e''}{e'} = \frac{e_r''}{e_r'} \tag{9.7}$$

The complex index of refraction n is equal to the square root of the relative complex dielectric constant, and is given by

$$n = \sqrt{e_r} = n' - jn'' \tag{9.8}$$

$$= \frac{1}{k}(\beta - j\alpha) \tag{9.9}$$

where

n' = real part of the complex index of refraction or simply the index of refraction;
n'' = imaginary part of the complex index of refraction;
k = $2\pi/\lambda$ = wave number in the dielectric material in m^{-1};
λ = wavelength in the dielectric material in m;
α = attenuation constant in Np/m;
β = phase constant.

Thus,

$$n^2 = e_r = \frac{e_c}{e_o} \tag{9.10}$$

and

$$n' = \left(\frac{c_o}{c}\right) \tag{9.11}$$

The extinction coefficient γ is the sum of the absorption and scattering coefficients as described in Chapter 3. If scattering is negligible, then

$$\gamma \approx K_\alpha = 2\alpha \tag{9.12}$$

The term 2α is also referred to as the power absorption coefficient [3].

When the transverse electric field (here assumed to have an x-component) propagates in the z-direction, it can be written as

$$E_x(z, t) = E_{xo} \exp[-\alpha z - j\beta z] \exp[j\omega t] \tag{9.13}$$

In a general lossy dielectric material,

$$\alpha = k_o|Im\sqrt{e_r}| = k\left[\frac{1}{2}\left(\sqrt{1+\tan^2\delta} - 1\right)\right]^{1/2} \tag{9.14}$$

and

$$\beta = k_o|Re\sqrt{e_r}| = k\left[\frac{1}{2}\left(\sqrt{1+\tan^2\delta} + 1\right)\right]^{1/2} \tag{9.15}$$

where

k_o = $2\pi/\lambda_0$ = free space wave number;
λ_0 = free space wavelength.

In a low-loss material, such as those desirable for use as windows or domes,

$$\alpha \approx k/2 \tan\delta = \eta_i\sigma_c/2 \tag{9.16}$$

and

$$\beta = k = \omega\sqrt{\mu e'} \tag{9.17}$$

where

$$\eta_i = \sqrt{\frac{\mu}{e'}} \text{ ohms} = \text{intrinsic impedance of the material.} \tag{9.18}$$

For the low-loss material, (9.14), (9.4), (9.18), and (9.9) are used to express the power absorption coefficient as

$$K_\alpha = 2\alpha \approx k \tan\delta = \eta_i\sigma_c = 4\pi n''/\lambda \text{ m}^{-1} \tag{9.19}$$

when λ is expressed in meters.

The index of refraction, and hence absorption coefficient, of anisotropic window materials is defined for the ordinary wave and the extraordinary wave. Crystalline window materials with symmetry lower than cubic, such as sapphire and quartz, are examples of anisotropic materials. An ordinary wave is one where the electric field exists purely in the direction of propagation (for example, the *z*-direction). An extraordinary wave is one where the electric field is purely transverse to the direction of propagation (for example, in the *xy*-plane). The symbol // is sometimes used to denote the value of a window property when the direction of propagation is parallel to the threefold rotational symmetry axis of an anisotropic crystal (i.e., the ordinary direction referred to by *o*-ray). The symbol ⊥ denotes the value for propagation parallel to the twofold rotational symmetry axis (i.e., the extraordinary direction referred to by *e*-ray). In the sapphire crystal, the three-fold rotational symmetry axis is designated as the *c*-axis or optical axis. The twofold rotational symmetry axis, perpendicular to the threefold axis, is designated the *a*-axis. The index of refraction in anisotropic materials varies continuously as a function of the propagation direction through the crystal. An anisotropic material with two limiting values of the index of refraction is called birefringent.

9.3 WINDOW AND RADOME WALL CONFIGURATIONS

Window and radome walls are designed to maximize the propagation of electromagnetic energy at the required incidence angles and frequencies. In addition to providing good electromagnetic performance, the material must possess sufficient strength with minimum weight. There are two basic classes of walls that can be designed to satisfy these requirements: (1) monolithic solid walls and (2) layered walls. These are described in the following sections.

9.3.1 Monolithic Walls

Monolithic walls are usually an integer multiple N of a half-wavelength (in the material) in thickness. This wall exhibits 100% transmission at normal incidence for lossless materials. The thickness can also be optimized for transmission at any desired incidence angle θ by choosing the thickness of the wall d as

$$d = \frac{N\lambda_o}{2}\left(\frac{1}{\sqrt{\epsilon' - \sin^2\theta}}\right), \qquad N > 0 \tag{9.20}$$

Thus, the required wall thickness for maximum transmission increases with increasing incidence angle and decreasing real part of the complex dielectric constant.

Thin walls, those whose thickness is less than approximately 1/20 of a wavelength (in the material), exhibit good broadband electromagnetic performance. The loss becomes large at frequencies where the thickness of the wall approaches 1/20 of a wavelength except when the thickness is an integer multiple of a half-wavelength in the wall. Thin walls are rarely used at millimeter wavelengths because of the resulting low strength of such thin structures. For example, a thin wall $\lambda/20$ configuration at 35 GHz, using a material whose real part of the relative complex dielectric constant is 5, results in a maximum thickness of only 7.55 thousandths of an inch (0.19 mm). At 95 GHz, the thickness is 2.78 thousandths of an inch (0.07 mm).

9.3.2 Layered Walls

Figure 9.2 shows an example of a common *A*-sandwich layered wall used for MMW radomes. This wall is composed of two outer skins, one on each side of a core material. The outer skins are selected from thin but strong materials represented by the relative dielectric constant e_{rs} and thickness *s*. The core is a low-loss, low-dielectric constant, and usually low-strength material represented by the relative dielectric constant e_{rc} and thickness *c*. The thickness of the inner core is approximately one-quarter wavelength in the material. This thickness provides good transmission at normal incidence with a low-dielectric constant ($<\approx 1.5$) material. The core is often composed of foam or a honeycomb material that is largely air and whose real part of the relative complex dielectric constant is typically less than 1.3. The skin materials are selected to withstand environmental conditions such as stress, temperature, and rain and other moisture erosion. The skin may have a higher dielectric constant or loss tangent than is usually desired for optimum electromagnetic transmission. If a large temperature change is anticipated, then the core and skin materials must have similar expansion rates to avoid delamination.

9.4 PHYSICAL PROPERTIES OF WINDOW MATERIALS

The physical properties of several dual-wavelength sensor candidate window materials are shown in Table 9.2 [5]. The properties listed for these crystalline materials

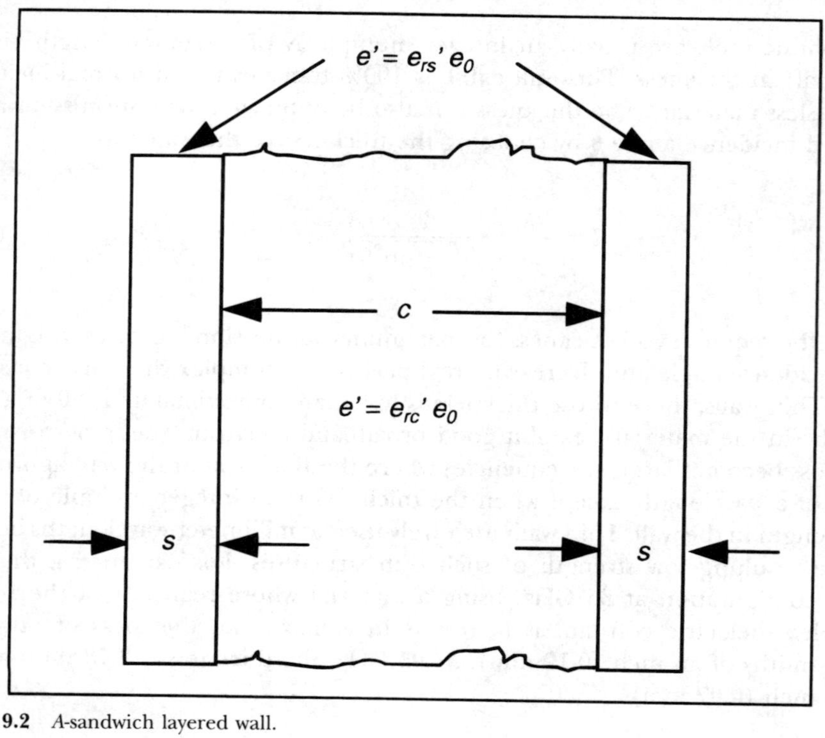

Figure 9.2 *A*-sandwich layered wall.

Table 9.2
Dual-Wavelength Window Material Properties at Room Temperature (295K) [5]

Material	*IR/visible Spectral Interval (μm)*	*MMW Spectral Interval (μm)*	*n′ (IR) (λ = 4 μm)*	*n′ (MMW) (f = 3 GHz)*	*dn′ (IR)/dT (K^{-1})*	*dn′ (MMW)/dT (K^{-1})*	*Flexure Strength (psi)*	*Melting Temp. (K)*	*Heat Capacity (cal/g·K)*
BeO (Beryllium oxide) o-ray	0.11–4.5	—	1.604	2.59	—	—	35,000	≈2,800	0.26
MgO (Magnesium oxide)	0.16–7	—	1.668	3.401	1.1×10^{-5}	1.3×10^{-4}	—	3,070	0.21
SiO_2 (Quartz)	0.16–4.2	>100 (<3000 GHz)			—	—	—	2,000	0.17
o-ray			1.466	2.105					
e-ray			1.474	2.153					
Fused silica	0.16–4.2	—	1.371	1.944	1.3×10^{-5}	—	—	—	—
$ZrO_2{:}Y_2O_3$ (Yttria-doped sapphire)	0.35–7	—	2.049	>3.73	—	—	—	≈3,000	0.10
Al_2O_3 (Sapphire)	0.15–5	>300 (<1000 GHz)					70,000	2,300	0.18
o-ray			1.675	3.056	1×10^{-5}	1.9×10^{-4}			
e-ray			1.668	3.400	1×10^{-5}	2.1×10^{-4}			
Y_2O_3 (Yttria)	0.2–7	>450 (<667 GHz)	1.859	3.43	4×10^{-5}	1×10^{-4}	24,000	2,650	0.115
$MgAl_2O_4$ (Spinel)	0.21–5.5	>450 (<667 GHz)	1.633	2.88	1×10^{-5}	2×10^{-5}	26,700	2,320	0.215
$Al_{23}O_{27}N_5$ (ALON)	0.24–4.8	>350 (<857 GHz)	1.662	3.046	2.7×10^{-5}	2.1×10^{-4}	43,600	2,410	0.20
ZnS (Zinc sulfide)	0.4–14	>800 (>375 GHz)	2.202*	2.89	5×10^{-5}	2.8×10^{-4}	14,000	2,100	0.12
Diamond	>0.22		2.38*	2.38	1×10^{-5}	4.6×10^{-5}	427,000	3,770	0.12

* Data for λ = 10 μm.

(normally associated with IR applications) include MMW and IR values for transparency range, index of refraction n', and change of index of refraction with temperature. Values for flexure strength, melting temperature, and specific heat are also given.

The flexure strength values provide a relative measure of the strength of the different materials. Since the strength of a ceramic is controlled by the presence of defects, it is plausible that a given stress operating over a large area is more likely to encounter a strength-limiting defect than the same stress operating over a small area. Therefore, large parts tend to fail at lower stress than small parts. Furthermore, the strength of a material depends on the type and quality of surface finish, material fabrication process, material purity, test method, and sample size. If the strength of a material for a particular application is marginal, it is best to obtain data from the manufacturer for the specific material with the specific surface finish to be used [6].

The specific heat of a material is the ratio of its heat capacity to that of water at 15°C and thus is a dimensionless quantity. Heat capacity is the amount of heat required to raise the temperature of a given mass of a material 1°C. If heat capacity is expressed in calories/(g°C) or equivalently calories/(g K), then the numerical values for heat capacity and specific heat are the same since the heat capacity of water is 1.00 calorie/(g°C). Heat capacity can also be expressed in units of joules/(g K), where 1 calorie = 4.1868J. It is desirable for window materials to have high heat capacities.

Quartz (SiO_2) has the lowest index of refraction in both the MMW and IR frequency regions and, therefore, the lowest reflection loss of the materials listed. Its index of refraction, unlike that of the other materials, decreases with increasing temperature. In the hostile environment of a missile nose cone, any coating to reduce reflection loss probably will not survive, making low index of refraction materials desirable for this application. The change of index of refraction with temperature, dn'/dT, affects the boresight error produced by a material and is higher at MMW than at IR because of the contributions of lattice vibrations to the index of refraction at MMW frequencies.

Measured absorption coefficients at 5 μm for spinel ($MgAl_2O_4$), sapphire (Al_2O_3), and ALON (aluminum oxynitride: $Al_{23}O_{27}N_5$) at 25°C, 250°C, and 500°C are given in Table 9.3 [7]. The absorption coefficients gradually increase with increasing temperature.

9.4.1 Heat

Heating causes window and radome materials to change their physical thickness, the real part of the complex dielectric constant, and the loss tangent. The combined effect of these changes causes the electromagnetic transmission and reflection properties of the material to vary. The variation in material properties can also have an effect on boresight error in seeker systems due to the temperature-induced change in the index of refraction. Most materials currently in use at IR wavelengths are ceramic, and some can withstand temperatures up to 2,200°C. These inorganic

Table 9.3
Absorption Coefficients of Spinel, Sapphire, and ALON at 25°C, 250°C, and 500°C as Measured at 5 μm [7]

	Absorption Coefficient (cm^{-1})		
Material	*25°C*	*250°C*	*500°C*
Spinel	0.4	0.7	1.3
Sapphire	0.8	1.3	2.1
ALON	1.6	2.4	3.7

materials have some potential for dual-wavelength MMW and IR sensor and seeker applications. However, as will be described later in the chapter, they exhibit a larger than desirable real part of the complex dielectric constant at MMW frequencies. Organic materials are usually large-molecule polymers that find application in MMW radomes where the temperature is less than 370°C. These include duroid, thermal plastics such as lexan and teflon, and laminates such as epoxy-E glass.

Oxides such as sapphire, spinel, ALON, and yttria are stable at temperatures above 1,000°C. On the other hand, materials such as zinc sulfide, zinc selenide, and gallium arsenide whose anions are not oxides, are less stable and have upper operating temperatures less than 1,000°C. For long-term use, such as days or years as opposed to seconds or minutes, lower operating temperature limits are required. For example, zinc sulfide is considered safe for indefinite use at 450°C and zinc selenide at 350°C.

Another consideration at high temperatures is the emission from the window material itself. Since the emitted energy is proportional to the product of the emissivity and the absolute temperature, an increase in either or both factors increases emission. This energy may be of sufficient magnitude to mask the radiation entering the window from the outside sources by decreasing the signal-to-noise ratio or by saturating the detector with photons so that it no longer responds to the outside signal. High temperatures are encountered, for example, when the stagnation temperature at the front of a dome on a missile reaches 350°C at Mach 3 and 1,060°C at Mach 5 at 10-km altitude. At sea level, the temperatures increase to approximately 520°C and 1,475°C, respectively [6]. Increased absorption (and hence emissivity) is also experienced at high temperatures at the absorption edges of a transmission band.

To understand the effect of high window temperatures and emissivity on signal detection, consider the energy of a 10-μm photon signal given by

$$E = hf \tag{9.21}$$

$$= 6.6256 \times 10^{-34}\ \text{J-sec} \times 3 \times 10^{13}\ \text{Hz} = 2 \times 10^{-20}\ \text{J} \tag{9.22}$$

where Planck's constant is denoted by h and frequency by f.

The competing emission at a temperature T of 500°C from a window with an emissivity ϵ of 0.5 is

$$E = k_B \epsilon T \tag{9.23}$$

$$= 1.38 \times 10^{-23}\,\text{J/K} \times 0.5 \times 773\text{K} = 0.53 \times 10^{-20}\,\text{J} \tag{9.24}$$

where k_B is Boltzmann's constant.

Thus, the energy in the signal is only a factor of 3.8 greater than the noise emitted by the window. If the window temperature increases to 1,000°C, then the ratio is reduced to approximately 2.3:1.

9.4.2 Thermal Shock

When a dome material is used in the nose of a missile, friction with the air heats the outside leading surface more rapidly than the rest of the dome at launch. The hot front surface expands more rapidly than the cooler inside and causes significant stress between the expanded and unexpanded material. If the stress is greater than the strength of some critical flaw, the dome shatters. Thermal shock failure may occur at the inside or outside surface of a dome, depending on the flight stage of the missile. Thermal shock failure typically occurs on the inside surface of the dome during the boost phase at a site where the maximum tensile stress is exceeded. After the boost stage ends and the dome reaches maximum temperature, the dome cools from the outside during the glide stage. Now, maximum tensile stress occurs on the outside surface at which failure may originate.

Thermal shock resistance is obtained by using materials with high strength, high thermal conductivity, low thermal expansion, and low Young's modulus, where Young's modulus is the ratio of stress to strain. Thinner domes and windows are more resistant to thermal shock than are thicker ones. Sapphire and gallium phosphide are candidate materials that transmit energy in the 3- to 5-μm region and are also resistant to thermal shock. Silicon is resistant to thermal shock, but cannot operate above temperatures of approximately 260°C because of the increase in free carrier absorption that increases the absorption coefficient. Free carrier absorption, also known as intraband absorption, typically occurs in materials with small bandgaps when the temperature is high enough to cause a thermally induced transition of electrons from the valance band into the conduction band. This creates electrons in the conduction band and holes in the valence band (described in Section 2.6) that can absorb small increments of energy that promote these carriers to a higher energy state within the same band. Potential 8- to 14-μm materials include zinc sulfide, gallium arsenide, and type IIa (nitrogen free) diamond. Thin diamond coatings (<100 μm) can also operate in the 3- to 5-μm region. Germanium, which can potentially span the 2- to 23-μm wavelength band, cannot be used above approximately 100°C at the longer wavelengths and above 200°C at midband because of the increase in free carrier absorption.

9.4.3 Moisture

Surface water and moisture within the material also have a significant effect on the performance of the window or radome. The real and imaginary components of the dielectric constant of water were described in Chapter 6. The large real part of the complex dielectric constant, and the sizable imaginary part, cause water to influence the propagation of electromagnetic radiation more than the conventional low dielectric constant, low-loss window and dome materials. Hydrophobic outer layers are therefore of value in repelling moisture from window and radome materials.

9.4.4 Absorption Coefficient Model for Far Infrared Wavelengths Approaching the MMW Spectrum

The absorption coefficient may be modeled in terms of lattice vibrations or, equivalently, phonon transitions that characterize the energy absorbed when a molecule of the window material changes vibrational energy states. The phonon is the quantum unit of crystal vibrational energy. The smallest amount of energy that the molecule can absorb upon going from vibrational energy level $v = 0$ to vibrational energy level $v = 1$ is $h\omega$, where v is the vibrational quantum number (0, 1, 2, 3, . . .), h is Planck's constant,

$$\omega = \frac{1}{2\pi}\sqrt{\frac{k_f}{\mu_{rm}}} \tag{9.25}$$

k_f = force constant as illustrated in the simple model of a diatomic molecule in which two atoms of masses m_1 and m_2 are connected by a spring of force constant k_f shown in Figure 9.3, and

$$\mu_{rm} = \text{reduced mass} = \frac{m_1 m_2}{m_1 + m_2} \tag{9.26}$$

If the molecule absorbs energy $h\omega$ in going from $v = 0$ to $v = 1$, the associated energy state transition is referred to as the fundamental transition. If the molecule absorbs energy capable of transitioning from $v = 0$ to $v = 2$, the transition is called

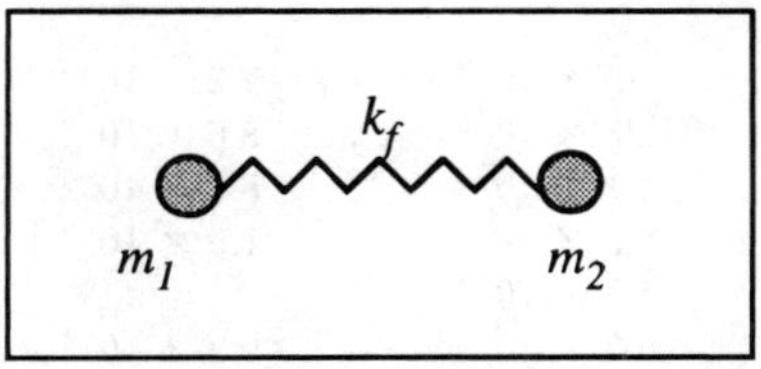

Figure 9.3 Vibration model of a diatomic molecule whose atoms are connected by a spring of force constant k_f.

the first overtone. Similarly, the $v = 0$ to $v = 3$ state transition is referred to as the second overtone.

Table 9.4 contains the coefficients for a model that predicts the absorption coefficient in the far infrared wavelength region as a function of wavenumber k and the vibrational energy states in a molecule of the material [5,8–10]. Far infrared in this context implies wavelengths of 100 to 200 μm or greater, depending on the material, and denotes the absorption at frequencies approaching the MMW spectrum.

The absorption coefficient at wavenumbers approaching the MMW spectrum is expressed as

$$K_\alpha = 2\alpha = A(T_O)k^2 \text{ cm}^{-1} \text{ at room temperature of 295K} \tag{9.27}$$

and as

$$K_\alpha = 2\alpha = [A_{1p} + A_{2p} \times T/T_o + A_{3p} \times (T/T_o)^2]k^2 \text{ cm}^{-1} \tag{9.28}$$

at other temperatures T.
where the coefficients A_{vp} are frequency-independent constants that relate the v-phonon vibrational energy state to the absorption coefficient, k is the wavenumber in cm^{-1}, and temperature T_o equals 295K. The model predicts the increase in absorption and, hence, emissivity that is observed as temperature increases.

Not all of the coefficients at the higher order phonon energy states have been measured as indicated by the gaps in the table. For the o-ray of sapphire, where

Table 9.4
Far-Infrared (Approaching MMW Spectrum) Absorption Coefficient Model Parameters [9]

Material	A (T_0 = 295 K) (cm)	A_{1p} (cm)	A_{2p} (cm)	A_{3p} (cm)
MgO (Magnesium oxide)	n/a	6.1×10^{-4}	n/a	n/a
SiO_2 (Quartz)				
o-ray	1.07×10^{-4}	8.9×10^{-5}	n/a	n/a
e-ray	8.2×10^{-5}	1.39×10^{-4}		
Al_2O_3 (Sapphire)				
o-ray	9.15×10^{-4}	2.2×10^{-4}	4.7×10^{-4}	2.2×10^{-4}
e-ray	1.28×10^{-3}	8.6×10^{-4}	n/a	n/a
Y_2O_3 (Yttria)	2.1×10^{-3}	8.4×10^{-4}	n/a	n/a
$MgAl_2O_4$ (Spinel)	2.2×10^{-3}	1.6×10^{-3}	n/a	n/a
ALON (Aluminum oxynitride)	1.1×10^{-3}	n/a	n/a	n/a
ZnS (Zinc sulfide)	5.8×10^{-4}	9.8×10^{-4}	n/a	n/a

Note: n/a = not available.

the complete three-phonon model is available, the sum of the A_{1p}, A_{2p}, and A_{3p} coefficients is equal to the coefficient A at $T = T_0$ as predicted by the theory. Extrapolation of the absorption coefficient, using (9.28), to MMW frequencies has been demonstrated for the ordinary ray of sapphire and quartz at room temperature [9].

9.4.5 Dielectric Properties

The dielectric properties of potential dual-wavelength IR and MMW crystalline window materials at 35 GHz are shown in Table 9.5 [6,11]. These materials are those commonly used for IR windows and domes. The frequency and temperature at which the values were measured are indicated. Reflection is reduced if the real part of the complex dielectric constant e' is low and absorption is reduced if the loss tangent is low. The MMW values of the dielectric constants for the IR ceramic materials are fairly high (e'_r = 8 to 11), but exhibit negligible losses. The dielectric constants are almost independent of MMW frequency, where the decrease in value between 35 and 95 GHz is ≤ 0.2% [11]. The measured loss tangents are lowest in

Table 9.5
Dielectric Properties of Potential Dual-Wavelength IR and MMW Crystalline Window Materials at Microwave and Millimeter-Wave Frequencies [6,11]

Material	*Frequency/ Temperature*	*Re|Relative Complex Dielectric Constant|* e'_r*	*Loss Tangent* e''/e'*
Zinc sulfide (ZnS)	35 GHz/296K	8.347(10)	0.0024(3)
Zinc selenide (ZnSe)	35 GHz/296K	8.976(10)	0.0017(2)
ALON (Aluminum oxynitride: $Al_{23}O_{27}N_5$)	35 GHz/296K	9.280(10)	0.00027(5)
Spinel (Magnesium Aluminate: $MgOAl_2O_3$)	35 GHz/296K	9.190(10)	0.00022(5)
Sapphire (Al_2O_3)	35 GHz/296K	9.392(5) (*o*-ray)	0.00005(2)
		11.582(5) (*e*-ray)	0.00006(2)
Yttria (Y_2O_3)	**	11.8	0.0005
Calcium fluoride (CaF_2)	**	6.5	0.00015
Magnesium fluoride (MgF_2)	**	5.1	0.0001
Silicon (10 ohm-m resistivity)	**	≈12	≈0.009
Diamond	**	5.7	<0.0004

* Numbers in parentheses are estimated uncertainties in the final digit.
** Applies over frequencies from hundreds of megahertz to tens of gigahertz. Loss tangents increase as the frequency is increased from the microwave to the far infrared.

single crystal sapphire (0.00005) and highest in zinc sulfide and zinc selenide (0.0024 and 0.0017, respectively).

A set of values for the relative dielectric constant and loss tangent is shown in Table 9.6 for zinc sulfide at frequencies from 8 to 94 GHz [12]. These values are consistent with those reported above. Corresponding dielectric properties of the IR crystalline materials at infrared wavelengths are given in Table 9.7 [6,13]. For comparison, the dielectric properties of inorganic and organic materials, typically used in microwave and millimeter-wave radome construction, are listed in Table 9.8 [14,15].

Effect of Temperature on Dielectric Constant

The real part of the dielectric constant significantly increases with temperature. Figure 9.4 shows measurements at 35 GHz for zinc sulfide, zinc selenide, spinel, sapphire, and ALON. The relative increases are largest for zinc selenide and zinc sulfide, and smallest for sapphire. The increase in MMW dielectric constant with temperature exhibited by these crystalline materials is directly proportional to the thermal expansion coefficient [11]. Therefore, the use of low coefficient of thermal expansion materials minimizes the sensitivity to temperature and resultant reflection losses and boresight errors.

Effect of Temperature on Loss Tangent

The measured increase in loss tangent at 35 GHz for the IR crystalline window materials is shown in Figure 9.5. The loss tangent can also be expressed as the power dissipated divided by the power stored per cycle. Thus it is a measure of the energy lost as heat when a wave propagates through the material. The loss tangent of the aluminum-based materials is typical of low-loss ceramics (i.e., a gradual

Table 9.6
Dielectric Properties of Zinc Sulfide (ZnS) in the Microwave and Millimeter-Wave Spectrum* [12]

Frequency (GHz)	*Re |Relative Complex Dielectric Constant|* e'_r	*Loss Tangent* e''/e'
8 to 12	8.02	—
12.5 to 18	8.23	—
35	8.39	0.0012
94	8.60	0.0024

* Sassoon Advanced Materials Limited of Dumbarton, Scotland quotes a constant refractive index for ZnS of 2.9 (e'_r = 8.41) at 10, 30, and 100 GHz. The absorption coefficient at these frequencies is given as 0.01, 0.03, and 0.1 cm^{-1}, respectively.

Table 9.7
Dielectric Properties of Potential Dual-Wavelength IR and MMW Crystalline Window Materials at Infrared Wavelengths [6,13]

Material	*Wavelength*	*Re* |*Refractive Index*| n'	*Re* |*Relative Complex Dielectric Constant*| ϵ'_r
Zinc sulfide (ZnS)	3 to 13 μm	2.26 to 2.15	5.10 to 4.64
Zinc selenide ($ZnSe$)	0.6 μm	2.61	6.81
	3.39 μm	2.44	5.95
	10.6 μm	2.40	5.76
ALON (Aluminum oxynitride: $Al_{23}O_{27}N_5$)	4 μm	1.66	2.76
Spinel (Magnesuim Aluminate: $MgOAl_2O_3$)	4 μm	1.63	2.67
Sapphire (Al_2O_3)	4 μm	1.68 (*o*-ray)	2.81 (*o*-ray)
		1.67 (*e*-ray)	2.78 (*e*-ray)
Yttria (Y_2O_3)	4 μm	1.86	3.46
Calcium fluoride (CaF_2)	3 to 9 μm	1.42 to 1.32	2.01 to 1.73
Magnesium fluoride (MgF_2)	1.08 μm	1.37 (*o*-ray)	1.89 (*o*-ray)
		1.39 (*e*-ray)	1.92 (*e*-ray)
Germanium	3 μm	4.04	16.36
Quartz	3 to 5 μm	1.50 to 1.42	2.25 to 2.01
Silicon (purity not specified)	1.4 to 11 μm	3.49 to 3.42	12.20 to 11.68
Diamond	10 μm	2.38	5.66

increase with temperature up to approximately 1,000°C, followed by a much more rapid rise at higher temperatures). By contrast, sapphire shows a large drop in loss at cryogenic temperatures, followed by a gradual rise up to 800°C. The zinc compounds show moderate absorption at MMW frequencies, exhibiting a more rapid increase when compared to the lower loss materials. The loss mechanism at the lower temperatures in the figure is attributed to localized doping of the primary and secondary crystalline phases that create isolated microscopic regions of finite electrical conductivity. These conductive regions are localized within the grains and along individual grain junctions and grain boundaries. The loss mechanism at the higher temperatures is attributed to the increase in electrical conductivity of the liquid phase as the softening temperatures of residual amorphous glassy phases are approached [11].

To minimize the reflective losses and boresight error, MMW window designs require the wavelength λ within the material to be an integral number of half-wavelengths. The wavelength λ is related to the free-space wavelength λ_0 by the real part of the index of refraction as

$$\lambda = \lambda_0 / n' \tag{9.29}$$

The reflection coefficient of a half-wavelength and full-wavelength ALON

Table 9.8
Dielectric Properties of Microwave and Millimeter-Wave Radome Materials [14,15]

Material	*Re |Relative Complex Dielectric Constant|* e'_r	*Loss Tangent* e''/e'
MMW Inorganic Materials		
Aluminum oxide	7.85	0.0005
Alumina, hot pressed	9.6	0.0005
Boron nitride, hot pressed	4.87	0.0005
Boron nitride, pyrolytic	5.12	0.0005
Magnesium aluminum silicate (cordierite ceramic)	4.75	0.002
Magnesium oxide	9.72	0.0005
Pyroceram 9606	5.58	0.0008
Silicon dioxide	3.82	0.0005
Slip-cast fused silica	3.33	0.001
Silicon nitride	5.50	0.003
MMW Organic Materials		
Lexan	2.86	0.006
Teflon	2.10	0.0005
Epoxy-E glass cloth	4.40	0.016
Polyester-E glass cloth	4.10	0.015
Polyester-quartz cloth	3.70	0.007
Quartz-reinforced polyimide	3.2	0.008
Duroid 5650™ (loaded PTFE)	2.65	0.003

window, optimized for room-temperature conditions, is shown in Table 9.9 for several operating temperatures. At 500°C, the reflection coefficient is 0.02, increasing to 0.13 at 1,000°C for the half-wavelength thickness. The full-wavelength window is designed for increased mechanical strength, but exhibits a reflection coefficient of 0.07 at 500°C. The loss tangent must be low enough at the operating temperatures so as not to significantly degrade the window properties. For example, absorption was measured at 3% at 35 GHz with the 2.8-mm full-wavelength-thick ALON at 1,000°C [11].

9.4.6 Rain Erosion

Collisions of an infrared window or dome with raindrops at sufficient velocity can damage or, in the extreme case, break the material. Long-wave materials such as ZnS and Ge are especially susceptible to damage that slowly accumulates during a lifetime of occasional encounters with clouds and rain at subsonic aircraft speeds. In ZnS, a single raindrop of 2.3 mm diameter can cause impact craters with damage rings surrounding a relatively undamaged central area. This circular damage pattern is characteristic of a polycrystalline material with small, randomly oriented grains.

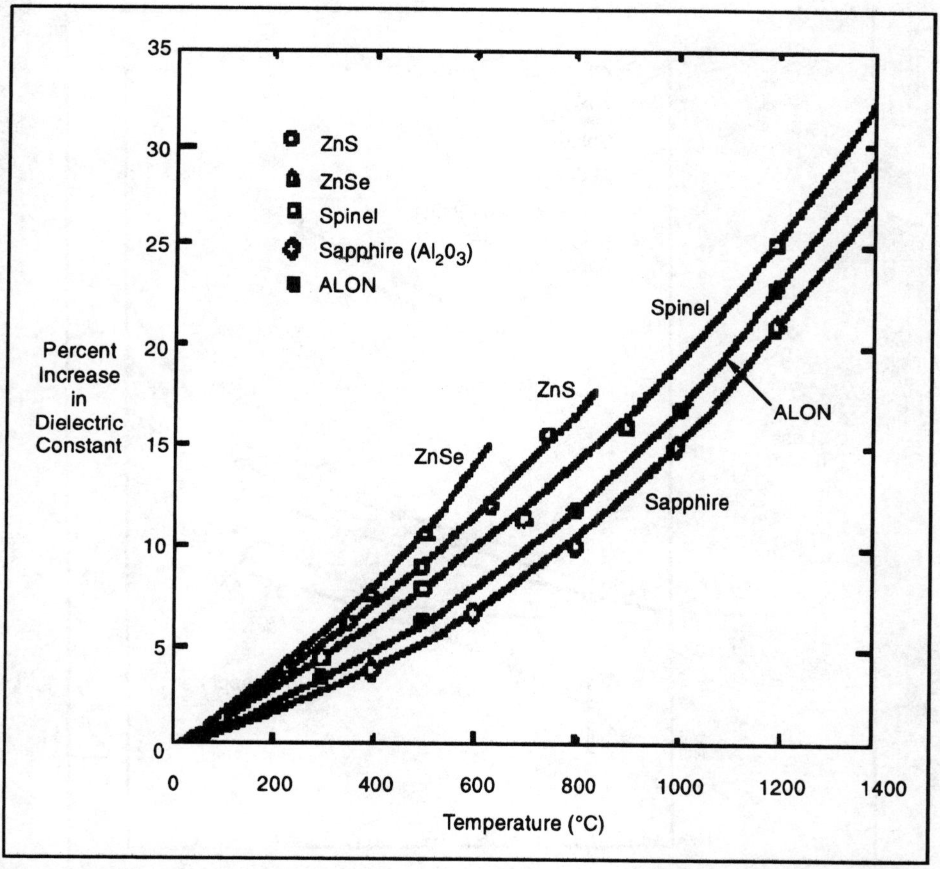

Figure 9.4 Measured increase in dielectric constant with temperature at 35 GHz. The increases are normalized with respect to room-temperature values. (*Source:* [11]. Reprinted with permission.)

Single-crystal materials suffer rain damage by forming straight-line cracks along crystal cleavage planes. Glasses may give ill-defined fracture patterns without the well-developed ring fractures that are characteristic of ZnS [6]. Layered zinc sulfide and zinc selenide windows are now being used in the U.S. Air Force Low Altitude Navigation and Targeting Infrared System for Night (LANTIRN) program. The windows are transmissive in the 1- to 14-μm range, and are rugged and resist erosion from rain, dust, and dirt, presumably through control of material purity, surface finish, and other manufacturing processes that control the formation of defects [16]. Fluoride glasses have the potential for use as windows in the visible through mid IR. They are non-hygroscopic and of moderate hardness that lends itself to achieving optical surface polishes [17].

The damage caused by impact with rain drops reduces the mechanical strength of window and dome materials. Collisions with ice at higher altitudes appears to

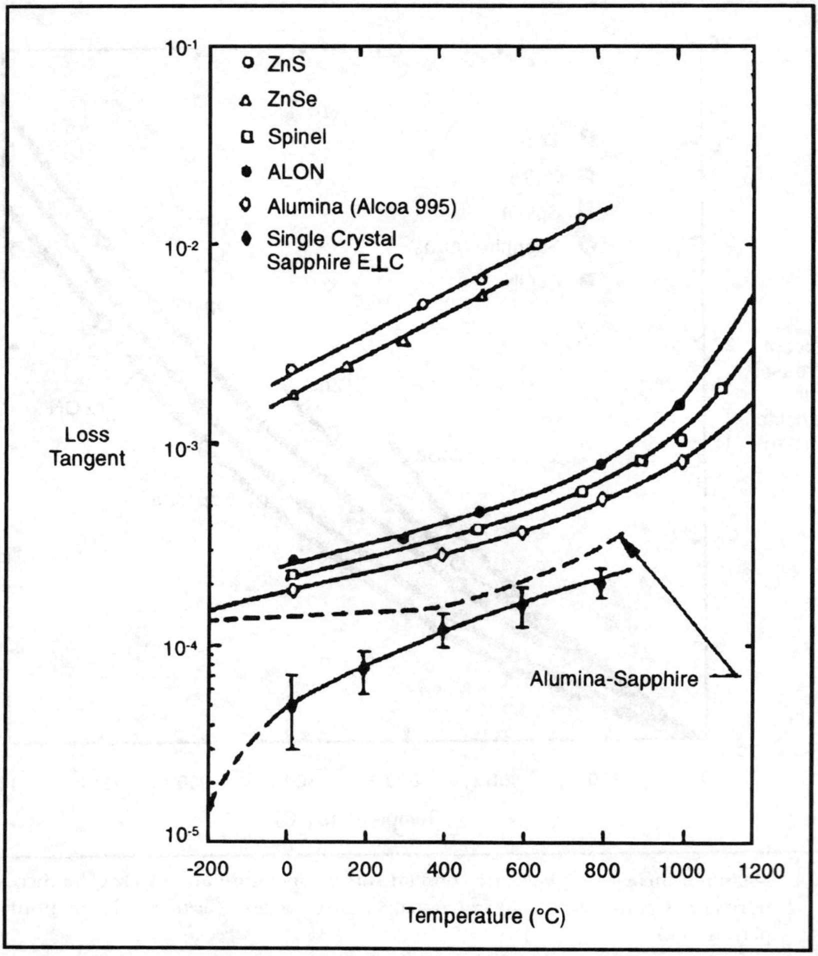

Figure 9.5 Measured increase in loss tangent with temperature at 35 Ghz. (*Source:* [11]. Reprinted with permission.)

cause less damage. In ZnS, for example, the mechanical strength of undamaged material is approximately 80 MPa[1], but is reduced to approximately 40 MPa after impact with a water jet at speeds between 150 and 600 m/sec. The damage threshold velocity decreases with increasing drop size by approximately the cube root of the drop diameter [6,18]. In addition to crack growth and loss of strength, rain impact damage increases transmission loss by increasing scattering of the electromagnetic energy. The transmission loss increases as the number of impacts increases and as the wavelength decreases.

Two types of protective coatings can prevent erosion of IR window and dome materials. These are a thin layer of a rigid material (those with a high Young's

[1]MPa = 10^6 Pascals, where 6.89×10^3 Pascals or Newton/square meter = 1 psi.

Table 9.9
Reflection Coefficient of a Half-Wavelength and Full-Wavelength ALON Window at Room Temperature, 500°C, and 1,000°C [11]

Window Thickness	*Reflection Coefficient* Room Temperature	500°C	1,000°C
Half-wavelength (1.4 mm) optimized at room temperature	0.0	0.02	0.13
Full-wavelength (2.8 mm) for increased mechanical strength	0.0	0.07	—

modulus) or a thin layer of a compliant elastomer [6]. Either must adhere strongly to the more brittle substrate.

The rigid material protects by allowing the impacting particles to produce less strain in the coating than in the unprotected substrate. The coating then transmits less strain into the substrate than direct particle impact on the substrate produces. Examples of rigid protective materials are germanium-carbon, having an index of refraction between 2 and 4 and absorption coefficient <10 cm^{-1} in the 3- to 12-μm region, and boron phosphide with an index of refraction of 2 at 10 μm. The increase in absorption coefficient above 10 μm limits the usefulness of boron phosphide to the 8- to 10-μm region, where the absorption coefficient varies from 20 to 50 cm^{-1}. Near 12 μm, it increases to approximately 200 cm^{-1} [19]. Boron phosphide is used for rain erosion protection of Ge and ZnS.

A compliant coating protects by spreading the stress over a greater area. Elastomers accommodate large strains by elastic deformation and need Poisson's ratios >0.4 to be effective. The Poisson ratio is the fractional change in diameter of a solid object divided by the fractional change in length after a tensile stress is applied. The thickness of the elastomer should be about 20% of the contact radius of the impacting particle. The development of elastomers for protection of IR windows has lagged that of the rigid materials [20].

9.5 SUMMARY

The available database on the dielectric properties of IR window materials at 35 GHz indicates that there is a substantial increase in both the real part of the dielectric constant and the loss tangent as a function of temperature for the materials studied by Ho [11]. This may limit the useful operating frequency range in dual MMW and IR window applications when excessive temperatures are encountered.

The increase appears lowest in single-crystal sapphire and highest in the polycrystalline zinc compounds. Since MMW window and radome designs require the optical path length in the material to be an integral number of half-wavelengths (at normal incidence) to minimize reflection losses, a change in the dielectric constant with temperature can result in substantial losses and increased boresight error. Although the loss tangents of these materials vary over two orders of magnitude, they are still small enough at most temperatures to not significantly degrade window properties. Traditional long-wavelength IR transmitting materials contain compounds with heavy atoms, such as yttrium and selenide, because these give rise to lower energy vibrations and, hence, lower absorption coefficients. However, the chemical bonds within these crystals are weaker, causing compounds made from them to have inferior mechanical properties than those made of lighter atoms. Hence, the longer wavelength materials are generally weaker, softer, and less tough than mid-wave window materials. Diamond is the notable exception to the material tradeoff between gaining long-wavelength transmission and losing desirable mechanical properties. It combines the best possible mechanical performance with excellent long-wave transmission. Other considerations in selecting a window material include manufacturing process control to ensure electromagnetic and mechanical performance repeatability, sufficient mechanical strength and shock resistance to withstand high speeds and launch stress, and resistance to erosion from rain and other environmental agents.

References

[1] Benser, E. T., J. A. Cox, and H. B. French, "IR/MMW dual-mode, common aperture optics," *Infrared Technology XIII*, SPIE Vol. 819, 1987, pp. 288–296.

[2] Cooper, B., "A New Generation of IR Windows," *Photonics Spectra*, June 1990, pp. 159–162.

[3] Ulaby, F. T., R. K. Moore, and A. K. Fung, *Microwave Remote Sensing: Active and Passive, Vol. I, Microwave Remote Sensing Fundamentals and Radiometry*, Norwood, MA: Artech House, 1981.

[4] Kruse, P. W., L. D. McGlauchlin, and R. B. McQuistan, *Elements of Infrared Technology: Generation, Transmission, and Detection*, New York, NY: John Wiley & Sons, 1962.

[5] Thomas, M. E., "Electromagnetic Properties of Dual-Mode Materials for IR/RF Domes," *Proc. Third Electromagnetic Windows Symposium*, Huntsville, AL, Nov. 1989.

[6] Harris, D. C., *Infrared Window and Dome Materials*, Vol. TT 10, SPIE Optical Engineering Press, Bellingham, WA, 1992.

[7] Cox, J. A., et al., "Comparative Study of Advanced IR Transmissive Materials," *Infrared and Optical Transmitting Materials*, SPIE Vol. 683, 1986.

[8] Thomas, M. E., et al., "Characterization of the Infrared and Far-Infrared Properties of Sapphire and Quartz," *13th International Conf. on Infrared and Millimeter Waves*, R. J. Temkin, (Ed.), SPIE Conf. Dig. Vol. 1039, Dec. 1988, pp. 339–341.

[9] Thomas, M. E., "Temperature Dependence of the Complex Index of Refraction," Chapter 8 in *Handbook of Optical Constants of Solids*, Vol. 2, E. Palik, (Ed.), San Diego, CA: Academic Press, 1991.

[10] Thomas, M. E., personal communication.

[11] Ho, W. W., "Millimeter-Wave Dielectric Properties of Infrared Window Materials," *Infrared Systems and Components*, SPIE Vol. 750, 1987, pp. 161–165.

[12] Dielectric Properties of Amber ZnS, Paper 7, British Aerospace, *5th European Electromagnetic Windows Conference*, Antibes-Juan-Les-Pins, France, Sept. 27–29, 1989.

[13] Wolfe, W. L., "Optical Materials," Chapter 7 in *The Infrared Handbook*, W. L. Wolfe and G. J. Zissis, (Eds.), 3rd Printing, The Infrared Information Analysis Center, Environmental Research Institute of Michigan, Ann Arbor, MI, 1989.

[14] Huddlestone, G. K. and H. L. Bassett, "Radomes," Chapter 44 in *Antenna Engineering Handbook*, R. C. Johnson and H. Jasik, (Eds.), New York, NY: McGraw-Hill, 1984.
[15] Joy, E. B., "MMW Radomes," Chapter 12 in *Principles and Applications of Millimeter-Wave Radar*, N. C. Currie and C. E. Brown, (Eds.), Norwood, MA: Artech House, 1987.
[16] "IR Windows Match Night-Vision Needs," *Laser Fous World*, Dec. 1989, p. 35.
[17] Klocek, P., and B. M. Kale, "Applications of infrared transmitting glasses in electro-optic systems," *Infrared Sensor Technology*, SPIE Vol. 344, 1982, pp. 98–105.
[18] Field, J. E., R. J. Hand, and C. J. Pickles, "Strength and Rain Erosion Studies of IR Materials," *Proc. SPIE, Vol. 1112*, 1989, pp. 306–315.
[19] Hand, R. J., J. E. Field, and S. van der Zwaag, "High Modulus Layers as Protective Coatings for Window Materials," *Proc. SPIE*, Vol. 1112, 1989, pp. 120–128.
[20] Matthewson, M. J., "The Effect of a Thin Compliant Protective Coating on Hertzian Contact Stress," *J. Phys. D*, Vol. 15, 1982, pp. 237–249.

CHAPTER 10

Millimeter-Wave/Infrared Dual-Sensor Design

Millimeter-wave and imaging infrared sensors have increased the capabilities of modern weapon systems, examples of which were described in Chapter 1. These multiple-sensor systems enhance the ability of a missile to detect and classify targets in inclement weather, clutter, and countermeasures because they operate in different portions of the electromagnetic spectrum and exploit both the active and passive signatures of targets. Sensor performance may be extended further by incorporating signal-processing algorithms that are tailored to the resolution and dimensionality of the output signals from the transducers. This chapter applies the sensor design and performance criteria discussed in past chapters to the design of a millimeter-wave (MMW) and passive infrared (IR) dual sensor system. The implications of the transducer hardware design on seeker gimbal requirements, signal-to-noise ratio, signal-to-clutter ratio, minimum detectable signals, and spatial resolution are analyzed. However, the signal-processing approaches that enhance the detection and classification of objects in the field of view of the transducers are not specifically addressed. The latter were discussed in previous chapters that dealt with each of the different sensor types.

10.1 MISSION SCENARIO

A typical short-to-medium range submissile application is postulated to derive the design requirements for the sensors. The submissile is assumed to be dispensed from a larger carrier rocket or munition at some distance from the target zone. The terminally guided submissile then autonomously flies to the target zone and searches for and acquires targets in an area defined in Figure 10.1. Different missions require different missile trajectory patterns, some with steep terminal descents, and thus also require different sensor designs as the detection and

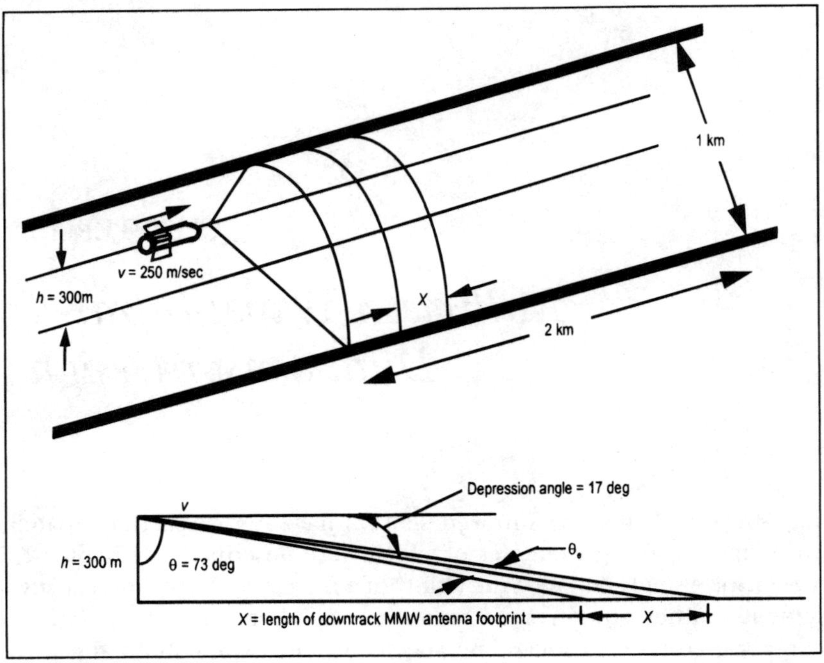

Figure 10.1 Mission scenario.

engagement ranges and geometries vary. In the terminally guided submissile application described in this chapter, the missile engages in horizontal flight while its sensors scan a 1-km by 2-km area for potential targets from a depression angle of 17 degrees. The submissile flies at an altitude of 300 m with a velocity of 250 m/sec. Based on the look-down angle and altitude, the slant range is 1,000 m. The required sensor system probability of detection is 90% when the target has the nominal radar cross section of 25 m^2 and the false target density is 0.5 false targets per square kilometer. Sensor system performance is to be maintained at some nominal level when operating in high-clutter backgrounds and against targets whose signatures are suppressed. The sensor system operates autonomously and has a lock-on-after-launch capability. The MMW and IR sensors operate continuously without a provision for hand-off to one or the other sensor when the missile is in close proximity to the target. The submissile diameter is 6 inches and its length is 26 inches. Possible targets encompass stationary and moving armor and surface-to-air missile (SAM) sites.

10.2 DUAL-SENSOR CONFIGURATION

Naturally it is desirable to design one low-cost sensor to satisfy all the mission requirements. By using a MMW/IR sensor combination, however, the potential exists to design and build two relatively inexpensive sensors that together satisfy the design goals. As will be shown, one sensor can satisfy the detection and tracking requirements in a nominal clutter background against a nominal target. However, single-sensor performance severely degrades when the target signature is suppressed or the background is one of high clutter. The candidate sensor system concept is shown in Figure 10.2. Gimbal-mounted, coherent, monopulse MMW radar and imaging IR sensors are postulated along with their respective digital signal processors. Feature- or sensor-level fusion in the data fusion processor further enhances the ability of the sensors to detect and classify targets, especially under inclement weather and countermeasure conditions.

A common aperture sensor configuration is produced by having a 2-inch clear aperture IR imaging array positioned in the center of the MMW antenna array, as shown in Figure 10.3. The common boresight of the sensors eliminates parallax, reduces parts, and leads to a compact form factor that minimizes packaging volume while maintaining adequate performance. The design specifications for the MMW radar are summarized in Table 10.1 and those for the imaging IR array in Table 10.2. These specifications are the results of the analyses described in the subsequent sections of this chapter.

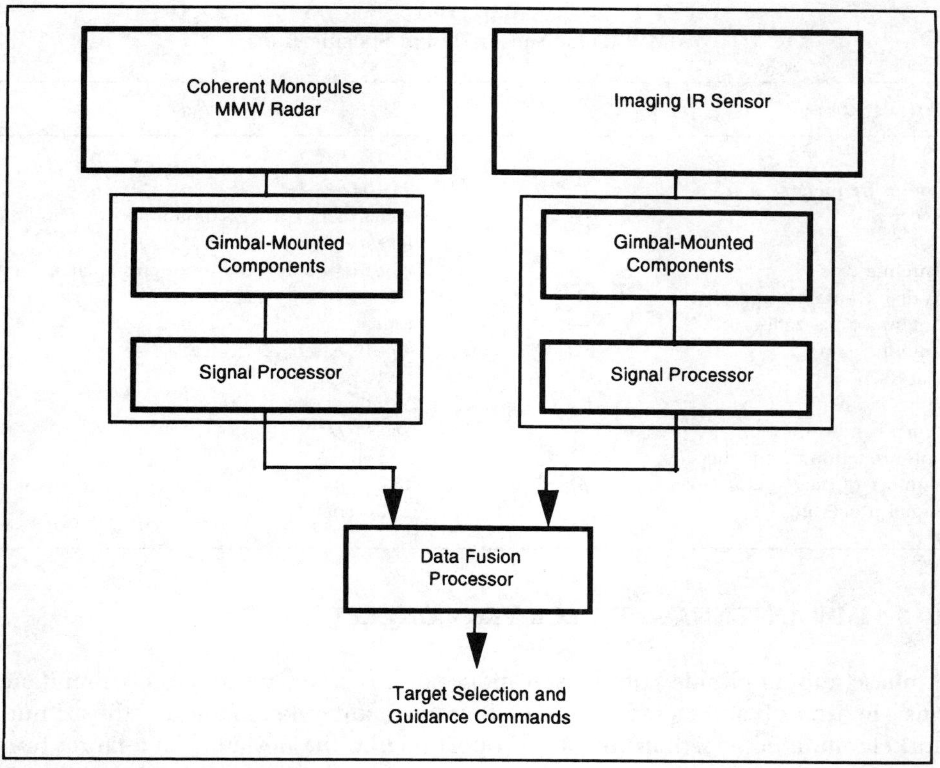

Figure 10.2 Dual-sensor concept.

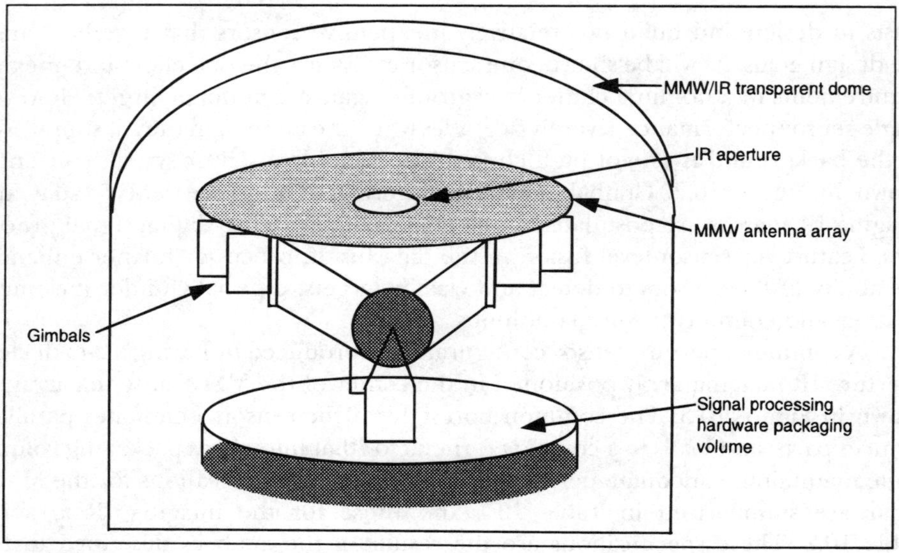

Figure 10.3 Common aperture MMW/IR seeker.

Table 10.1
MMW Radar Sensor Design Specifications

Design Parameter	*Symbol*	*Value*
Center frequency	f	94 GHz
Receiver	—	2 channel to process monopulse sum and difference signals
Antenna type	—	2-beam, multiplexed monopulse planar array
Available antenna aperture	D_o	5.5 in (14 cm) maximum
Antenna polarization	—	Linear
Antenna gain	G_T, G_R	45 dB search, 39 dB track
Output power	P_T	20 W peak
Pulse width	τ	50 nsec
Chirp bandwidth	—	250 MHz
Pulse repetition frequency	f_r	50 kHz
Number of range gates	n_r	18
Signal processing	—	Coherent

10.3 MMW ANTENNA AND RECEIVER CONCEPT

A phase and amplitude comparison monopulse receiver tracks targets simultaneously in azimuth and elevation using a four-beam antenna to generate the azimuth and elevation error signals that are proportional to the deviation of a target from the antenna boresight axis [1,2]. The error signals are shown in the three-channel

Table 10.2
Imaging IR Sensor Design Specifications

Design Parameter	*Symbol*	*Value*
Wavelength band	λ_L to λ_U	3 to 5 μm (nominal)
Optical clear aperture	D_o	2 in (5 cm)
Detector material	—	Indium antimonide (InSb)
Specific detector detectivity	D_d^*	2×10^{11} cm Hz$^{1/2}$ W^{-1}
Focal plane temperature	T_f	77K
Cooler type	—	Joule-Thompson
Number of detectors	n_d	132 in two columns of 66
Focal number	$F/\#$	2
Detector size	$a \times b$	≈0.0019 in by ≈0.0019 in (0.048 mm × 0.048 mm)
Instantaneous field of view	IFOV	0.48 mrad
Resolvable spatial frequency	f_T	1.0 cycle/mrad
Noise equivalent temperature difference	NEΔT	0.062K
Minimum resolvable temperature difference	MRTD	0.23K
Signal processing	—	Image processing

monopulse direction-finding radar of Figure 10.4 as ΔA and ΔE, respectively, along with the Σ signal produced by the sum of the four beams. In another monopulse radar concept, the antenna can measure the azimuth and elevation positions on alternate seeker scans by designing the available aperture to form two narrower beams using the mechanical or electrical beam-forming network. During the first scan cycle, for example, two left-to-right beams are used to form the resultant sum

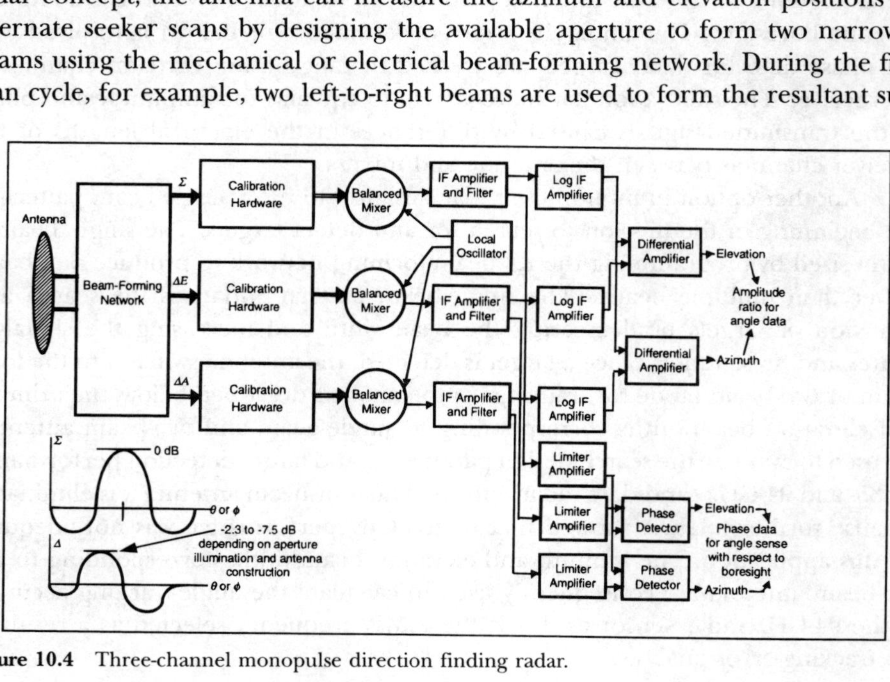

Figure 10.4 Three-channel monopulse direction finding radar.

and difference antenna patterns that sense azimuth position. On the next scan, the beams are formed up and down to create the sum and difference patterns that sense elevation. When the time-shared tracking approach is used, the azimuth and elevation position errors are computed by multiplexing the input signals in either radio frequency (RF) components before the mixer or in the intermediate frequency (IF) stage that follows the mixer [1]. With this radar configuration one of the difference receiver channels in Figure 10.4 is eliminated. Although detailed analysis of the angle-tracking error sources is beyond the scope of this chapter, the errors in the two-channel and three-channel receivers are due to polarization and antenna pattern effects, RF beam-forming network errors, thermal noise, calibrated receiver imbalance, and mechanical misalignment [2].

The sum and difference antenna radiation patterns, Σ and Δ, produce the signals processed by the sum and difference receiver channels. The sum pattern contains a single lobe and the difference pattern two lobes as shown in the lower left of Figure 10.4. Because the difference radiation pattern is in phase with the sum radiation pattern on one side of the antenna axis and out of phase on the other, bearing angle information (in terms of the target being to the right or left of, or above and below, the antenna axis) is available using a phase detector. When the Σ and Δ signals are processed through the log IF amplifiers and differential amplifiers, the output voltage is proportional to the amplitude ratio of the difference to sum signals [1,2]. The amplitude ratio supplies information about the magnitude of the azimuth and elevation tracking error angles on alternate scans. These angles are ambiguous because the antenna patterns are rotationally symmetric about the antenna boresight. The ambiguity (right or left, above or below) is resolved with the phase detector data. Targets are tracked in range using the sum channel of the receiver. The calibration hardware is used to equalize the amplitude and phase of the transmitted signals caused by differences in the electrical lengths of the receiver channels between the antenna and mixers.

Another option in monopulse radar design is to use a single-beam pattern at the beginning of the mission to search for and detect targets. The single beam is synthesized by programming the RF beam-forming network to produce one beam rather than multiple beams. The single-beam pattern enhances the search and detection of targets by decreasing the beamwidth and increasing the signal-to-clutter and noise ratios. Once a target is detected, the antenna switches to the four-beam or two-beam mode for tracking. In the analysis described below, the azimuth and elevation beamwidths corresponding to single-beam and two-beam antennas are used to evaluate the search design parameters and target detection performance for 35- and 94-GHz candidate radar sensors. The four-beam antenna was eliminated because further analysis showed that its tracking performance was not adequate for this application. The azimuth and elevation beamwidths corresponding to the two-beam antenna are consequently used to calculate the angle-tracking accuracy of the 94-GHz radar sensor as this is the MMW frequency selected as a result of the tracking-error analysis.

The two-beam pattern is formed by dividing the aperture into quadrants as shown in Figure 10.5. On the scan that acquires azimuth angle track data, the beam-forming network generates the two azimuth beams using the left and right halves of the array. On the next scan where elevation data are needed, the network generates the two elevation beams from the upper and lower halves of the array. In the single-beam search and detect mode, the entire aperture area is used to generate one forward-scanning beam.

Since the radar transmits and receives backscattered energy, the two-way 3-dB beamwidth is of interest. To calculate the beamwidth of each of the beams, the effective diameter D_e of the aperture is found, which is related to the two-way azimuth and elevation 3-dB beamwidths $\theta_{a,e}$ through

$$\theta_{a,e} = \frac{\lambda}{\sqrt{2}D_e} \text{ rad} \tag{10.1}$$

where λ is the wavelength corresponding to the transmitted frequency.

The effective antenna aperture is shown by the cross-hatched area on the left of Figure 10.5 and is given by

$$A_e = \frac{\pi D_o^2}{4} - \frac{\pi D_i^2}{4} \tag{10.2}$$

where D_o and D_i are 5.5 inches or 14 cm (the portion of the submissile diameter available for sensor design) and 2 inches or 5 cm (the diameter set aside for the coboresighted optical clear aperture), respectively.

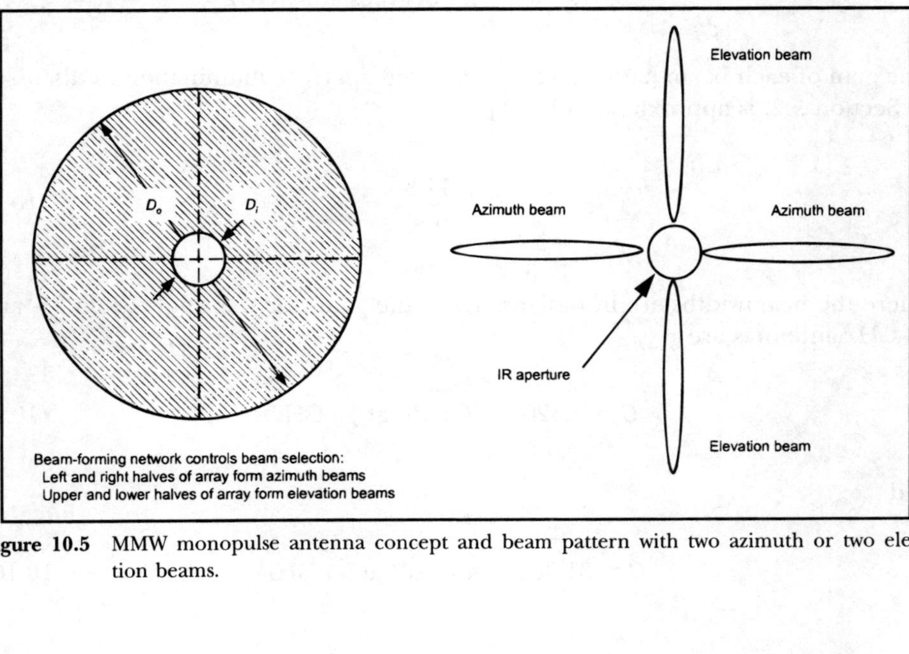

Figure 10.5 MMW monopulse antenna concept and beam pattern with two azimuth or two elevation beams.

Thus,

$$A_e = 20.6 \text{ in}^2 = 133 \text{ cm}^2 \tag{10.3}$$

and the effective aperture diameter becomes

$$D_e = \sqrt{\frac{4A_e}{\pi}} = 13 \text{ cm} \tag{10.4}$$

If the optical aperture was not present, the available diameter would be 14 cm. Therefore, the presence of the IR optics does not significantly reduce the available antenna aperture.

The two-way azimuth and elevation 3-dB beamwidths for the single-beam search and detection mode are calculated as

$$\theta_a = \theta_e = \frac{\lambda}{\sqrt{2}D_e} \tag{10.5}$$

$$= 0.050 \text{ rad} = 2.9 \text{ degrees at 35 GHz} \tag{10.6}$$

and

$$= 0.019 \text{ rad} = 1.1 \text{ degrees at 94 GHz} \tag{10.7}$$

The gain of each beam pattern, using a tapered aperture illumination as discussed in Section 5.2, is approximated by [1]

$$G = \frac{11.3}{\theta_a \theta_e} \tag{10.8}$$

where the beamwidths are in radians. Thus, the gains for the single-beam 35- and 94-GHz antennas are

$$G = 4{,}520 = 36.6 \text{ dB at 35 GHz} \tag{10.9}$$

and

$$G = 31{,}300 = 45.0 \text{ dB at 94 GHz} \tag{10.10}$$

When two beams are formed for tracking, D_e is divided by 2 and each beam is allocated a maximum effective diameter of

$$D_e/2 = 6.5 \text{ cm} \tag{10.11}$$

Using 6 cm as the antenna design goal for the effective aperture D_e' of each beam, the two-way azimuth and elevation 3-dB beamwidths for the two-beam antenna become

$$\theta_a = \theta_e = \frac{\lambda}{\sqrt{2}D_e'} \tag{10.12}$$

Upon evaluating equation (10.12) at 35 and 94 GHz, the beamwidths for the two-beam antenna are found to be

$$\theta_a = \theta_e = 0.101 \text{ rad} = 5.8 \text{ degrees at 35 GHz} \tag{10.13}$$

and

$$\theta_a = \theta_e = 0.038 \text{ rad} = 2.2 \text{ degrees at 94 GHz} \tag{10.14}$$

Therefore, the gains for the two-beam 35- and 94-GHz antennas are

$$G = 1{,}110 = 30.5 \text{ dB at 35 GHz} \tag{10.15}$$

and

$$G = 7{,}830 = 38.9 \text{ dB at 94 GHz} \tag{10.16}$$

The downtrack and crosstrack antenna footprints are computed using the antenna beamwidths and the incidence angle of the seeker. The downtrack antenna footprint X, shown in Figure 10.6 with respect to the total search area, is calculated as

$$X = h\left[\tan\left(\theta + \frac{\theta_e}{2}\right) - \tan\left(\theta - \frac{\theta_e}{2}\right)\right] \tag{10.17}$$

and the azimuth or crosstrack footprint Y as

$$Y = 2R\sin\frac{\theta_a}{2} = \frac{2h}{\cos\theta}\sin\frac{\theta_a}{2} \tag{10.18}$$

$$= \frac{h\theta_a}{\cos\theta} \quad \text{for} \quad \theta_a << 1 \tag{10.19}$$

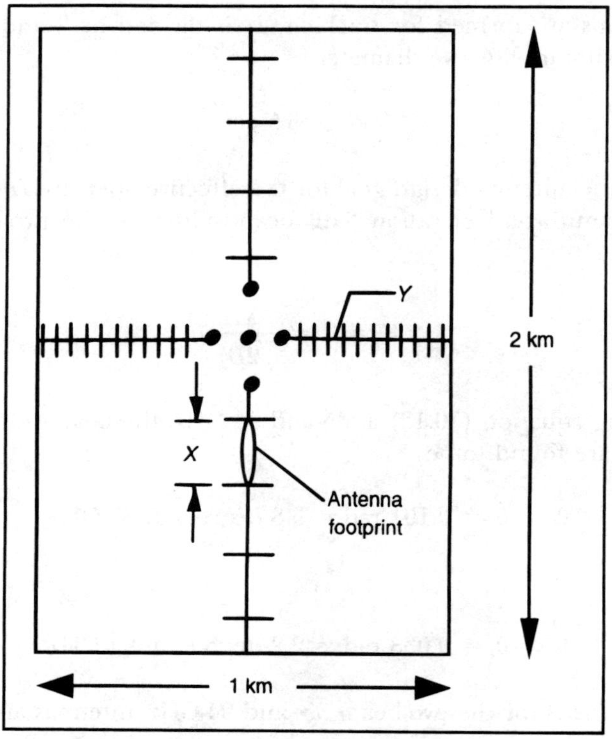

Figure 10.6 Downtrack *X* and crosstrack *Y* antenna footprints.

where

h = missile altitude = 300 m;
R = slant range from the sensor to the target = 1 km;
θ = sensor incidence angle with respect to nadir = 73 degrees;

and the beamwidths θ_a and θ_e are as calculated above.

The antenna beamwidths, gains, and footprint values are summarized in Table 10.3 for the single-beam and two-beam configurations.

10.4 SEEKER SCAN RATE AND ACCELERATION

The scan rate and acceleration of the gimbal-mounted seeker head are a function of the downtrack search area, missile velocity, antenna footprint, and scan efficiency. The total azimuthal field of view searched, θ_S, is given by

$$\theta_S = \frac{\text{1-km crosstrack}}{\text{1-km range}} = 1 \text{ rad} \tag{10.20}$$

Table 10.3
Antenna Beamwidths, Gains, and Footprints at 35 and 94 GHz

Design Parameter	*Symbol*	*35 GHz*	*94 GHz*
Wavelength	λ	0.857 cm	0.319 cm
Effective antenna diameter	D_e		
• Single beam		13 cm	13 cm
• Two beams		6 cm	6 cm
Two-way 3-dB azimuth beamwidth	θ_a		
• Single beam		0.050 rad = 2.9 degrees	0.019 rad = 1.1 degrees
• Two beams		0.101 rad = 5.8 degrees	0.038 rad = 2.2 degrees
Two-way 3-dB elevation beamwidth	θ_e		
• Single beam		0.050 rad = 2.9 degrees	0.019 rad = 1.1 degrees
• Two beams		0.101 rad = 5.8 degrees	0.038 rad = 2.2 degrees
Antenna beam gain	G		
• Single beam		4,520 = 36.6 dB	31,300 = 45.0 dB
• Two beams		1,110 = 30.5 dB	7,830 = 38.9 dB
Downtrack footprint	X		
• Single beam		177 m	69 m
• Two beams		366 m	135 m
Crosstrack (azimuth) footprint	Y		
• Single beam		51 m	20 m
• Two beams		104 m	39 m

The scan rate is found by referring to Figure 10.7, which shows the azimuth scan position and scan rate as a function of time. The scan rate $\dot{\theta}$ is given by

$$\dot{\theta} = \frac{\theta_S}{t_s} \text{ rad/sec,} \tag{10.21}$$

where t_s is the time required to search θ_S.

The scan rate may be expressed in terms of the scan efficiency η_s, where η_s is the fraction of the downtrack travel time available for scanning the azimuthal 1-km distance. As a design requirement, two side-to-side azimuth scans occur during the time the missile travels forward a distance equal to the downtrack footprint X of the antenna. Hence

$$\eta_S = \frac{t_s}{t_x/2} = \frac{2t_s}{t_x} \tag{10.22}$$

where t_x is the time for the missile to travel a distance corresponding to the downtrack footprint X of the antenna. The value t_x is equal to

$$t_x = X/v \tag{10.23}$$

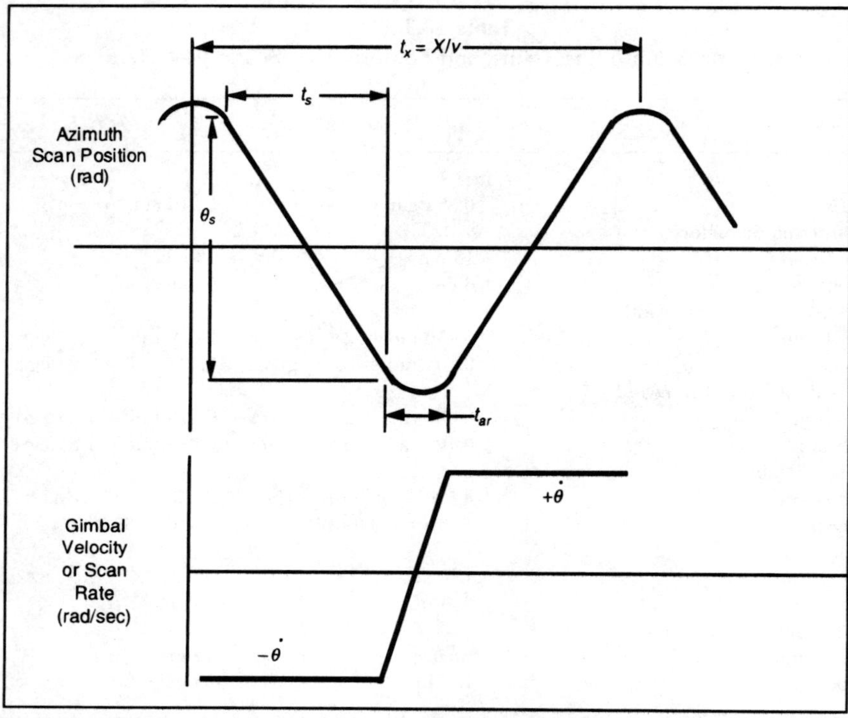

Figure 10.7 Azimuth scan position and scan rate as a function of scan time.

where v = missile velocity = 250 m/sec.

In terms of η_s, the scan rate becomes

$$\dot{\theta} = \frac{2\theta_s}{\eta_s t_x} \text{ rad/sec} \tag{10.24}$$

The time t_s available for scanning the 1-km azimuth distance is not equal to $t_x/2$ because of the time required by the gimbal to reverse its scan direction. In terms of this turnaround time, t_{ar}, t_s becomes

$$t_s = \frac{t_x}{2} - t_{ar} = \frac{\eta_s t_{ar}}{1 - \eta_s} \tag{10.25}$$

In this example, the turnaround time is set equal to 40 msec. In terms of t_{ar} and t_x, the scan efficiency is expressed as

$$\eta_s = 1 - \frac{2t_{ar}}{t_x} \tag{10.26}$$

The acceleration a of the gimbal is given by

$$a = \frac{2\dot{\theta}}{t_{ar}} = \frac{4\dot{\theta}}{t_x(1 - \eta_s)} = \frac{8\theta_S}{t_x^2\eta_s(1 - \eta_s)} \text{ rad/sec}^2 \tag{10.27}$$

where the gimbal velocity or scan rate $\dot{\theta}$ was depicted in Figure 10.7.

Table 10.4 shows the search and gimbal parameters based on the single-beam (search and detect) and two-beam (track) antenna designs. As the beamwidth decreases, the downtrack footprint and the corresponding time to travel that distance decrease. Hence, the gimbal must scan the azimuth distance at a faster rate.

10.5 MMW SENSOR DESIGN AND PERFORMANCE CALCULATIONS

The following sections describe the analysis procedures for determining the available integration time, false alarm probability, raw signal-to-clutter ratio, diversity gain, integration gain, effective signal-to-clutter ratio, noise-limited detection probability, and angle-tracking error that apply to the MMW sensor.

Table 10.4
Search and Gimbal Parameters for Search and Detect Mode (Single-Beam) and for Track Mode (Two Beams)

Design Parameter	*Symbol*	*35 GHz*	*94 GHz*
Azimuth field of view	θ_S	1 rad	1 rad
Turnaround time	t_{ar}	40 msec	40 msec
Downtrack footprint travel time	t_x		
• Single beam		0.71 sec	0.28 sec
• Two beams		1.46 sec	0.54 sec
Scan efficiency	η_s		
• Single beam		0.89	0.71
• Two beams		0.94	0.85
Azimuth search time	t_s		
• Single beam		0.32 sec	0.10 sec
• Two beams		0.69 sec	0.23 sec
Scan rate	$\dot{\theta}$		
• Single beam		3.13 rad/sec	10.0 rad/sec
• Two beams		1.45 rad/sec	4.35 rad/sec
Gimbal acceleration	a		
• Single beam		156 rad/sec^2	500 rad/sec^2
• Two beams		72.5 rad/sec^2	218 rad/sec^2

10.5.1 Integration Time

The integration time is formulated in terms of the azimuth scan time t_s and the number of resolution cells in the antenna footprint. The design parameter that determines the number of azimuth resolution cells is the two-way 3-dB azimuth beamwidth of the antenna, while the number of elevation cells is governed by the range gate resolution, which is a function of the transmitted pulse width. The integration time t_i is written as

$$t_i = \frac{(t_x/2) - t_{ar}}{n_a n_r} = \frac{t_s}{n_a n_r} \tag{10.28}$$

where

n_a = number of footprints or resolution elements in the azimuth (crosstrack) search distance

$$= \frac{\text{Azimuth search distance}}{(\text{Slant range})\,(\text{Azimuth 3-dB antenna beamwidth})} \tag{10.29}$$

$$= \frac{1{,}000 \text{ m}}{R\theta_a} \tag{10.30}$$

n_r = number of range gates in a downtrack footprint

$$= \frac{\text{Downtrack antenna footprint}}{\text{Transmitted pulse width footprint}} \tag{10.31}$$

$$= \frac{X}{c\tau/2} \tag{10.32}$$

where pulse width τ = 50 nsec and the other parameters are as defined above.

The integration time calculations are summarized in Table 10.5 for the 35- and 94-GHz MMW sensors using the single-beam and two-beam antenna parameters.

10.5.2 False Alarm Probability

The Fehlner false alarm number n_F is easily computed from the available data. As shown in Section 4.9.1, it is defined as

$$n_F = f_r n_r t_{fa}/N \tag{10.33}$$

Table 10.5
Integration Time Parameters for Single-Beam and Two-Beam Antenna Patterns

Design Parameter	*Symbol*	*35 GHz*	*94 GHz*
Turnaround time	t_{ar}	40 msec	40 msec
Downtrack footprint travel time	t_x		
• Single beam		0.71 sec	0.28 sec
• Two beams		1.46 sec	0.54 sec
Number of footprints in azimuthal search distance	n_a		
• Single beam		20	53
• Two beams		10	26
Number of range gates	n_r		
• Single beam		24	9
• Two beams		48	18
Integration time	t_i		
• Single beam		0.67 msec	0.21 msec
• Two beams		1.44 msec	0.49 msec

where

f_r = pulse repetition frequency;

n_r = number of range gates per downtrack footprint distance;

t_{fa} = average time interval between false alarms;

N = number of target return pulses or samples that are noncoherently integrated

$$= t_i f_r \tag{10.34}$$

t_i = integration time.

The average time interval between false alarms is expressed as

$$t_{fa} = (\text{False alarm density}) \times (\text{Total field of regard}) \times \left(\frac{\text{Total downtrack distance searched}}{\text{Forward missile velocity}}\right) \tag{10.35}$$

$$= (0.5/\text{km}^2)\,(1 \text{ km} \times 2 \text{ km})\left(\frac{2 \text{ km}}{0.25 \text{ km/sec}}\right) = 8 \text{ sec} \tag{10.36}$$

As discussed in Section 4.9.1, the false alarm probability P_{fa} is calculated from the Fehlner false alarm number as

$$P_{fa} = \frac{0.693}{n_F} \tag{10.37}$$

A summary of the false alarm probability calculations for single-beam and two-beam 35- and 94-GHz sensor operation is given in Table 10.6.

10.5.3 Raw Signal-to-Clutter Ratio

When operating in clutter-limited conditions, the signal-to-clutter ratio of the unprocessed backscattered signal is given by

$$\text{Raw } S/C = \frac{\sigma_T}{\sigma^o A_a L} = \frac{\sigma_T}{\sigma^o (R\theta_a)(c\tau/2) L_{atm} L_{beam} L_{clutter}} \tag{10.38}$$

where

σ_T = target radar cross section in m^2;
σ^o = clutter backscatter coefficient = σ_c/A_a in m^2/m^2;
σ_c = clutter radar cross section in m^2;
A_a = two-way 3 dB antenna footprint area in m^2;
L = dimensionless loss factor that accounts for all losses experienced by the sensor;

Table 10.6
False Alarm Probability Parameters for Single-Beam and Two-Beam Antenna Patterns

Design Parameter	*Symbol*	*35 GHz*	*94 GHz*
Pulse repetition frequency	f_r	50 kHz	50 kHz
Average time between false alarms	t_{fa}	8 sec	8 sec
Integration time	t_i		
• Single beam		0.67 msec	0.21 msec
• Two beams		1.44 msec	0.49 msec
Number of target samples noncoherently integrated	N		
• Single beam		33	10
• Two beams		72	24
Number of range gates per downtrack antenna footprint	n_r		
• Single beam		24	9
• Two beams		48	18
Fehlner false alarm number	n_F		
• Single beam		291,000	360,000
• Two beams		267,000	300,000
False alarm probability	P_{fa}		
• Single beam		2.38×10^{-6}	1.93×10^{-6}
• Two beams		2.60×10^{-6}	2.31×10^{-6}

L_{atm} = atmospheric loss due to absorption and scattering over the propagation path;

L_{beam} = antenna beam loss due to antenna pattern variation from the peak of the beam;

$L_{clutter}$ = clutter loss from non-Rayleigh clutter.

The antenna beam loss represents the loss in signal energy received from an average target during a scan compared with the energy that would be received if the beam were rectangular in shape or were to point at the target for the full integration time. The clutter loss occurs when target detection probability in non-Rayleigh clutter distributions, such as Weibull or log normal, are calculated from statistics that apply only in Rayleigh clutter. Clutter loss decreases (1) as the number of samples increases because the clutter statistics become more Gaussian due to the central limit theorem, (2) in Weibull clutter when the skewness parameter α_W approaches 2, and (3) in log normal clutter when the mean-to-median ratio decreases. For $N = 30$ samples, the clutter loss is approximately 4 dB in Weibull tree clutter ($\alpha_W = 1.2$) and approximately 5 dB in log normal clutter with a mean-to-median ratio of 1.

Table 10.7 contains the design example values used to calculate the raw signal-to-clutter ratio with the single-beam and two-beam antenna patterns. The raw signal-to-clutter ratio is enhanced by diversity and integration gain. Diversity gain is provided by the wide bandwidth of the transmitted chirp pulse. The effect of the chirp is to reduce the target fluctuation loss through the increased number of independent target returns brought about by frequency decorrelation. Integration

Table 10.7

Raw Signal-to-Clutter Ratio Parameters for Single-Beam and Two-Beam Antenna Patterns

Design Parameter	*35 GHz*	*94 GHz*
σ_T(nominal)	25 m^2	25 m^2
σ^o based on tree clutter at 17-degrees depression angle	−16 dB m^2/m^2	−13 dB m^2/m^2
R	1 km	1 km
τ	50 nsec	50 nsec
L_{atm} over 1 km (2-way) at an elevation of 0.25 km and water vapor concentration of 7.5 g/m^3	0.2 dB	0.6 dB
L_{beam}	2.5 dB	2.5 dB
$L_{clutter}$	4 dB	4 dB
θ_a		
• Single beam	0.050 rad	0.019 rad
• Two beams	0.101 rad	0.038 rad
Raw signal-to-clutter ratio		
• Single beam	0.62 = −2.1 dB	0.82 = −0.9 dB
• Two beams	0.31 = −5.1 dB	0.41 = −3.9 dB

gain increases the signal-to-noise ratio when the fluctuation rate of the target is less than that of the clutter.

10.5.4 Effective Signal-to-Clutter Ratio

Two types of diversity gain may be present to increase the signal-to-clutter ratio. The first reduces the target fluctuation loss by providing independent target samples through frequency decorrelation. The second mechanism reduces fluctuation loss by providing independent samples at intervals equal to the decorrelation time of the target. In this example, only the frequency decorrelation component is present since the decorrelation time of the target is assumed to be larger than the available integration time.

The number of independent target samples N_d produced by diversity gain from frequency decorrelation is expressed as

$$N_d = 1 + \frac{\Delta F}{\Delta f_F} \tag{10.39}$$

where ΔF is the chirp bandwidth (250 MHz) and Δf_F is the decorrelation frequency of the target as discussed in Chapter 5. The decorrelation frequency of the target is modeled as

$$\Delta f_F = \frac{c}{2\pi L} \tag{10.40}$$

when L is the target length in the downtrack direction and c equals the speed of light.

Since the smallest value of L gives the smallest number of independent samples (the conservative assumption), L is set equal to 3.3m, the approximate width of a tank target. Thus,

$$\Delta f_F = 14.5 \text{ MHz} \tag{10.41}$$

and

$$N_d = 1 + \frac{\Delta F}{\Delta f_F} = 1 + \frac{250 \text{ MHz}}{14.5 \text{ MHz}} = 18 \text{ independent target samples} \tag{10.42}$$

The diversity gain $G_d(N_d)$ in dB is given by (5.24) as

$$G_d(N_d) = [1 - 1/N_d][L_f(1)]_{\text{dB}} \tag{10.43}$$

where $L_f(1)$ is the fluctuation loss for a Swerling 1 target.

Figure 5.4 shows that $L_f(1)$ is at least 8 dB when the design goals for detection probability of at least 0.9 and false alarm probability of less than or equal to 10^{-6} are achieved. Thus the diversity gain is

$$G_d(N_d) = 5.96 = 7.7 \text{ dB} \tag{10.44}$$

In addition to diversity gain, integration gain is realized since it is assumed that the target fluctuates at a lower rate than does the clutter. Hence, some of the target-masking effects or competition from the clutter is removed when the target and clutter returns are integrated following detection. Equation (10.39) is also used to calculate the number of independent clutter samples produced from the relatively large chirp frequency. The clutter decorrelation frequency Δf_F is conservatively modeled by assuming a uniform distribution for the clutter scatterers. Thus from Table 5.6,

$$\Delta f_F = 0.44c/D = 17.6 \text{ MHz} \tag{10.45}$$

for the clutter decorrelation frequency where D is set equal to the range gate depth of 7.5m.

The number of independent clutter samples that is integrated becomes

$$N_c = 1 + \frac{250 \text{ MHz}}{17.6 \text{ MHz}} = 15 \text{ independent clutter samples} \tag{10.46}$$

The integration gain is given by (5.31) as

$$G_i(N_c) = \frac{N_c}{L_i(N_c)} \tag{10.47}$$

where L_i is the integration loss found from the curves in Figure 4.26.

For small signal-to-clutter ratios,

$$G_i(N_c) = \frac{N_c}{\sqrt{N_c}} \tag{10.48}$$

$$= (15)^{1/2} = 3.9 = 5.9 \text{ dB} \tag{10.49}$$

The integration loss decreases with large signal-to-clutter ratios, as shown in Figure 4.26.

Finally, the presence of both diversity gain and integration gain combine to produce an effective signal-to-clutter ratio of

$$(S/C)_e = G_d(N_d)\, G_i(N_c)\, (S/C)_{\text{single sample}} \tag{10.50}$$

$$= (5.96)\,(3.9)\,(S/C)_{\text{single sample}} = 21.9\ (S/C)_{\text{single sample}} \tag{10.51}$$

$$= 13.6 \text{ dB} + (S/C)_{\text{single sample dB}} \quad (10.52)$$

Using the raw signal-to-clutter ratio values from Table 10.7 and the combined diversity and integration gain of 13.6 dB produces the effective signal-to-clutter ratios shown in Table 10.8.

10.5.5 Detection Probability Summary for Nominal and Suppressed Targets in Nominal Background Clutter and High Background Clutter (Snow)

Table 10.9 presents a summary of the detection performance of the candidate 35- and 94-GHz sensors that use (1) a wide-bandwidth transmission pulse to provide diversity gain and (2) integration of target returns over a number of clutter samples

Table 10.8
Effective Signal-to-Clutter Ratios for Single-Beam and Two-Beam Antenna Patterns

Design Parameter	*Symbol*	*35 GHz*	*94 GHz*
Raw signal-to-clutter ratio	$(S/C)_{\text{single sample}}$		
• Single beam		0.62 = −2.1 dB	0.82 = −0.9 dB
• Two beams		0.31 = −5.1 dB	0.41 = −3.9 dB
Combined diversity and integration gains	$G_d(N_d)\, G_i(N_c)$	13.6 dB	13.6 dB
Effective signal-to-clutter ratio	$(S/C)_e$		
• Single beam		11.5 dB	12.7 dB
• Two beams		8.5 dB	9.7 dB

Table 10.9
Clutter-Limited Detection Probability of Candidate Single-Beam Antenna 35- and 94-GHz MMW Sensors With a Nominal Target

Design Parameter	*Symbol*	*35 GHz*	*94 GHz*
Target radar cross-section	σ_T	25 m^2	25 m^2
Clutter backscatter coefficient	σ^o	−16 dB m^2/m^2	−13 dB m^2/m^2
Range	R	1 km	1 km
Two-way 3-dB single-beam antenna beamwidth	θ_a, θ_e	50 mrad (2.9 degrees)	19 mrad (1.1 degrees)
Effective signal-to-clutter ratio	$(S/C)_e$	11.5 dB	12.7 dB
False alarm probability	P_{fa}	2.38×10^{-6}	1.93×10^{-6}
Number of signal pulses integrated	N	33	10
Detection probability with Swerling 1 target model*	P_d	92%	88%

* The final detection probability and false alarm probability are dependent on the specific signal-processing algorithms that are implemented for the MMW sensor and the data fusion processor.

to provide integration gain. The value of the target radar cross-section is the nominal $25m^2$. The added benefits of digital, neural network, or other type of signal processing in the MMW sensor and data fusion processing to further eliminate false alarms and false targets are not included in this analysis.

Table 10.10 shows the effect on detection probability when the target radar cross-section is reduced to $5m^2$ from $25m^2$. Table 10.11 shows the effect of a high-

Table 10.10

Clutter-Limited Detection Probability of Candidate Single-Beam Antenna 35- and 94-GHz MMW Sensors With a Suppressed Target

Design Parameter	*Symbol*	*35 GHz*	*94 GHz*
Target radar cross-section	σ_T	5 m^2	5 m^2
Clutter backscatter coefficient	σ^o	−16 dB m^2/m^2	−13 dB m^2/m^2
Range	R	1 km	1 km
Two-way 3-dB antenna beamwidth	θ_a, θ_e	50 mrad (2.9 degrees)	19 mrad (1.1 degrees)
Effective signal-to-clutter ratio	$(S/C)_e$	4.5 dB	5.7 dB
False alarm probability	P_{fa}	2.38×10^{-6}	1.93×10^{-6}
Number of signal pulses integrated	N	33	10
Detection probability with Swerling 1 target model*	P_d	66%	53%

* The final detection probability and false alarm probability are dependent on the specific signal-processing algorithms that are implemented for the MMW sensor and the data fusion processor.

Table 10.11

Clutter-Limited Detection Probability of Candidate Single-Beam Antenna 35- and 94-GHz MMW Sensors With Nominal and Suppressed Targets in a High-Clutter Snow Background

Design Parameter	*Symbol*	*35 GHz*		*94 GHz*	
Clutter backscatter coefficient	σ^o	−5 dB m^2/m^2		−5 dB m^2/m^2	
Range	R	1 km		1 km	
Two-way 3-dB antenna beamwidth	θ_a, θ_e	50 mrad (2.9 degrees)		19 mrad (1.1 degrees)	
False alarm probability	P_{fa}	2.38×10^{-6}		1.93×10^{-6}	
Number of signal pulses integrated	N	33		10	
Target radar cross-section	σ_T	25 m^2	5 m^2	25 m^2	5 m^2
Effective signal-to-clutter ratio	$(S/C)_e$	0.5 dB	−6.5 dB	4.7 dB	−2.3 dB
Detection probability with Swerling 1 target model*	P_d	38%	2.5%	45%	3.5%

* The final detection probability and false alarm probability are dependent on the specific signal-processing algorithms that are implemented for the MMW sensor and the data fusion processor.

clutter snow background on the detection probability of $5m^2$ and $25m^2$ targets. The added benefits of MMW sensor signal processing and data fusion processing are not included in the detection performance listed in the table. Without these, the combination of suppressed target and high clutter make the proposed MMW sensor design relatively ineffective when acting alone. Single-sensor performance such as this provides the rationale for the multiple-sensor suite, improving system performance under conditions including inclement weather, high clutter, and countermeasures.

The diversity and integration gains have a large effect on the detection probability when target signatures are suppressed to $5m^2$ from their nominal value of $25m^2$. With a false alarm probability of $\approx 2 \times 10^{-6}$, the 13.6-dB increase in signal-to-clutter ratio from diversity and integration gains leads to an increase in the detection probability of the 94-GHz sensor to 53% from the 0.04% achieved without the effects of diversity and integration gains.

10.5.6 Raw Signal-to-Noise Ratio

When the MMW sensor operates under noise-limited conditions, the noise power at the receiver is the dominant interference. It is calculated from the system input noise temperature and the receiver bandwidth as determined by the transmitted pulse width. The signal power backscattered from the target is calculated using the radar range equation. The ratio of the signal power to noise power gives the raw signal-to-noise ratio. As with clutter-limited operation, diversity and integration gain combine to create an effective signal-to-noise ratio that is larger than the raw value and hence improves the detection probability and decreases the angle-tracking error.

Because of the need for accurate angle tracking with the MMW sensor under the imposed aperture constraints, the calculations in this section reflect only 94-GHz operation. Corresponding calculations at 35 GHz would give a somewhat lower noise contribution, but a larger angle tracking error because of the larger antenna beamwidth.

Noise Power at 94 GHz

The equivalent system input noise temperature is given by

$$T_{SYS} = T'_A + T'_{REC} \tag{10.53}$$

where

T'_A = apparent temperature = sum of antenna temperature produced by the scene under observation and the temperature due to antenna self emission
= 290K

T'_{REC} = equivalent input noise temperature of the transmission line-receiver combination

$$= (F_{SSB} - 1)290L + (L - 1)T_L \text{ K} \tag{10.54}$$

F_{SSB} = single-sideband receiver noise figure = 9 dB = 8.0;
L = RF losses = 12.4 dB = 17.4;
T_L = physical temperature of lossy components = 290K.

With these values,

$$T'_{REC} = (7)(290)(17.4) + (16.4)(290) = 40{,}078\text{K} \tag{10.55}$$

and

$$T_{SYS} = 290\text{K} + 40{,}078\text{K} = 40{,}368\text{K} \tag{10.56}$$

The noise power is calculated as

$$P_N = k_B T_{SYS} B \tag{10.57}$$

where

k_B = Boltzmann's constant = 1.380662×10^{-23} J/K;
B = noise bandwidth.

Using a conservative formula for noise bandwidth in terms of the transmitted pulse width τ gives

$$B = (\text{pulse width})^{-1} = 1/\tau \tag{10.58}$$

$$= (50 \text{ nsec})^{-1} = 20 \text{ MHz} \tag{10.59}$$

With these values, the noise power becomes

$$P_N = (1.38 \times 10^{-23})(40{,}368)(20 \times 10^6) = 1.11 \times 10^{-11} \text{ W} = -79.6 \text{ dBm}. \tag{10.60}$$

Signal Power and Raw Signal-to-Noise Ratio at 94 GHz

The received signal power is calculated from the radar range equation as

$$P_R = \frac{P_T \tau f_r G_T G_R \lambda^2 \sigma_T}{(4\pi)^3 R^4 L_{\text{atm}} L_{\text{beam}}} \tag{10.61}$$

where

P_T = transmitted power in W;
f_r = pulse repetition frequency in Hz;
G_T, G_R = transmit, receive antenna gain;
λ = wavelength of transmitted frequency in m;
R = range to the target in m;

and the other parameters are as defined above.

Substituting the single-beam 94-GHz antenna parameters from Tables 10.1, 10.3, and 10.7 gives

$$P_R = \frac{(20)(50 \times 10^{-9})(50 \times 10^{3})(10^{4.5})(10^{4.5})(0.0032)^2(25)}{(1984.4)(10^{12})(10^{0.06})(10^{0.25})}$$

$$= 3.16 \times 10^{-12} \text{ W} = -85.0 \text{ dBm} \qquad (10.62)$$

as the received signal power during search.

The raw signal-to-noise ratio during search is calculated using the received power from (10.62) and the noise power from (10.60) as

$$\text{Raw } (S/N)_{\text{per pulse}} = \frac{P_R}{P_N} \qquad (10.63)$$

$$= \frac{3.16 \times 10^{-12}}{1.11 \times 10^{-11}} = 0.285 = -5.4 \text{ dB} \qquad (10.64)$$

The raw signal-to-noise ratio for the corresponding two-beam track mode is shown later in Table 10.12.

10.5.7 Effective Signal-to-Noise Ratio

As with clutter-limited operation, diversity and integration gain increase the raw signal-to-noise ratio. Diversity gain occurs from the creation of independent target returns due to the decorrelation of the returns. Integration gain is produced from the integration of the slower fluctuating target returns and the faster fluctuating noise pulses during the integration interval.

Diversity Gain

The diversity gain reduces the target fluctuation loss through the action of the wide-bandwidth chirp pulse. The number of independent target samples from frequency decorrelation was found in (10.42) as

$$N_d = 18 \text{ samples} \tag{10.65}$$

and the diversity gain $G_d(N_d)$ in (10.44) as

$$G_d(N_d) = 5.96 = 7.7 \text{ dB} \tag{10.66}$$

Integration Gain

The integration gain arising from the integration of the target signal over a number of independent noise samples N_n is given by (10.47) when N_c is replaced by N_n. The number of independent noise samples is equal to

$$N_n = (\text{Intermediate frequency predetection bandwidth})(\text{Integration time}) \tag{10.67}$$

$$= (20 \times 10^6)(0.21 \times 10^{-3}) = 4{,}200 \tag{10.68}$$

using values from Tables 10.1 and 10.5 for the single-beam 94 GHz antenna. For small S/N the integration gain equals $\sqrt{N_n}$. Thus

$$G_i(N_n) = 64.8 = 18.1 \text{ dB} \tag{10.69}$$

Effective Signal-to-Noise Ratio and Detection Probability at 94 GHz

The effective signal-to-noise ratio is

$$(S/N)_e = G_d(N_d)\, G_i(N_n)\, (S/N)_{\text{per pulse}} \tag{10.70}$$

$$= (5.96)(64.8)(S/N)_{\text{per pulse}} = 386\ (S/N)_{\text{per pulse}} \tag{10.71}$$

$$= 25.8 \text{ dB} + (S/N)_{\text{per pulse dB}} \tag{10.72}$$

Using the raw signal-to-noise ratio of −5.4 dB and the combined gain of 25.8 dB, the effective signal-to-noise ratio during search becomes

$$(S/N)_e = 20.4 \text{ dB at 94 GHz} \tag{10.73}$$

With the number of target backscattered pulses integrated equal to

$$N = f_r t_i \tag{10.74}$$

$$= (50 \times 10^3)(0.21 \times 10^{-3}) = 10 \tag{10.75}$$

and the false alarm probability equal to $\approx 2 \times 10^{-6}$, the detection probability for a Swerling 1 fluctuating target is

$$P_d \approx 97\% \tag{10.76}$$

under noise-limited conditions using the single-beam antenna.

A summary of the single-beam search and two-beam track signal-to-noise ratio and noise-limited detection probability calculations is provided in Table 10.12.

Table 10.12

Signal-to-Noise Ratios and Noise-Limited Detection Probabilities for 94-GHz Single-Beam and Two-Beam Antenna Patterns

Design Parameter	*Symbol*	*94 GHz*
Target	σ_T	25 m^2, Swerling 1
Received power	P_R	
• Single beam		−85.0 dBm
• Two beams		−97.2 dBm
Noise power	P_N	−79.6 dBm
Raw signal-to-noise ratio	$(S/N)_{\text{per pulse}}$	
• Single beam		−5.4 dB
• Two beams		−17.6 dB
Combined diversity and integration gains	$G_d(N_d)\, G_i(N_n)$	
• Single beam		25.9 dB
• Two beams		27.8 dB
Effective signal-to-noise ratio	$(S/N)_e$	
• Single beam		20.5 dB
• Two beams		10.2 dB
Number of signal pulses integrated	N	
• Single beam		10
• Two beams		24
False alarm probability	P_{fa}	
• Single beam		1.93×10^{-6}
• Two beams		2.31×10^{-6}
Detection probability	P_d	
• Single beam		≈97%
• Two beams		≈88%

10.5.8 Noise-Limited Angle-Tracking Error

The one-sigma noise-limited angle-tracking error is found by modifying (4.71) to include the diversity and integration gains as

$$\sigma_\theta = 0.49\theta_a / \sqrt{(S/N)_e} \tag{10.77}$$

where

θ_a = two-way 3-dB antenna beamwidth in rad;
$(S/N)_e$ = effective signal-to-noise ratio (includes diversity and integration gains).

If detection probability curves based on Swerling fluctuating target models (these include a fluctuation loss) are used, then the effective signal-to-noise ratio is the raw signal-to-noise ratio multiplied by the diversity gain and the integration gain. If detection probability curves based on a nonfluctuating target model (that do not include a fluctuation loss) are used to estimate the performance of the sensor against a fluctuating target, then the effective signal-to-noise ratio is simply the raw signal-to-noise ratio multiplied by the integration gain. The diversity gain in this example nullifies the fluctuation loss that would otherwise have to be applied to signal-to-noise reduction when nonfluctuating detection probability curves are used with a fluctuating target.

Since as small an angle-tracking error as possible is desired from the MMW sensor in case it alone must be relied on for tracking, 94 GHz is chosen as the operating frequency of the MMW sensor. The diversity gain that applies to the angle-tracking error calculation is the same value as given by (10.66). The integration gain, however, now is based on the number of backscattered target pulses N_T received as the area between the half-power points of the *one-way* antenna pattern is scanned. Using the design parameters that apply to the two-beam 94-GHz antenna, the number of received target pulses is found as

$$N_T = (\text{Time to scan one-way 3 dB beamwidth})(\text{Pulse repetition frequency}) \tag{10.78}$$

$$= \frac{\text{one-way 3-dB beamwidth}}{\text{Scan rate}} f_r \tag{10.79}$$

$$= \frac{(0.054 \text{ rad})(50 \times 10^3 \text{ Hz})}{4.35 \text{ rad/sec}} = 620 \text{ samples} \tag{10.80}$$

Therefore, the integration gain is

$$G_i(N_T) = \sqrt{620} = 24.9 = 14.0 \text{ dB} \tag{10.81}$$

The effective signal-to-noise ratio is then

$$(S/N)_e = G_d(N_d)\,G_i(N_T)\,(S/N)_{\text{per pulse}} \tag{10.82}$$

$$= (5.96)(24.9)(S/N)_{\text{per pulse}} = 148(S/N)_{\text{per pulse}} \tag{10.83}$$

$$= 21.7 \text{ dB} + (S/N)_{\text{per pulse dB}} \tag{10.84}$$

Using the raw signal-to-noise ratio of −17.6 dB for the two-beam antenna and the combined gain of 21.7 dB, the effective signal-to-noise ratio for the angle-tracking error calculation becomes

$$(S/N)_e = 2.6 = 4.1 \text{ dB at 94 GHz} \tag{10.85}$$

Using (10.77), the one-sigma value of the angle tracking error is

$$\sigma_\theta = (0.49)(0.038)(2.6)^{-1/2} = 11.5 \text{ mrad} = 0.66 \text{ degrees} \tag{10.86}$$

10.6 IR SENSOR DESIGN AND PERFORMANCE CALCULATIONS

The infrared sensor is designed to act alone in the event of a failure of the MMW sensor or when the target signature in the MMW spectrum is suppressed or the targets are in a high-clutter background. When both the MMW and IR sensors are able to provide information about target location and classification, their data are fed into a fusion processor, which makes the final decision as to target selection and point of missile impact. In this section, the instantaneous field of view, detector dimensions, spatial frequency, number of detectors, array configuration, sample frequency, NEΔT, and MRTD of the IR sensor are calculated. Since the 94-GHz operating frequency option has been selected for the MMW sensor, its beamwidth parameters are used in the design of the IR sensor where applicable.

10.6.1 Instantaneous Field of View

The diffraction-limited instantaneous field of view IFOV is expressed as

$$\text{IFOV} = 2.44\,\frac{\lambda}{D_o} \tag{10.87}$$

where

λ = wavelength of the incident IR energy;
D_o = optical clear aperture = 2 inches (5 cm).

To ensure that the detector operates in the diffraction-limited state at all wavelengths in the selected 3- to 5-μm energy band, the IFOV is sized at the longest wavelength at which energy is received, namely 5 μm. Thus,

$$\text{IFOV} = (2.44)(5 \times 10^{-6})/(0.05) = 0.24 \text{ mrad} \tag{10.88}$$

A square detector is desired, whose linear dimensions *a* and *b* are calculated from the IFOV as

$$a = b = (\text{IFOV})(\text{Focal length}) \tag{10.89}$$

$$= \left(2.44 \frac{\lambda}{D_o}\right)(F/\#)(D_o) = 2.44\,\lambda F/\# \tag{10.90}$$

$$= 2.44 \times 10^{-5}\text{ m}(\approx 0.001 \text{ inch}) \text{ when } F/\# = 2.0 \tag{10.91}$$

In order to maximize the energy collected by the detectors in the presence of image broadening from optical aberrations, the linear dimensions of the detector are doubled to

$$a = b = 4.88 \times 10^{-5}\text{ m}(\approx 0.0019 \text{ inch}) \text{ with } F/\# = 2.0 \tag{10.92}$$

Then, the IFOV becomes

$$\text{IFOV} = \alpha = \beta = a/(\text{Focal length}) \tag{10.93}$$

$$= \frac{4.88 \times 10^{-5}}{(2)(0.05)} = 0.48 \text{ mrad} \tag{10.94}$$

where α and β are the angles that the detector subtends.

10.6.2 Spatial Frequency

In order for the IR detector array to have sufficient resolution to classify objects in its field of regard, Johnson's criteria requires that the sensor's spatial frequency support placing at least four cycles across the minimum target dimension. With this resolution, the probability of classification is 50% when the image signal-to-noise ratio is 2.8. More cycles or a larger signal-to-noise ratio increase the probability of classification, as was shown in Figures 7.5 and 7.6.

In order to calculate the spatial frequency, the angular subtense α_T across the target minimum dimension is first calculated as

$$\alpha_T = h/R \tag{10.95}$$

where

h = target minimum dimension in m;
R = range to the target in m.

When h is 3.3m, corresponding to the width of a tank target, and R is 1 km,

$$\alpha_T = 3.3 \text{ mrad} \tag{10.96}$$

The target spatial frequency f_T is then

$$f_T = N_t/\alpha_T \tag{10.97}$$

where N_t is the desired number of spatial cycles, here equal to 4. Thus

$$f_T = 1.2 \text{ cycles/mrad} \tag{10.98}$$

A measure of the imaging quality of the sensor design can be found by comparing the maximum spatial frequency resolvable by the sensor with (10.98). The maximum spatial frequency resolvable by the sensor is given by

$$f_{T_{\max}} = \frac{1}{(2)(\text{IFOV})} \tag{10.99}$$

$$= \frac{1}{(2)(0.48)} = 1.0 \text{ cycle/mrad} \tag{10.100}$$

Therefore, the 2-inch clear aperture with an $F/\#$ of 2 appears adequate for the IR sensor design. Another trade that can be made is to increase D_o to reduce the IFOV and thereby increase the number of pixels across the minimum target dimension and the probability of correct target classification. However, such a design change will reduce the gain and increase the beamwidth of the MMW antenna, and thus adversely affect the detection and tracking performance of the MMW sensor. The final discrimination capability of the IR sensor can only be determined by a measurement of its minimum resolvable temperature difference versus spatial frequency, followed by application of the Center for Night Vision and Electro-Optics (CNVEO) Laboratory target discrimination model discussed in Chapter 7.

10.6.3 Number of Detectors

The number of detectors is governed by the desire to have the IR detector array view the same area as the 3-dB MMW antenna beamwidth and to include allowances for minimum interdetector gaps required for IR array manufacture. This concept is illustrated in Figure 10.8. The area the IR array must view is defined by the 94-GHz operating frequency of the MMW sensor and the two-beam antenna since it has the larger beamwidth (compared to the one-beam antenna pattern used for the search and detect mode). With two columns of detectors, there are no search area coverage gaps that would exist from the interdetector spaces if only one column were used. The two-column design also extends the modulation transfer function (MTF) useful spatial frequency limit closer to one-half the sample frequency, as was shown in Figure 7.23. Furthermore, multiple columns allow the azimuth distance to be covered by the search pattern in less time than is needed by a single column detector and thus eases the scan rate requirement.

Referring to Figure 10.8, the number of detectors n_d in the two-column design is given by

$$n_d = (\text{Number of columns})\left(\frac{\text{3-dB antenna beamwidth}}{\text{Detector FOV + Gap FOV}}\right) \tag{10.101}$$

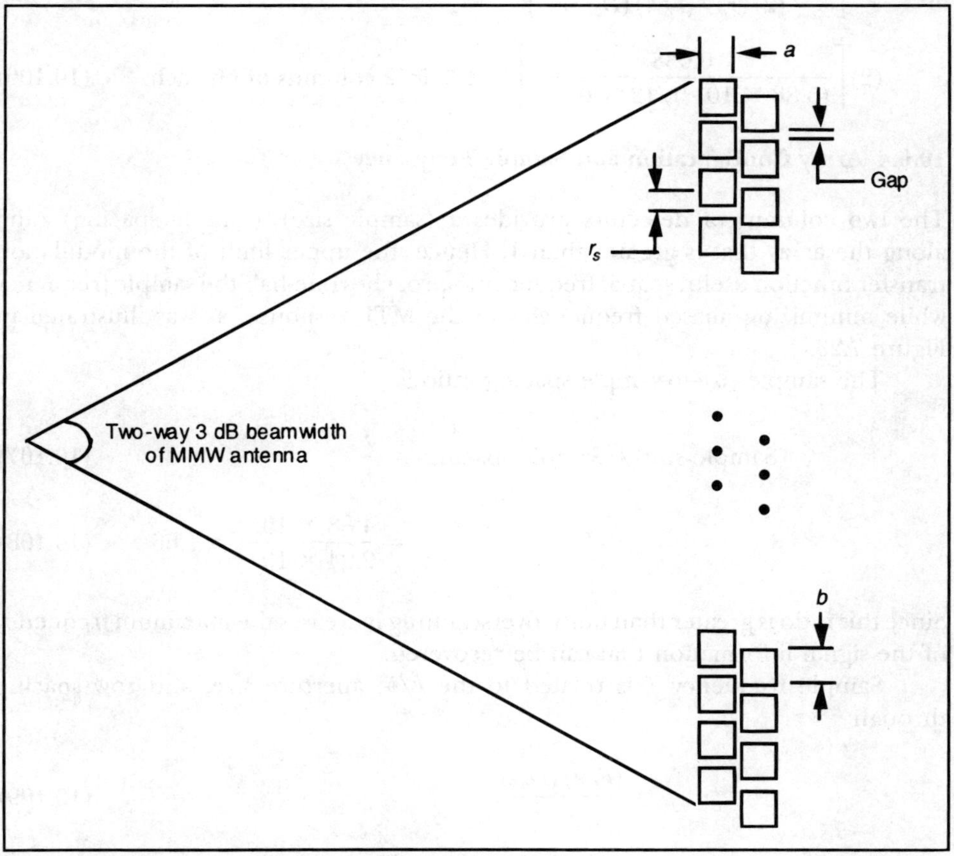

Figure 10.8 IR detector array configuration.

where

$$\text{Detector FOV} + \text{Gap FOV} = \frac{\text{Detector length} + \text{Gap length}}{(F/\#)(D_o)} = \frac{(2)(r_s)}{(F/\#)(D_o)} \tag{10.102}$$

and

$$r_s = \text{row spacing} = b/2 + \text{Gap}/2 \tag{10.103}$$

$$= 2.44 \times 10^{-5} + 5.0 \times 10^{-6} = 2.94 \times 10^{-5}\ \text{m} \approx 0.0012\ \text{inch} \tag{10.104}$$

The expression for the required number of detectors becomes

$$n_d = (2)\left[\frac{\text{3-dB antenna beamwidth}}{(2)(r_s)/(F/\#)(D_o)}\right] \tag{10.105}$$

$$= (2)\left[\frac{0.038}{(5.88 \times 10^{-5})/(2 \times 0.05)}\right] = 132 \text{ in 2 columns of 66 each} \tag{10.106}$$

10.6.4 Array Configuration and Sample Frequency

The two columns of detectors provides a (sample size)/(sample spacing) ratio along the array that is greater than 1. Hence, the upper limit of the modulation transfer function useful spatial frequency approaches one-half the sample frequency while minimizing aliased frequencies in the MTF response, as was illustrated in Figure 7.23.

The sample size-to-sample spacing ratio is

$$\text{(Sample size)/(Sample spacing)} = \frac{b}{r_s} \tag{10.107}$$

$$= \frac{4.88 \times 10^{-5}}{2.94 \times 10^{-5}} = 1.66 \tag{10.108}$$

Since this ratio is greater than unity, overscanning increases the maximum frequency of the signal information that can be recovered.

Sample frequency f_s is related to the $F/\#$, aperture size, and row spacing through

$$f_s = \frac{(F/\#)(D_o)}{r_s} \tag{10.109}$$

$$= \frac{(2)(0.05)}{2.94 \times 10^{-5}} = 3.5\ \text{cycles/mrad} \tag{10.110}$$

Thus, the required spatial frequency of at least 1 cycle/mrad can be realized by the sensor as the usable portion of the modulation transfer function spectrum extends to spatial frequencies approaching one-half the sample frequency (namely, 1.75 cycles/mrad) with the two-column detector design.

10.6.5 Noise-Equivalent Temperature Difference

As discussed in Section 7.5.3, the noise equivalent temperature difference (NEΔT) provides a performance measure for tracking sensors. It is given by

$$\text{NE}\Delta\text{T} = \frac{4}{D_o^2}\frac{(ab\Delta F)^{1/2}}{\alpha\beta}\frac{1}{\displaystyle\int_{\lambda_L}^{\lambda_U}\tau_o\tau_a\frac{\partial W_b}{\partial T}D_d^*\,d\lambda} \tag{10.111}$$

where

D_o = optical clear aperture in cm;
a,b = linear detector dimensions in cm;
α,β = detector angle subtense in rad;
ΔF = electrical noise bandwidth referenced to an RC-filter in Hz;
τ_o = optical spectral transmission;
τ_a = atmospheric spectral transmission during the NEΔT measurement;
$\dfrac{\partial W_b}{\partial T}$ = derivative of blackbody spectral emittance with respect to background temperature in $\text{W}/(\text{cm}^2\ \text{K})$;
D_d^* = specific detector detectivity in cm $\text{Hz}^{1/2}/\text{W}$;
λ_L to λ_U = lower and upper bounds of the spectral wavelength band of interest.

Substituting the design values from Table 10.2 into (10.111) gives

$$\begin{aligned}\text{NE}\Delta\text{T} &= \frac{4}{25.8}\frac{(4.88\times10^{-3})(\Delta F)^{1/2}}{2.30\times10^{-7}}\frac{1}{(0.9)(1.0)(2.1\times10^{-5})(2\times10^{11})}\\ &= 0.00087(\Delta F)^{1/2}\ \text{K}\end{aligned} \tag{10.112}$$

where the value used for the derivative of the blackbody spectral emittance is extracted from Table 7.5 and applies to the 3- to 5-μm band when the background temperature is 300K. The values used for the optical spectral transmission and the atmospheric spectral transmission were 0.9 and 1, respectively.

The bandwidth of the RC-filter is calculated as

$$\Delta F = f_T \text{ (cycles/mrad)} \times \text{Scan rate (mrad/sec)} \tag{10.113}$$

$$= (1.2)(4.3 \times 10^3) = 5{,}160 \text{ Hz} \tag{10.114}$$

Thus,

$$\text{NE}\Delta\text{T} = 0.062\text{K} \tag{10.115}$$

which is small enough to detect the temperature differences between areas on targets and between targets and the background, even when measures are in place to diffuse the heat emitted by the target. Recall from Section 7.5.3 that NEΔT is not constant at all temperatures because the value of the integral of the derivative of spectral emittance is smaller at colder temperatures. This results in NEΔT increasing in value as the temperature of the object under observation decreases.

Detector material trades can be made at this point. For example, lead selenide (PbSe) with a lower D_d^* will increase NEΔT, but will operate at a lower focal plane temperature. Also, the frequency response of PbSe is lower than that of InSb, but may still be adequate for this particular application. Other considerations include cost and availability of the detector array.

10.6.6 Minimum Resolvable Temperature Difference

The minimum resolvable temperature difference (MRTD) is given by

$$\text{MRTD} = \frac{12}{D_o} \frac{F/\# f_T}{\tilde{r}_s \tau_d^{1/2}} \frac{1}{\int_{\lambda_L}^{\lambda_U} \tau_o \tau_a \frac{\partial W_b}{\partial T} D_d^* \, d\lambda} \tag{10.116}$$

where

$$\tau_d = \text{time to scan IFOV}$$

$$= \frac{\text{IFOV}}{\text{Scan rate}} \tag{10.117}$$

$$= \frac{0.48 \text{ mrad}}{4.3 \times 10^3 \text{ mrad/sec}} = 1.1 \times 10^{-4} \text{ sec} \tag{10.118}$$

and the other parameters are as defined above.

By substituting the values from Table 10.13 into (10.116), the MRTD is computed to be

$$\text{MRTD} = 0.23\text{K} \tag{10.119}$$

when f_T equals 1.2 cycles/mrad.

10.6.7 Detection Probability

The detection probability and resultant false alarm probability of the imaging IR sensor are dependent on the specific image-processing algorithms that are implemented. The design and evaluation of these algorithms are beyond the scope of this chapter. The approach to signal processing of multipixel imagery was discussed in Section 7.11.

10.7 SUMMARY

The trajectory and search area for a terminally guided submissile application were used to develop a system-level design for a dual MMW/IR sensor that provides target detection and tracking under nominal conditions, in high clutter, and against suppressed signature targets. Many of the design and performance enhancement principles developed in the preceding chapters were applied to this sensor system design example. Although the signal-processing algorithms for each of the sensors were not explicitly described in this chapter, they have been discussed elsewhere and their functions were accounted for in the sensor concept diagrams and in the data fusion processor.

Table 10.13
Parameter Values Used in MRTD Calculation

Design Parameter	*Symbol*	*Value*
Lower and upper limits of wavelength band	λ_L to λ_U	3 to 5 μm
Optical clear aperture	D_o	5 cm
F-number	$F/\#$	2
Target spatial frequency	f_T	1.2×10^3 cycles/rad
Modulation transfer function	$\tilde{r}_s$	0.6
Time to scan IFOV	τ_d	1.1×10^{-4} sec
Optical spectral transmission	τ_o	0.9
Atmospheric spectral transmission	τ_a	1.0
Derivative of blackbody spectral emittance	$\partial W_b/\partial T$	2.1×10^{-5} W/(cm^2 K)
Specific detector detectivity	D_d^*	2×10^{11} cm Hz$^{1/2}$/ W
Minimum resolvable temperature difference	MRTD	0.23K

The design criteria for search area coverage and optimum sensor performance conflict, leading to compromises in the seeker design. For example, a long integration time is desirable to increase the number of samples, implying a small number of azimuth resolution elements. However, the number of azimuth resolution elements is fixed by the desire to scan the 1-km azimuth distance in the time it takes the missile to travel a distance equal to the downtrack antenna footprint. If the number of azimuth resolution elements is decreased, there will be gaps in the area searched for targets unless the antenna beamwidth is increased. However, the latter option will degrade the MMW sensor resolution and angle-tracking capability. The trade in the operating frequency of the MMW sensor was between the somewhat larger detection probability of the 35-GHz sensor and the better angle-tracking accuracy of the 94-GHz sensor. With respect to the IR sensor, the major constraint was the aperture size, which limited the IFOV and, hence, the spatial resolution. Although detailed performance calculations were only given for the MMW sensor under nominal conditions, reduced target signature, and high clutter to show the degradation in sensor operation, similar illustrations can also be used to show the nominal and degraded performance of the IR sensor. Thus, a dual-sensor design was selected, utilizing different portions of the electromagnetic spectrum and both the active and passive signature characteristics of potential targets to enhance the capability of the sensor suite and ultimately the missile.

The detailed trades developed as part of the MMW sensor design process included operating frequency and antenna design. These impacted the integration time, false alarm probability, number of independent target return pulses, signal-to-clutter and signal-to-noise ratios, and detection probability. A wide-bandwidth chirp was used to provide frequency decorrelation of target returns that, in turn, reduced the target fluctuation loss and increased detection probability and tracking performance. The integration gain that improved signal-to-clutter and signal-to-noise ratios also increased the detection probability and tracking performance of the sensor. The 94-GHz frequency was selected for the MMW sensor because of the smaller angle-tracking error. The two-beam, multiplexed monopulse receiver was chosen so that the antenna beamwidth could be decreased, as compared to the four-beam approach, to again improve angle tracking and yet provide sum and difference channel information for range and angle tracking.

In the IR sensor, the trades were among optical clear aperture, detector size, spatial resolution, detector array configuration, and detector material (D^*_d, focal plane temperature, cost, availability). The 3- to 5-μm wavelength was chosen for the IR sensor primarily because of the availability of the detector material and its cost. This band also has a higher radiation contrast than the long-wavelength IR band and is thus more effective in performing thermal imaging. The two-column detector design extends the modulation transfer function useful spatial frequency limit closer to one-half the sample frequency without aliasing effects. Furthermore, multiple columns allow the azimuth distance to be covered by the search pattern in less time than is needed by a single-column detector. The resulting NEΔT and MRTD are adequate to satisfy the mission requirements.

References

[1] Barton, D. K., *Modern Radar System Analysis,* Norwood, MA: Artech House, 1988.

[2] Bullock, L. G., G. R. Oeh, and J. J. Sparagna, "An Analysis of Wide-Band Microwave Monopulse Direction-Finding Techniques," *IEEE Trans. on Aerospace and Electronic Systems,* Vol. AES-7, No. 1, Jan. 1971, pp. 188–203.

List of Symbols, Abbreviations, and Acronyms

a	detector x-dimension, area of detected image, acceleration
$a(S, T)$	polynomial used to define the static dielectric constant of seawater as a function of salinity and temperature
A	area of video display
A_a	two-way 3-dB antenna ground footprint area
A_B	ground-projected antenna beam footprint area
A_c	clutter cell area intercepted by a pulse radar
A_e	effective area of antenna aperture
A_p	two-way 3-dB antenna footprint as projected normal to the viewing direction
A_R	effective area of receiving antenna aperture
A_D	area of a detector
A_i	area of pixel i
A_o	area of optical clear aperture
A_T	target area
ac	alternating current
A/D	analog-to-digital (converter)
ALON	aluminum oxynitride ($Al_{23}O_{27}N_5$)
AM	amplitude modulation
AMRAAM	Advanced Medium Range Air-to-Air Missile
AVHRR	Advanced Very High Resolution Radiometer
b	detector y-dimension
$b(S, T)$	polynomial used to define the relaxation time of seawater as a function of salinity and temperature
B	bandwidth
B_i	shape or roughness of pixel i
B_{IF}	predetection IF bandwidth
B_{3dB}	bandwidth measured at 3-dB (half-power) points
BAT	Brilliant Antiarmor Submunition
BFF	antenna beamfill factor

BLIP	background-limited infrared photodetector performance
c	speed of light in free space; also used to denote speed of light in a dielectric material
c_o	speed of light in free space = 3×10^8 m/sec; used when dielectric materials are discussed
°C	degrees centigrade
C_R	radiation contrast
C/N	clutter-to-noise ratio
$(C/N)_e$	effective clutter-to-noise ratio
CCD	charge coupled device
CD	calendar day
CFAR	constant false alarm rate
CFE	Conventional Forces Europe (Treaty)
cm	centimeter
CNVEO	Center for Night Vision and Electro-Optics
CO_2	carbon dioxide
CRREL	Cold Regions Research and Engineering Laboratories
CW	continuous wave
d	snow depth, surface dimensionality
d	value of the d-test statistic
dA	illuminated target area
deg	degree
D	$1/e$ penetration depth of an electromagnetic wave through a particular distribution of scatterers
D_{BLIP}	detector detectivity for background-limited performance
D^*_{BLIP}	specific detector detectivity for background-limited performance
D_d	detector detectivity
D^*_d	specific detector detectivity
D_o	effective diameter of antenna aperture, optical clear aperture
dc	direct current
$e = e'$	real part of the complex dielectric constant
e''	imaginary part of the complex dielectric constant
e_c	complex dielectric constant
e_i, e_o	mixer input and output signals
e_0	permittivity of free space = 8.854×10^{-12} farad/m
e_r	relative complex dielectric constant
e'_r	real part of the relative complex dielectric constant
e''_r	imaginary part of the relative complex dielectric constant
e_s	static value of the dielectric constant
e_∞	dielectric constant at infinite frequency
E	energy
E_L	amplitude of local oscillator component of mixer input signal
E_B	background irradiance
E_S	amplitude of RF component of mixer input signal

E_s	signal irradiance
E_x	component of electric field vector in *x*-direction
ESM	electronic support measures
eV	electron volt
f	frequency of operation, focal length of optics
$\dot{f}$	rate of change of frequency with respect to time
f_a	aliased frequency
f_D	Doppler frequency shift
f_e	effective spatial frequency
f_h	horizontal dimension spatial frequency
f_{IF}	intermediate frequency at mixer output
f_m	RF modulation frequency, maximum frequency content of a signal
f_0	instantaneous frequency of a signal
f_r	pulse repetition frequency
f_s	sampling frequency, switching frequency
f_T	target spatial frequency
f_v	vertical dimension spatial frequency
$f(\theta,\phi)$	radiation pattern of an antenna in θ and ϕ coordinates over 4π steradians
F	noise figure
F_{DSB}	double-sideband noise figure
F_{SSB}	single-sideband noise figure
$\dot{F}$	frame rate
$F/\#$	*f*-number
$F(\omega)$	Fourier transform of the delay function $P(t)$
FET	field-effect transistor
FLIR	forward-looking infrared (sensor)
FMCW	frequency modulated continuous wave
FOR	field of regard
FOV	field of view
fractal	fractal dimension
G	power gain, antenna gain
G_d	diversity gain
G_i	integration gain
G_0	peak value of antenna gain
G_R	gain of receiving antenna
G_T	gain of transmitting antenna
GaAs	gallium arsenide
GaAlAs	gallium aluminum arsenide
GEP	generalized evidence processing
GHz	gigahertz
h	Planck's constant, minimum target dimension, height of object above ground
h_1	parameter used in Parzen window nonparametric classification

h_N	window width used in Parzen window nonparametric classification
H	relative humidity
H	horizontal (polarization)
HBT	heterojunction bipolar transistor
HE	high explosive
HEMT	high electron mobility transistor
HgCdTe	mercury cadmium telluride
Hz	hertz
i_N	noise current
i_s	signal current
i_T	threshold current
IF	intermediate frequency
IIR	imaging infrared
IMPATT	impact avalanche and transit time (diode)
InP	indium phosphide
InSb	indium antimonide
IR	infrared
IFOV	instantaneous field of view
IRST	infrared search and track (sensor)
j	$\sqrt{-1}$
J	joule
k	integer, phase constant, wave number
k_a	aperture illumination taper factor
k_B	Boltzmann's constant = 1.380662×10^{-23} J/K
k_f	force constant
k_o	free space wave number
K	number of nearest neighbors selected for a *K*th nearest neighbor algorithm
K_F	video filter constant
K_e	electrical efficiency
K	kelvin
kg	kilogram
km	kilometer
kn	knot (nautical mile per hour)
kHz	kilohertz
l, m	lengths of edges of retroreflectors
lb	pound
L	length, dimensionless loss factor
L_a	system losses
L_{ant}	antenna loss factor
L_{atm}	atmospheric loss
L_{beam}	antenna beam loss
L_{clutter}	clutter loss
L_f	length of the aperture in a focused SAR
$L_f(1)$	fluctuation loss for a Swerling 1 target

L_i	integration loss
L_s	synthetic aperture length in a spotlight SAR
L_u	length of the aperture in an unfocused SAR
LHC	left hand circular (polarization)
LO	local oscillator
LWIR	long-wavelength infrared
m	mean value of a random variable
$\overline{m}$	estimate of mean value of a random variable
m	meter, mass
mm	millimeter
m_v	volume percentage of liquid water in a unit volume of snow
M	spectral radiant exitance
M_s	number of range cells that contain scatterers
M_w	mass of condensed water per unit volume of air
M	mega
MIMIC	microwave/millimeter-wave monolithic integrated circuit
MMIC	monolithic microwave integrated circuit
MESFET	metal semiconductor field effect transistor
MMW	millimeter-wave
MRTD	minimum resolvable temperature difference
MDTD	minimum detectable temperature difference
MMT	manufacturing methods and techniques
MSS	Multispectral Scanner
MDT	minimum detectable temperature
MTF	modulation transfer function
MWIR	mid-wavelength infrared
n	integer, complex index of refraction
n'	real part of the complex index of refraction or simply the index of refraction
n''	imaginary part of the complex index of refraction
n_a	number of resolution elements contained in the crosstrack search distance
nm	nanometer
n_b	average background and dark-current photon count
n_d	number of detectors
n_F	Fehlner false alarm number
n_M	Marcum false alarm number
n_p	number of target return pulses received per integration time
n_s	average signal photon count
n_r	number of range gates
N	number of cells, frequency cycles, samples, resolution elements
N_c	number of independent clutter samples
N_d	number of independent samples produced by temporal and frequency decorrelation
N_F	number of independent samples from frequency decorrelation

N_n	number of independent noise samples
N_e	effective number of spatial cycles
NEP	noise-equivalent power
$(NEP)_{PL}$	noise-equivalent power under conditions of photon-limited detection
N_t	number of line pairs resolved across a target
N_T	number of target pulses received, number of independent samples from temporal decorrelation
NATO	North Atlantic Treaty Organization
NEΔT	noise-equivalent temperature difference
NOAA	National Oceanic and Atmospheric Administration
OTF	optical transfer function
p	probability density function
P	power, number of pixels enclosed by a target
P_B	power incident on a detector from background radiation
P_C	probability of classification
PC	photoconductive
P_d	detection probability
P_{fa}	false alarm probability
P_s	incident signal power
P_{LO}	local oscillator output power
P_{ni}	input noise power
P_{no}	output noise power
P_N	noise power
P_R	received power
P_s	scattered power
P_{si}	input signal power
P_{so}	output signal power
$P(t)$	delay function produced by depth distribution of scatterers
P_T	transmitted power, threshold input optical power
Pa	pascals
PbSe	lead selenide
PbS	lead sulfide
pixels	picture elements
PRF	pulse repetition frequency
PTF	phase transfer function
PV	photovoltaic
q	electron charge = 1.6021×10^{-19} coulombs
r	resolution
$\bar{r}$	average rate of production of electrons from optical energy striking a detector
r_a	down track resolution
rad	radian
r_g	radius of gyration

r_i	distance between scatterers on an object located at position x_i and the position of a SAR
r_I	image resolution
r_O	sensor optics resolution
r_r	cross-track resolution
r_s	row spacing in a detector array
$\tilde{r}_S$	system modulation transfer function
r_{tgt}	target radius
R	precipitation rate, radar-to-target slant range
$\|R\|^2$	square of Fresnel reflection coefficient at boundary between two different layers of matter
R_{bkg}	Fresnel reflection coefficient at the air-background interface
R_c	current responsivity of an optical detector
R_{fa}	false alarm rate
R_i	output of a binary integrator detector, range between an observer and an image on a video monitor
R_{mf}	maximum operating range for a focused SAR
R_{mu}	maximum operating range for an unfocused SAR
R_S	Fresnel reflection coefficient at the air-ground surface interface
R_T	range between sensor and target
R_{tgt}	Fresnel reflection coefficient at the air-target interface
R_u	unambiguous range
RAID	ram air-inflatable device
RF	radio frequency
RHC	right-hand circular (polarization)
rms	root mean square
rss	root sum squared
RV	Gaussian distributed random variable
s_o	number of samples appended to a median filter to reduce edge effects
S	salinity
S_L	length of swath scanned by a sensor
S/C	signal-to-clutter ratio
$(S/C)_e$	effective signal-to-clutter ratio
S/I	signal-to-interference ratio
S/N	signal-to-noise ratio
$(S/N)_e$	effective signal-to-noise ratio
$(S/N)_I$	image signal-to-noise ratio
$(S/N)_V$	video signal-to-noise ratio
S/NC	signal-to-noise plus clutter ratio
SADARM	Sense and Destroy Armor (Munition)
SAM	surface-to-air missile
SAR	synthetic aperture radar
Si	silicon
SiO_2	silicon dioxide or quartz

SFM	sensor-fuzed munition
SFW	Sensor-Fused Weapon
SLAM	Standoff Land Attack Missile
SMMR	Scanning Multichannel Microwave Radiometer
SWIR	short-wavelength infrared
t	time
$t(\cdot)$	value of the log t-test statistic
t_{ar}	turnaround time of seeker gimbal
tan δ	loss tangent
t_d	decorrelation time
t_f	frame scan time
t_{fa}	average time interval between false alarms
t_i	integration time
t_r	pulse repetition interval
t_s	time to scan the azimuthal field of view
t_x	time for missile to travel a distance equal to the downtrack antenna footprint
T	temperature, time between samples
T_A	antenna temperature
T'_A	apparent temperature
T_B	brightness temperature
T_{bkg}	background physical temperature
T_{BS}	ground surface component of brightness temperature
$T_{B_{bkd}}$	brightness temperature of background
$T_{B_{tgt}}$	brightness temperature of target
T_c	sum of cosmic and galactic brightness temperatures
T_D	total downward or downwelling atmospheric emission
T_{DSB}	equivalent input double-sideband noise temperature
T_E	equivalent input noise temperature
T_e	integration time of the human eye
T_f	physical temperature of an antenna feed
T_i	physical temperature of a ferrite switch and isolator assembly, physical temperature in the region enclosed by pixel i
T_L	physical temperature of lossy RF components
T_m	time duration of an FMCW RF-modulation cycle
T'_N	noise added in a noise injection radiometer
T_0	reference temperature, usually 290K
T_p	physical (thermometric) temperature of a lossy component
T_{REC}	equivalent input noise temperature of receiver
T'_{REC}	equivalent input noise temperature of receiver and lossy transmission line
T_S	physical temperature of the ground surface
T_{sky}	sum of cosmic and galactic brightness temperatures and downwelling atmospheric emission in the specular direction

T_{SSB}	equivalent input single-sideband noise temperature
T_{SYS}	equivalent system input noise temperature
T_{tgt}	target physical temperature
T_U	total upward or upwelling atmospheric emission
T_w	physical temperature of the waveguide that connects an antenna to a ferrite switch
TDI	time delay and integration
TGM	terminally-guided munition
TOT	time on target
TOW	Tub-Launched, Optically Tracked, Wire Guided (Missile)
TR	transmit/receive (switch)
TTPF	target transform probability function
TWT	traveling wave tube
u	hyperbolically distributed random variable
v	velocity
$\overline{v}$	average scan velocity
v_b	blind velocity
v_i	voltage at output of an envelope detector
v_m	maximum scan velocity
v_t	target velocity
V	resolution cell volume of a sensor
V_c	volume concentration of snow
V_i	visibility
V	vertical (polarization)
VCO	voltage controlled oscillator
W	antenna length in cross-track direction, snowpack water equivalent
W_M	window size in a median filter
$W(\omega)$	power spectrum
W	watt
W_b	background spectral emittance
W_t	target spectral emittance
WAM	Wide Area Mine
x	downtrack dimension
x_a	downtrack resolution distance corresponding to a pulse width τ
X	downtrack antenna footprint
y	crosstrack dimension
Y	crosstrack antenna footprint
z	altitude dimension
ZnS	zinc sulfide
ZnSe	zinc selenide
α	detector angular subtense in x direction, attenuation constant
α_B	antenna beamshape factor
α_e	effective angular subtense of the target
α_T	angular subtense of the target as seen by a sensor

α_τ	relaxation time distribution parameter
α_W	Weibull clutter skewness parameter
β	depression angle, detector angular subtense in y direction, phase constant
$\beta(\Delta)$	polynomial used to model the ionic conductivity of seawater as a function of temperature and salinity
δ	diameter of central Airy disk in diffraction limited resolution
δf_d	difference frequency produced by FMCW downsweep waveform when a Doppler shift is present
δf_u	difference frequency produced by FMCW upsweep waveform when a Doppler shift is present
δR	range resolution per intermediate frequency cycle
Δ	temperature offset parameter used to model the ionic conductivity of seawater as a function of temperature and salinity
ΔA	azimuth error signal from a monopulse tracker
ΔE	elevation error signal from a monopulse tracker
Δf_D	Doppler bandwidth
Δf_F	decorrelation frequency of a target, clutter, or other background object
Δf_s	frequency spread produced by a mechanically or electrically scanned antenna
Δf_T	rss value of frequency spread produced by antenna scanning and Doppler-induced signal fading
ΔF	RF modulation bandwidth, bandwidth
ΔG	gain fluctuation
ΔR	range resolution per count per intermediate frequency cycle
Δt	time delay
ΔT	target-to-background temperature difference
ΔT_{avg}	average target-to-background temperature difference
ΔT_B	background-to-target brightness temperature contrast
ΔT_D	minimum detectable temperature difference
ΔT_S	target-to-background temperature difference as observed at the sensor
Δx	distance between pixels
$\Delta \omega_F$	half-power bandwidth of the power spectrum of a function
ϵ	emissivity
ϵ_{bkg}	emissivity of the background
ϵ_H	emissivity for horizontally polarized emission
ϵ_S	emissivity of the ground surface
ϵ_{tgt}	emissivity of the target
ϵ_V	emissivity for vertically polarized emission
ϕ	azimuth angle in x-y plane
γ	extinction or attenuation coefficient

γ^o	clutter backscatter coefficient used when antenna footprint is projected normal to viewing direction
Γ_f	power reflection coefficient of an antenna feed
Γ_s	power reflection coefficient of a ferrite switch
η	radar backscatter coefficient, quantum efficiency
η_B	antenna beam efficiency equal to fraction of antenna radiation pattern solid angle enclosed by the antenna mainlobe
η_i	intrinsic impedance of a dielectric material
η_l	antenna radiation efficiency formed by the ratio of the power at the interface of the antenna and the RF hardware component to which it is attached, to the incident power at the antenna aperture
η_{os}	overscan ratio
η_s	scanning efficiency
K_α	absorption coefficient
K_σ	scattering coefficient
λ	wavelength
λ_L	lower wavelength limit to spectral band of interest
λ_m	wavelength at which maximum power is emitted from a blackbody
λ_o	free space wavelength
λ_U	upper wavelength limit to spectral band of interest
μ	magnetic permeability
μ_o	permeability of free space $= 4\pi \times 10^{-7}$ henry/m
μ_r	relative magnetic permeability
μ_{rm}	reduced mass
μm	micrometer
ν	frequency of radiation striking an optical detector, vibrational quantum number
π	3.14
Π_f	power transmission coefficient through an antenna feed
Π_i	power transmission coefficient through a ferrite switch and isolator assembly
Π_w	power transmission coefficient through a waveguide that connects an antenna to a ferrite switch
θ	incidence angle with respect to nadir
$\dot{\theta}$	scan rate
θ'	parameter that partially defines the antenna mainlobe shape when a Gaussian model is used for the mainlobe
θ_a	3-dB azimuth beamwidth of antenna
θ_b	3-dB antenna beamwidth
θ_e	3-dB elevation beamwidth of antenna
θ_i	angle between incident power vector and the surface normal
θ_M	angle between maximum energy reflection from a retroreflector and the direction in which it decreases by a factor of two
θ_0	antenna boresight angle as measured from nadir

θ_p	one-way 3-dB antenna beamwidth
θ_r	3-dB beamwidth of antenna pattern generated by a real aperture of length L
θ_s	3-dB beamwidth of synthetic aperture, angle between scattered power vector and the surface normal
θ_S	total azimuthal field of view in scan direction
ρ	mean-to-median ratio of log normal distributed clutter, reflectivity
ρ_s	snow density
σ	radar cross section, standard deviation of a random variable
$\overline{\sigma}$	estimate of standard deviation of a random variable
σ_a	one-sigma value of angle tracking error from two-way antenna beamwidth and glint sources
σ_{bkd}	one-sigma background temperature variation
σ_g	angle-tracking error component due to glint from target scintillation
σ_T	target radar cross-section
σ_c	clutter radar cross-section, conductivity
σ_v	standard deviation of clutter velocity spectrum
$\sigma(z)$	distribution of scattering cross sections from scatterers at a distance z from the sensor
σ^o	clutter backscatter coefficient used when antenna footprint is projected onto the ground
σ_θ	noise-limited angle-tracking error governed by two-way antenna beamwidth
Σ	signal produced by sum of returns from antenna patterns in a monopulse receiver
τ	measurement interval, time duration of transmitted radar pulse, relaxation time
τ_a	atmospheric spectral transmission
τ_d	dwell time
τ_h	one-way opacity from the ground to a height h
τ_o	optical spectral transmission
τ_0	total one-way opacity through the atmosphere
τ_p	time duration of noise injected pulse
ω	radian frequency
ω_{IF}	radian frequency corresponding to intermediate frequency signal
ω_L	local oscillator radian frequency
ω_S	radian frequency corresponding to sideband signals
Ω	target scattering solid angle
ψ	arbitrary phase angle
ψ_i	phase shift in SAR signal from scatterers at position x_i
ψ_{mf}	phase shift associated with maximum operating range of a focused SAR
ψ_{mu}	phase shift associated with maximum operating range of an unfocused SAR

About the Author

Lawrence A. Klein is a consultant in millimeter-wave and infrared sensor design and data fusion. He develops multiple sensor system, communications, and system architecture concepts for intelligent transportation systems and defense applications. He received the B.E.E. degree from the City College of New York, a masters degree in electrical engineering from the University of Rochester, and a Ph.D. in electrical engineering from New York University. His early research at NASA-Langley Research Center explored theoretical and experimental aspects of remote sensing of terrain and weather phenomena. Later work in the defense industry included development of polarimetric radars, millimeter-wave radiometers, and multiple sensor and signal processing systems for tactical weapons and space-based surveillance of the Earth. He has authored the book *Sensor and Data Fusion Concepts and Applications* for the SPIE. Dr. Klein is member of Eta Kappa Nu and is listed in *Who's Who in Science and Engineering* (1996/1997).

Index

The Artech House Radar Library

David K. Barton, *Series Editor*

Active Radar Electronic Countermeasures, Edward J. Chrzanowski

Airborne Early Warning System Concepts, Maurice Long, *et al.*

Airborne Pulsed Doppler Radar, Second Edition, Guy V. Morris and Linda Harkness, editors

Analog Automatic Control Loops in Radar and EW, Richard S. Hughes

CARPET (Computer-Aided Radar Performance Evaluation Tool): Radar Performance Analysis Software and User's Manual, developed at the TNO Physics and Electronic Laboratory by Albert G. Huizing and Arne Theil

Coherent Radar Performance Estimation, James A. Scheer and James L. Kurtz, editors

DETPROB: Probability of Detection Calculation Software and User's Manual, William A. Skillman

Digital Techniques for Wideband Receivers, James Tsui

Electronic Homing Systems, M. V. Maksimov and G. I. Gorgonov

Electronic Intelligence: The Analysis of Radar Signals, Second Edition, Richard G. Wiley

Electronic Intelligence: The Interception of Radar Signals, Richard G. Wiley

Estimation and Tracking: Principles, Techniques, and Software, Yaakov Bar-Shalom and Xiao-Rong-Li

High Resolution Radar, Second Edition, Donald R. Wehner

Introduction to Electronic Defense Systems, Fillippo Neri

Introduction to Electronic Warfare, D. Curtis Schleher

Introduction to Multisensor Data Fusion: Multimedia Software and User's Guide, TECH REACH, Inc.

IONOPROP: Ionospheric Propagation Assessment Software and Documentation, Herbert Hitney

Kalman-Bucy Filters, Karl Brammer

Millimeter-Wave and Infrared Multisensor Design and Signal Processing, Lawrence A. Klein

Millimeter-Wave Radar Clutter, Nicholas Currie, Robert Hayes, and Robert Trebits

Modern Radar System Analysis, David K. Barton

Modern Radar System Analysis Software and User's Manual, David K. Barton and William F. Barton

Multifunction Array Radar Design, Dale R. Billetter

Multifunctional Array Radar: System Design and Analysis, Sergio Sabatini and Marco Tarantino

Multisensor Data Fusion, Edward L. Waltz and James Llinas

Photonic Aspects of Modern Radar, Henry Zmuda and Edward N. Toughlian

Principles and Applications of Millimeter-Wave Radar, Charles E. Brown and Nicholas C. Currie, editors

Principles of Data Fusion Automation, Richard T. Antony

Principles of High-Resolution Radar, August W. Rihaczek

Principles of Traveling Wave Tubes, A. S. Gilmour

Radar Cross Section Analysis and Control, A. K. Bhattacharyya and D. L. Sengupta

Radar Cross Section, Second Edition, Eugene F. Knott, *et al.*

Radar Evalution Handbook, David K. Barton, *et al.*

Radar Meteorolgy, Henri Sauvageot

Radar Polarimetry for Geoscience Applications, Fawwaz Ulaby

Radar Reflectivity Measurement: Techniques and Applications, Nicholas C. Currie, editor

Radar Resolution and Complex-Image Analysis, August W. Rihaczek

Radar Signals: An Introduction to Theory and Application, Charles E. Cook and Marvin Bernfield

Radar Technology Encyclopedia, David K. Barton and Sergey A. Leonov, editors

Radar Vulnerability to Jamming, Robert N. Lothes, Michael B. Szymanski, and Richard G. Wiley

RGCALC: Radar Range Detection Software and User's Manual, John E. Fielding and Gary D. Reynolds

Russian-English and English-Russian Dictionary of Radar and Electronics, Sergey A. Leonov and William F. Barton

Small-Aperture Radio Direction Finding, Herndon Jenkins

Space-Based Radar Handbook, Leopold J. Cantafio, editor

Spaceborne Weather Radar, Robert M. Meneghini and Toshiaki Kozu

Statistical Signal Characterization, Herbert Hirsch

Statistical Signal Characterization: Algorithms and Analysis Programs Software and User's Manual, Herbert Hirsch

Synthetic-Aperture Radar and Electronic Warfare, Walter W. Goj

For further information on these and other Artech House titles, contact:

Artech House
685 Canton Street
Norwood, MA 02062
617-769-9750
Fax: 617-769-6334
Telex: 951-659
e-mail: artech@artech-house.com

Artech House
Portland House - Stag Place
London SW1E 5XA England
+44 (0) 171-973-8077
Fax: +44 (0) 171-630-0166
Telex: 951-659
e-mail: artech-uk@artech-house.com

WWW: http://www.artech-house.com